Semiconductor Equations

P. A. Markowich
C. A. Ringhofer
C. Schmeiser

Springer-Verlag Wien New York

Peter A. Markowich
Fachbereich Mathematik
Technische Universität, Berlin

Christian A. Ringhofer
Department of Mathematics
Arizona State University, Tempe, Arizona, USA

Christian Schmeiser
Institut für Angewandte und Numerische Mathematik
Technische Universität, Wien, Austria

Typeset by Asco Trade Typesetting Ltd., Hong Kong
Printed in Austria by Novographic, Ing. W. Schmid, A-1238 Wien

Printed on acid-free paper

With 33 Figures

Library of Congress Cataloging-in-Publication Data. Markowich, Peter A., 1956– Semiconductor equations / Peter A. Markowich, Christian Ringhofer, Christian Schmeiser. p. cm. Includes bibliographical references. ISBN 0-387-82157-0 (U.S.) 1. Semiconductors—Mathematical models. I. Ringhofer, Christian, 1957– . II. Schmeiser, Christian, 1958– . III. Title. TK7871.85.M339. 1989. 621.381'52—dc20

ISBN 3-211-82157-0 Springer-Verlag Wien-New York
ISBN 0-387-82157-0 Springer-Verlag New York-Wien

Preface

In recent years the mathematical modeling of charge transport in semiconductors has become a thriving area in applied mathematics. The drift diffusion equations, which constitute the most popular model for the simulation of the electrical behavior of semiconductor devices, are by now mathematically quite well understood. As a consequence numerical methods have been developed, which allow for reasonably efficient computer simulations in many cases of practical relevance. Nowadays, research on the drift diffusion model is of a highly specialized nature. It concentrates on the exploration of possibly more efficient discretization methods (e.g. mixed finite elements, streamline diffusion), on the improvement of the performance of nonlinear iteration and linear equation solvers, and on three dimensional applications.

The ongoing miniaturization of semiconductor devices has prompted a shift of the focus of the modeling research lately, since the drift diffusion model does not account well for charge transport in ultra integrated devices. Extensions of the drift diffusion model (so called hydrodynamic models) are under investigation for the modeling of hot electron effects in submicron MOS-transistors, and supercomputer technology has made it possible to employ kinetic models (semiclassical Boltzmann-Poisson and Wigner-Poisson equations) for the simulation of certain highly integrated devices.

The focus of this book is the presentation of the hierarchy of semiconductor models ranging from kinetic transport equations to the drift diffusion equations. Particular emphasis is given to the derivation of the models and the physical and mathematical assumptions used therefore. We do not go into the mathematical technicalities necessary for a detailed analysis of the models but rather sacrifice rigour for the sake of conveying the basic properties and features of the model equations. The mathematically interested reader is encouraged to consult the references for in-depth investigations of specific subjects.

We address applied mathematicians, electrical engineers and solid state physicists. The exposition is accessible to graduate students in each of the three fields. In particular, we hope that this book will be useful as a text for advanced graduate courses in this area and we urge students to work the problems, which can be found at the end of each chapter, for a deeper penetration of the material.

We are grateful to our colleagues U. Ascher, J. Batt, F. Brezzi, P. Degond, P. Deuflhard, D. Ferry, W. Fichtner, L. Gastaldi, P. Kasperkovitz, N. Kluksdahl, D. Marini, H. Neunzert, F. Nier, R. O'Malley, F. Poupaud, T. Seidman, S. Selberherr, H. Steinrück, P. Szmolyan, and T. Taylor for many hours of stimulating discussions. In particular we are indebted to A. Arnold, N. Mauser, P. Pietra, and R. Weiß for reading large parts of the manuscript. We thank U. Schweigler for skillfully typing a portion of the manuscript. We are indebted to the Centre des Mathématiques Appliquées of the École Polytéchnique, Palaiseau, France, and to the Institut für Technische Elektronik of the Technische Universität Hamburg-Harburg where a part of the manuscript was written. The first and the third author acknowledge support from the "Österreichischer Fonds zur Förderung der wissenschaftlichen Forschung" under the grant numbers P6771 and P4919, respectively. The second author acknowledges support by the National Science Foundation, USA under grant no. PMS 880-1153.

Paris, 1989

Peter A. Markowich

Fachbereich Mathematik
Technische Universität Berlin
Straße des 17. Juni 136
D-1000 Berlin 12

Christian A. Ringhofer

Department of Mathematics
Arizona State University
Tempe, AZ 85287, USA

Christian Schmeiser

Institut für Angewandte und Numerische Mathematik
Technische Universität Wien
Wiedner Hauptstraße 8–10
A-1040 Wien, Austria

Contents

Introduction

Semiconductor device modeling spans a wide range of areas in solid state physics, applied and computational mathematics. The involved topics range from the most basic principles of kinetic transport in solids over statistical mechanics to complicated bifurcation problems in the mathematical description of certain devices and to numerical methods for partial differential equations.

This book tries to give an overview of the involved models and their mathematical treatment. It addresses, on one hand, the engineer and the physicist interested in the mathematical background of semiconductor device modeling. On the other hand it can be used by the applied mathematician to familiarize himself (herself) with a field which has immediate and technologically relevant applications and gives rise to a whole variety of interesting mathematical problems. The scope of semiconductor device modeling is clearly interdisciplinary. Quantitative answers are needed to describe devices and these answers can be obtained from a variety of different physical models.

We start from the most basic physical principles for kinetic transport of charged particles. Then we discuss a hierarchy of simplified model equations culminating in the drift diffusion equations which are the most widely used model today.

In order to make this book accessible to as wide a range of readers as possible the emphasis has been placed on concepts, and mathematical details have been replaced by references to the corresponding literature.

In the first Chapter the classical and quantum mechanical transport models in ensemble phase space and single particle phase space are discussed. Furthermore it is shown how the quantum mechanical models can be incorporated into the classical transport picture via the so called semiclassical models. The solution of transport equations in phase space is a very complex task. Therefore, simplified equations for integral quantities, such as particle and energy densities, are frequently used. These simplified equations are partial differential equations in position space only. The derivation of these equations, i.e. the hydrodynamic models and finally the drift diffusion equations, is the subject of Chapter 2. Chapter 3 is devoted to a mathematical discussion of the drift diffusion equations which are the underlying model for the bulk of the simulations performed today. Chapter 4 is concerned with the analysis of specific device structures. Here the analytical tools developed in Chapter 3, mainly asymptotic analysis, are used to approximately calculate current flows and to study the qualitative behaviour of voltage-current characteristics. Each Chapter is self contained, making it possible to use parts of the book as a text in seminars or courses. At the ends of the Chapters we have collected selections of problems (referenced in the text) which should make it easier for the student in such a course to reflect on the presented material.

Kinetic Transport Models for Semiconductors 1

1.1 Introduction

In this Chapter we shall derive and discuss transport equations, which model the flow of charge carriers in semiconductors. The common feature of these equations is that they describe the evolution of the phase space (position-momentum space) density function of the ensemble of negatively charged conduction electrons or, resp., positively charged holes, which are responsible for the current flow in semiconductor crystals.

The kinetic equations are the starting point for the derivation of the drift diffusion semiconductor model (often referred to as the Basic Semiconductor Equations or the van Roosbroeck Equations), which, together with its extensions (hydrodynamic models), constitutes the core of state-of-the-art semiconductor device simulation programs. This already necessitates a close scrutiny of kinetic transport models. Another reason is provided by the fact that the mathematical assumptions, which allow the derivation of the drift diffusion model from the kinetic models and which guarantee its validity, are—particularly for highly integrated devices—not satisfied. Thus, kinetic models must be used for the simulation of such devices. Until recently this approach was generally not taken since the numerical solution of the kinetic equations requires a lot of computing power in real life applications. However, with the reduction of cost of supercomputer technology, which was at least partly prompted by efficient VLSI-simulation, the numerical treatment of kinetic models for semiconductors was facilitated for at least some realistic applications. We expect the trend towards the kinetic equations to continue in the near future and, thus, we encourage simulation oriented researchers to become acquainted with these models.

Principally, the kinetic equations split into quantum mechanical, semi-classical and classical models. The quantum mechanical models are based on the many-body Schrödinger equation or, equivalently, on the quantum Liouville equation obtained from the Schrödinger equation by performing the Wigner transformation. The starting point for the classical models is the description of the motion of particle ensembles based on Newton's second law. A probabilistic reformulation of these canonical equations of motion immediately gives the classical Liouville equation, which describes the evolution of the phase space distribution function of the particle ensemble. The quantum Liouville equation is consistent with its classical counterpart in the sense that in the (formal) classical limit $\hbar \to 0$, where $\hbar$ denotes the Planck constant scaled by 2π, the quantum Liouville equation reduces to the classical Liouville equation.

The semi-classical Liouville equation can be regarded as a modification of the classical Liouville equation, which incorporates the quantum effects of the semiconductor crystal lattice via the band-diagram of the material.

The Liouville equations contain many-body effects in the sense that they are posed on the usually high-dimensional ensemble phase-space, whose coordinates are the position and momentum coordinates of all particles of the ensemble. The interaction force field, which appears in these equations, generally depends on all these coordinates. Thus, it is desirable (and, in fact, necessary in order to facilitate a numerical solution) to reduce the dimension of the Liouville equations.

The procedure for the reduction of the dimension of the Liouville equations is based on postulating properties of the interaction force field. Two cases are usually considered: When only long range forces (like the Coulomb force) are considered, then the Vlasov or collisionless Boltzmann equation is obtained in either the classical or semiclassical formulation and the quantum Vlasov equation in the quantum mechanical case. These equations have the form of single particle Liouville equations supplemented by an effective field equation, which depends on the position space number density of the particles. The effective field equation represents the (averaged) effect of the many-body physics.

If, in addition to the long range forces, short range forces are included, then the (semi-) classical and, resp., quantum Boltzmann equations are obtained. These equations contain collision integrals, which model the short range interactions (scatterings) of the particles with each other and/or with their environment. The specific form of the kernel of the collision operator, which is nonlocal in the momentum direction, is determined by the considered short range interaction mechanisms.

In Section 1.2 we discuss the classical and semi-classical Liouville equations and in Section 1.4 their quantum mechanical counterparts. Section 1.3 is concerned with the classical Vlasov and Boltzmann equations and Section 1.5 with the corresponding quantum equations. In Section 1.6 we discuss the applications of kinetic transport models to semiconductor physics and modeling.

1.2 The (Semi-)Classical Liouville Equation

In this Section we shall derive the basic equation, which governs the motion of an ensemble of charged particles under the action of a driving force assuming that the particles obey the laws of classical mechanics. Since,

usually, it is not possible to obtain enough data to determine the initial state of the ensemble exactly, we shall take a probabilistic point of view and reformulate the equations for the trajectory of the ensemble as a deterministic equation for the probability density of the ensemble in the position-momentum space. This microscopic equation is referred to as classical Liouville equation.
We start out by considering

Particle Trajectories

We shall at first analyze the motion of a single electron in a vacuum under the action of an electric field E. We associate the position vector $x \in \mathbb{R}^3$ and the velocity vector $v \in \mathbb{R}^3$—both assumed to be functions of the time t—with the electron. Then, in the absence of a magnetic field, the force $\mathscr{F}$, which acts on the electron, is given by

$$\mathscr{F} = -qE \tag{1.2.1}$$

(see, e.g., [1.31]). Here $q(>0)$ denotes the *elementary charge*, i.e. the charge of the electron is $-q$.
Newton's second law reads:

$$\mathscr{F} = m\dot{v}, \tag{1.2.2}$$

where m stands for the mass of the electron and '$\dot{}$' denotes differentiation with respect to the time t ($\dot{v}$ is the acceleration vector). By inserting (1.2.1) into (1.2.2) we obtain the system of ordinary differential equations

$$\dot{x} = v \tag{1.2.3}$$

$$\dot{v} = -\frac{q}{m}E \tag{1.2.4}$$

for the trajectories of the electron in the position-velocity space. Together with a given initial state

$$x(t=0) = x_0, \qquad v(t=0) = v_0 \tag{1.2.5}$$

the system (1.2.3), (1.2.4) constitutes an initial value problem for the trajectory $w(t; x_0, v_0) = (x(t), v(t))$, which passes through (x_0, v_0) at time $t = 0$. Note that, generally, the electric field E depends on the position vector x and on the time t, i.e. $E = E(x, t)$.

A Potential Barrier

As an example and for future reference we consider the one-dimensional motion of an electron across a thin and high potential barrier. The static potential V is depicted in Fig. 1.2.1. The corresponding electric field

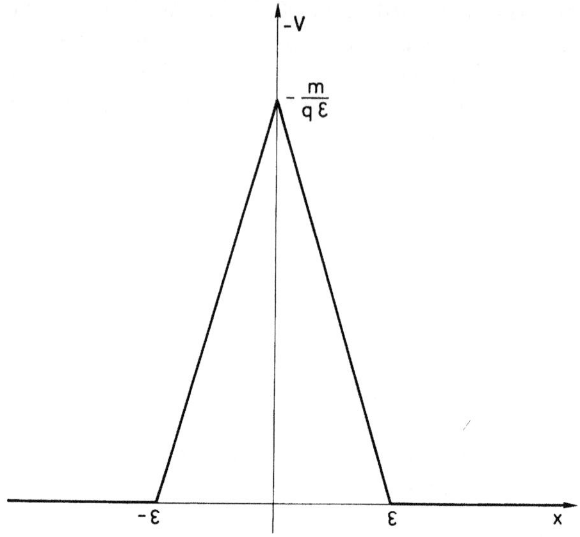

Fig. 1.2.1 Potential barrier

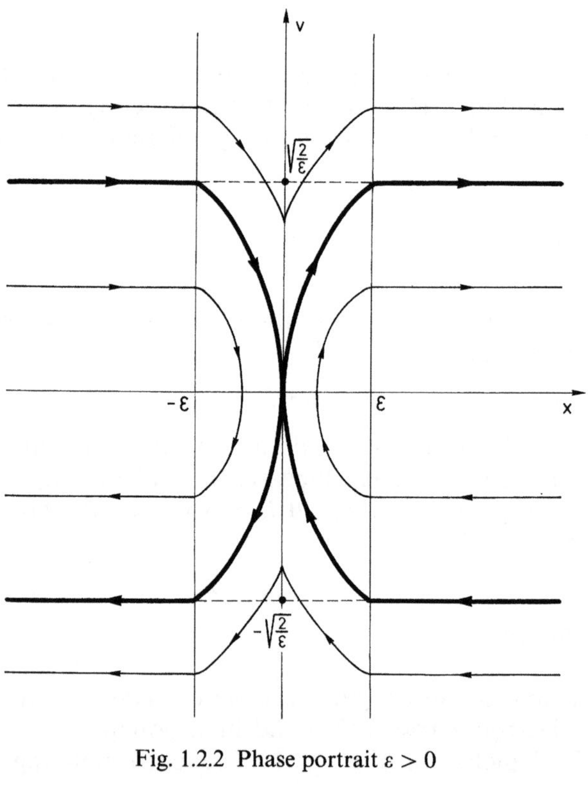

Fig. 1.2.2 Phase portrait $\varepsilon > 0$

$E = -V_x$ is given by

$$E(x) = \frac{m}{q}\begin{cases} 0, & |x| > \varepsilon \\ \dfrac{1}{\varepsilon^2}, & -\varepsilon < x < 0. \\ -\dfrac{1}{\varepsilon^2}, & 0 < x < \varepsilon \end{cases}$$

ε is a small positive parameter. The equations (1.2.3), (1.2.4) for the trajectories are easily integrated and we obtain the phase portrait shown in Fig. 1.2.2. The two thickly drawn curves are 'limiting' characteristics. A particle with velocity $|v| < \sqrt{2/\varepsilon}$ cannot cross the barrier, it is reflected. Only particles with $|v| > \sqrt{2/\varepsilon}$ cross over. As $\varepsilon \to 0+$ the barrier becomes thinner and higher, precisely speaking $V \xrightarrow{\varepsilon \to 0+} -(m/q)\delta(x)$.
The limiting phase portrait ($\varepsilon = 0$) is depicted in Fig. 1.2.3. All particles, no matter how big their velocity, are reflected. In Section 1.4 we shall consider the corresponding quantum mechanical model, which behaves totally different.

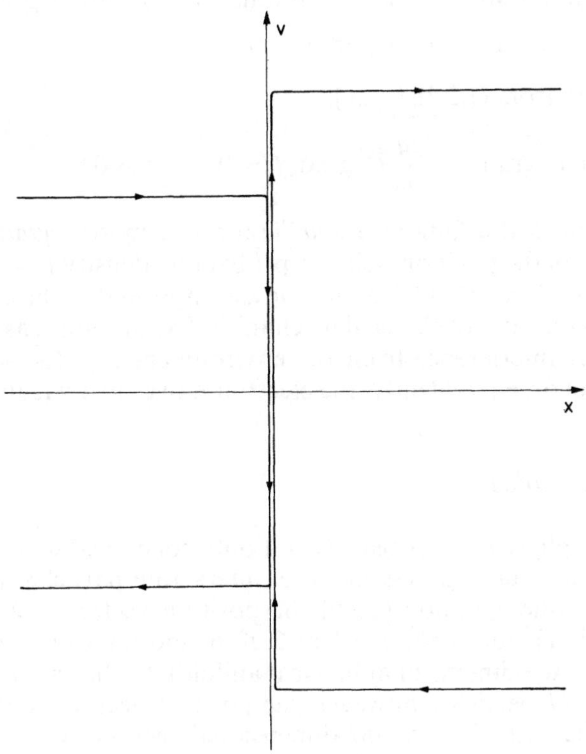

Fig. 1.2.3 Phase portrait $\varepsilon = 0$

The Transport Equation

Assume now that instead of the precise initial position x_0 and initial velocity v_0 of the electron we are given the joint probability density $f_I = f_I(x, v)$ of the initial position and velocity of the electron. f_I has the properties

$$f_I(x, v) \geqslant 0, \qquad \int\int f_I(x, v)\, dx\, dv = 1, \tag{1.2.6}$$

where the integration is performed over the whole (x, v)-space. Then

$$P(B) := \int\int_B f_I(x, v)\, dx\, dv$$

is the probability to find the electron in the subset B of the (x, v)-space at time $t = 0$. It is our goal now to derive a continuum equation for the probability density $f = f(x, v, t)$, which evolves from $f_I = f(x, v, t = 0)$.
It is reasonable to postulate that f does not change along the trajectories w, i.e. we require

$$f(w(t; x, v), t) = f_I(x, v) \tag{1.2.7}$$

for all x, v and for all $t \geqslant 0$. Differentiating (1.2.7) with respect to t gives

$$\partial_t f + \dot{x} \cdot \operatorname{grad}_x f + \dot{v} \cdot \operatorname{grad}_v f = 0 \tag{1.2.8}$$

and we obtain from (1.2.3), (1.2.4):

$$\partial_t f + v \cdot \operatorname{grad}_x f - \frac{q}{m} E \cdot \operatorname{grad}_v f = 0, \qquad t > 0. \tag{1.2.9}$$

This equation is the famous *Liouville* (or transport) *equation* governing the evolution of the position-velocity probability density $f = f(x, v, t)$ of the electron in the electric field E under the assumption that the electron moves according to the laws of classical mechanics. The motion is assumed to take place without interference from the environment (e.g. the semiconductor crystal lattice) or, equivalently, the electron moves in a vacuum.

Particle Ensembles

In solid state physics one is usually not only concerned with the motion of a single particle but of an ensemble of interacting particles. For the single electron Liouville equation (1.2.9) the position vector x and the velocity vector v are in $\mathbb{R}^3$ (or in $\mathbb{R}^d$, $d = 1$ or 2, if the motion can be restricted to a one- or resp., two-dimensional linear manifold). In the case of an ensemble consisting of M particles, however, the position vector x and the velocity vector v of the ensemble are $3M$-dimensional vectors, i.e. $x = (x_1, \ldots, x_M)$, $v = (v_1, \ldots, v_M)$ where $x_i, v_i \in \mathbb{R}^3$ represent the position and, resp., velocity vector of the i-th particle of the ensemble. Also, the force field $\mathscr{F} =$

$(\mathcal{F}_1, \ldots, \mathcal{F}_M)$ is a $3M$-dimensional vector, which in general depends on all $6M$ position and velocity coordinates and on the time t. $\mathcal{F}_i = \mathcal{F}_i(x, v, t)$ denotes the force acting on the i-th particle.

If all the particles of the ensemble have equal mass m (which we shall assume henceforth) then the trajectories of the ensemble satisfy the system of ordinary differential equations in the $6M$-dimensional ensemble position-velocity space:

$$\left.\begin{aligned} \dot{x}_i &= v_i \qquad (1.2.10)\\ \dot{v}_i &= \frac{1}{m}\mathcal{F}_i \qquad (1.2.11) \end{aligned}\right\} \quad i = 1, \ldots, M.$$

As above, we denote the ensemble trajectory, which passes through the initial state (x_0, v_0), by $w(t; x_0, v_0) = (x(t), v(t))$.

The classical (ensemble) Liouville equation

$$\partial_t f + v \cdot \mathrm{grad}_x f + \frac{1}{m}\mathcal{F} \cdot \mathrm{grad}_v f = 0, \tag{1.2.12}$$

now posed for $x \in \mathbb{R}^{3M}$, $v \in \mathbb{R}^{3M}$ is derived from (1.2.10), (1.2.11) as in the single electron case. Here $f = f(x, v, t)$ denotes the joint position-velocity probability density of the M-particle ensemble at time t, i.e.

$$P_M(B, t) = \iint_B f(x, v, t)\, dx\, dv$$

denotes the probability to find the particle ensemble in the subset B of the $6M$-dimensional ensemble position-velocity space at the time t (the preservation of the nonnegativity of f and the conservation of the integral of f over $\mathbb{R}^{6M}$ will be shown below under appropriate assumptions on the force field $\mathcal{F}$).

The Liouville equation (1.2.12) is linear and hyperbolic, its characteristics are the ensemble trajectories satisfying (1.2.10), (1.2.11). It has to be supplemented by the initial condition

$$f(x, v, t = 0) = f_I(x, v). \tag{1.2.13}$$

The Initial Value Problem

We consider the Liouville equation (1.2.12) subject to the initial condition (1.2.13) for $x \in \mathbb{R}^{3M}$, $v \in \mathbb{R}^{3M}$. In order to distinguish between the position and velocity spaces we shall in the sequel often write $x \in \mathbb{R}_x^{3M}$, $v \in \mathbb{R}_v^{3M}$. From (1.2.7) we conlude $f(x, v, t) \geqslant 0$, $x \in \mathbb{R}_x^{3M}$, $v \in \mathbb{R}_v^{3M}$ for all $t \geqslant 0$ for which a solution exists, if $f_I(x, v) \geqslant 0$, $x \in \mathbb{R}_x^{3M}$, $v \in \mathbb{R}_x^{3M}$. Thus, the nonnegativity of f is preserved by the evolution process generated by the Liouville equation.

For the following we shall assume that the force field $\mathcal{F}$ is divergence-free with respect to the velocity:

$$\operatorname{div}_v \mathscr{F} = 0, \qquad x \in \mathbb{R}_x^{3M}, \qquad v \in \mathbb{R}_v^{3M}, \qquad t \geqslant 0. \tag{1.2.14}$$

We integrate (1.2.12) over $\mathbb{R}_x^{3M} \times \mathbb{R}_v^{3M}$, assume that the solution decays to zero sufficiently fast as $|x| \to \infty$, $|v| \to \infty$ and calculate using (1.2.14):

$$\int_{\mathbb{R}_v^{3M}} \mathscr{F} \cdot \operatorname{grad}_v f \, dv = -\int_{\mathbb{R}_v^{3M}} f \operatorname{div}_v \mathscr{F} \, dv = 0.$$

We obtain

$$\frac{d}{dt} \int_{\mathbb{R}_x^{3M}} \int_{\mathbb{R}_v^{3M}} f(x, v, t) \, dv \, dx = 0$$

and conclude that the integral of f over the whole position-velocity space is conserved in time:

$$\int_{\mathbb{R}_x^{3M}} \int_{\mathbb{R}_v^{3M}} f(x, v, t) \, dv \, dx = \int_{\mathbb{R}_x^{3M}} \int_{\mathbb{R}_v^{3M}} f_I(x, v) \, dv \, dx = 1, \qquad t \geqslant 0. \tag{1.2.15}$$

The preservation of the nonnegativity of f and the conservation of the whole-space integral (1.2.15), both directly implied by the derivation of the Liouville equation and by the assumption (1.2.14) on $\mathscr{F}$, allow the full probabilistic interpretation of the solution of the initial value problem for the Liouville equation (cf. Liouville's Theorem [1.13]).

For the following the *moments* of the probability density f with respect to the velocity will be of importance. At this point we introduce the zeroth order moment

$$n_{\text{class}}(x, t) := \int_{\mathbb{R}_v^{3M}} f(x, v, t) \, dv \tag{1.2.16}$$

and the (negative) first order moment

$$J_{\text{class}}(x, t) := -q \int_{\mathbb{R}_v^{3M}} v f(x, v, t) \, dv. \tag{1.2.17}$$

The function $n_{\text{class}} = n_{\text{class}}(x, t)$ is the position probability density of the particle ensemble, i.e.

$$P_{M,x}(A, t) = \int_A n_{\text{class}}(x, t) \, dx$$

is the probability to find the ensemble in the subset A of the position space $\mathbb{R}_x^{3M}$ at the time t. n_{class} is called classical microscopic particle position density.

J_{class} represents a flux density, it is called classical microscopic particle current density.

The conservation property (1.2.15) can now be restated as

$$\int_{\mathbb{R}_x^{3M}} n_{\text{class}}(x, t) \, dx = \int_{\mathbb{R}_x^{3M}} n_{\text{class},I}(x) \, dx, \qquad t \geqslant 0 \tag{1.2.18}$$

with $n_{\text{class},I}(x) = \int_{\mathbb{R}_v^{3M}} f_I(x, v, t) \, dv$.

By formally integrating the Liouville equation (1.2.12) over $\mathbb{R}_v^{3M}$ we obtain the conservation law

$$q\partial_t n_{\text{class}} - \operatorname{div}_x J_{\text{class}} = 0, \tag{1.2.19}$$

which is referred to as macroscopic particle continuity equation.
The solvability of the initial value problem (1.2.12), (1.2.13) for the Liouville equation is closely related to the global (in time) existence of the characteristics $w(t; x, v)$ for all $(x, v) \in \mathbb{R}_x^{3M} \times \mathbb{R}_v^{3M}$, which in turn is related to the regularity and growth properties of the force field $\mathscr{F}$. If the maps

$$w(t; \cdot, \cdot)\colon \mathbb{R}_x^{3M} \times \mathbb{R}_v^{3M} \to \mathbb{R}_x^{3M} \times \mathbb{R}_v^{3M}, \qquad t \geqslant 0 \tag{1.2.20}$$

are sufficiently smooth and one-to-one, and if f_I is sufficiently differentiable, then the unique solution f of (1.2.12), (1.2.13) is given by

$$f(x, v, t) = f_I(w^{-1}(t; x, v)), \quad x \in \mathbb{R}_x^{3M}, \quad v \in \mathbb{R}_v^{3M}, \quad t \geqslant 0. \tag{1.2.21}$$

The invertibility requirement of the maps $w(t; \cdot, \cdot)$ excludes the intersection of trajectories ('collisions' of ensembles). Mathematically it prohibits certain strong singularities of the force field $\mathscr{F}$ at finite x, v and t.
An L^2-semigroup analysis of the classical Liouville equation (1.2.12) for v-independent, static gradient force fields can be found in [1.46].

The Classical Hamiltonian

We consider the motion of an electron ensemble under the action of a velocity-independent force field $\mathscr{F} = \mathscr{F}(x, t)$ and denote (as in the single electron case) the negative force per particle charge by E:

$$E = -\frac{1}{q}\mathscr{F}. \tag{1.2.22}$$

Assuming that $E = E(x, t)$ is a gradient field

$$E = -\operatorname{grad}_x V, \tag{1.2.23}$$

we can write the total energy of the ensemble as sum of the kinetic and potential energies

$$\varepsilon_{\text{tot}} = \frac{m|v|^2}{2} - qV(x, t). \tag{1.2.24}$$

When the total energy ε_{tot} is expressed in terms of the momentum vector

$$p = mv, \tag{1.2.25}$$

then we obtain the classical Hamiltonian of the ensemble

$$H(x, p, t) = \frac{|p|^2}{2m} - qV(x, t). \tag{1.2.26}$$

The equations (1.2.10), (1.2.11) for the ensemble trajectories are now equivalent to the so-called canonical equations of motion (or Hamiltonian equations):

$$\dot{x} = \operatorname{grad}_p H \tag{1.2.27}$$

$$\dot{p} = -\operatorname{grad}_x H \tag{1.2.28}$$

(see, e.g., [1.35]). Note that $\dot{H} = 0$ holds along the trajectories for static fields $E = E(x)$. In the static case the energy (1.2.26) is conserved by the motion. For time dependent fields $E = E(x, t)$ we have $\dot{H} = \partial_t H$.
The Liouville equation expressed in the (x, p)-coordinates takes the form

$$\partial_t f + \frac{p}{m} \cdot \operatorname{grad}_x f - qE \cdot \operatorname{grad}_p f = 0, \quad x \in \mathbb{R}_x^{3M}, \quad p \in \mathbb{R}_p^{3M}, \quad t > 0. \tag{1.2.29}$$

The $6M$-dimensional (x, p)-space is usually referred to as (ensemble) phase space.

The Semi-Classical Liouville Equation

So far the particles were assumed to move without interference from their environment, or equivalently, in a vacuum. In a semiconductor, however, the ions in the crystal lattice induce a lattice-periodic potential, which has a significant effect on the motion of the charged particles. Since the period of the lattice potential is very small (typically of the order of magnitude 10^{-8} cm), it is necessary to use quantum mechanics to model its impact on the transport of charged carriers. For precisely this reason the Liouville equation, which has incorporated the quantum effects of the crystal lattice, is referred to as semi-classical transport equation.
We start out with the mathematical set-up of the crystal lattice structure. We denote the (infinite periodic) crystal lattice by

$$L = \{ia_{(1)} + ja_{(2)} + la_{(3)} | i, j, l \in \mathbb{Z}\}. \tag{1.2.30}$$

$a_{(1)}, a_{(2)}, a_{(3)}$ are the primitive lattice vectors. The corresponding reciprocal lattice is given by

$$\hat{L} := \{ia^{(1)} + ja^{(2)} + la^{(3)} | i, j, l \in \mathbb{Z}\}, \tag{1.2.31}$$

where the reciprocal primitive lattice vectors $a^{(1)}, a^{(2)}, a^{(3)} \in \mathbb{R}^3$ satisfy

$$a_{(i)} \cdot a^{(j)} = 2\pi\delta_i^j. \tag{1.2.32}$$

A connected subset $Z \subseteq \mathbb{R}^3$ is called a primitive cell of the lattice L, if it satisfies the following two conditions:

(a) The volume of Z equals $|a_{(1)} \cdot (a_{(2)} \times a_{(3)})|$, which is the volume of the parallelepiped spanned by the primitive lattice vectors of L.
(b) $\mathbb{R}^3 = \bigcup_{x \in L} T_x Z$, where $T_x Z$ denotes the translate of Z by the lattice vector x. This means that the whole space $\mathbb{R}^3$ is covered by the union of translates of Z by the lattice vectors.

Primitive cells of the reciprocal lattice $\hat{L}$ are defined accordingly.

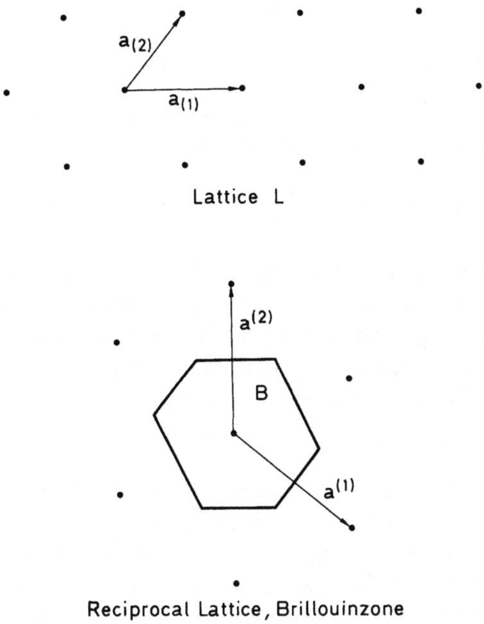

Fig. 1.2.4

The (first) *Brillouin zone* B is defined as that primitive cell of the reciprocal lattice $\hat{L}$, which consists of those points, which are closer to the origin than to any other point of $\hat{L}$ (see, e.g., [1.4], [1.31] for details). It is easy to show that the Brillouin zone B is point symmetric to the origin, i.e. $k \in B$ iff $-k \in B$ holds.

Fig. 1.2.4 shows a two-dimensional lattice, its reciprocal lattice and the Brillouin zone.

Consider now an electron whose motion is governed by the potential V_L generated by the ions located at the points of the crystal lattice L. Clearly, V_L is L-periodic, i.e.

$$V_L(x + X) = V_L(x), \qquad x \in \mathbb{R}_x^3, \qquad X \in L. \tag{1.2.33}$$

The steady state energies ε of the electron are the spectral values of the Schrödinger equation

$$H_L \psi = \varepsilon \psi \tag{1.2.34}$$

with the quantum mechanical Hamiltonian

$$H_L = -\frac{\hbar^2}{2m} \Delta - q V_L \tag{1.2.35}$$

(see Section 1.4 for details). Bloch's Theorem (see, e.g., [1.4], [1.31]) asserts that the bounded eigenstates ψ can be chosen to have the form of a plane wave $e^{ik \cdot x}$ times a function with the periodicity of the lattice L:

$$\psi(x) = e^{ik\cdot x}u_k(x), \qquad x \in \mathbb{R}_x^3 \tag{1.2.36}$$

$$u_k(x+X) = u_k(x), \qquad x \in \mathbb{R}_x^3, \qquad X \in L, \tag{1.2.37}$$

where, principally, k is an arbitrary (wave) vector in $\mathbb{R}_k^3$. Inserting (1.2.36) into (1.2.34) gives the eigenvalue equation

$$-\frac{\hbar^2}{2m}(\Delta u_k + 2ik\cdot\nabla u_k) + \left(\frac{\hbar^2}{2m}|k|^2 - qV_L(x)\right)u_k = \varepsilon u_k \tag{1.2.38}$$

subject to the periodicity condition (1.2.37). For given $k \in \mathbb{R}_k^3$ the problem (1.2.38), (1.2.37) constitutes a second order self-adjoint elliptic eigenvalue problem posed on a primitive cell of the crystal lattice L. Thus, we may expect an infinite sequence of eigenpairs $\varepsilon = \varepsilon_l(k)$, $u_k(x) = u_{k,l}(x)$, $l \in \mathbb{N}$. Note that (1.2.36), (1.2.37) can be reformulated as

$$\psi(x+X) = e^{ik\cdot X}\psi(x), \qquad x \in \mathbb{R}_x^3, \qquad X \in L.$$

Since $e^{ik\cdot X} = 1$ for all $k \in \hat{L}$, $X \in L$ we conclude that the set of wave functions ψ and the energies ε are identical for any two wave vectors which differ by a reciprocal lattice vector. Thus, we can assign the indices $l \in \mathbb{N}$ in such a way that the energy levels $\varepsilon_l(k)$ and the corresponding wave functions $\psi_{k,l}(x) = e^{ik\cdot x}u_{k,l}(x)$ are periodic on the reciprocal lattice $\hat{L}$:

$$\varepsilon_l(k+K) = \varepsilon_l(k), \qquad k \in B, \qquad K \in \hat{L}, \qquad l \in \mathbb{N} \tag{1.2.39}$$

$$\psi_{k+K,l} = \psi_{k,l}, \qquad k \in B, \qquad K \in \hat{L}, \qquad l \in \mathbb{N}. \tag{1.2.40}$$

Obviously, no information is lost when the wave-vector k is constrained to the Brillouin zone B.

For a thorough mathematical analysis of the spectral properties of the Schrödinger equation with a periodic potential we refer to [1.47].

The function $\varepsilon_l = \varepsilon_l(k)$, continuous on the Brillouin zone B, is called l-th energy band of the crystal. The corresponding mean electron velocity is given by

$$v_l(k) = \frac{1}{\hbar}\operatorname{grad}_k \varepsilon_l(k) \tag{1.2.41}$$

(see [1.4]).

In practice the lattice potential V_L is not known exactly, and therefore approximation methods have to be used to compute the band diagram $\{\varepsilon_l(k) | k \in B\}_{l\in\mathbb{N}}$ for a given material. For the technologically most important semiconductors the band diagrams can be found in the literature (see [1.31], [1.4]) and we shall henceforth assume that the energy bands and, thus, the velocities (1.2.41) are known functions of k.

Consider now the motion of an electron ensemble, where all M electrons are 'located' in the same energy band ε_l, i.e. the wave-function $\psi_{(i)} = \psi_{(i)}(x,t)$ of the i-th electron is represented by a 'linear combination' of eigenstates $\psi_{k,l}(x)$ over $k \in B$:

$$\psi_{(i)}(x,t) = \int_B c_{(i)}(k,t)\psi_{k,l}(x)\,dk, \qquad i = 1,\ldots,M.$$

In the sequel we denote the wave-vector of the i-th electron by $k_i \in \mathbb{R}^3$, $k = (k_1, \dots, k_M) \in \mathbb{R}^{3M}$, and its position vector by $x_i \in \mathbb{R}^3$, $x = (x_1, \dots, x_M) \in \mathbb{R}^{3M}$. Also, from now on we shall drop the band index l assuming that we are dealing with a specific band.

It is well known that in the presence of a driving force $\mathscr{F} = \mathscr{F}(x, k, t)$, $\mathscr{F} = (\mathscr{F}_1, \dots, \mathscr{F}_M)$, periodic in k_i for $i = 1, \dots, M$, the semi-classical equations of motion in the (x, k)-space read (see [1.31], [1.4]):

$$\left.\begin{aligned} \dot{x}_i &= v(k_i) \qquad (1.2.42)\\ \hbar\dot{k}_i &= \mathscr{F}_i \qquad (1.2.43) \end{aligned}\right\} \quad i = 1, \dots, M.$$

Note that band transitions are excluded since the band index is fixed in the equations.

If the driving force is independent of the wave vector k, we again write $E = E(x, t)$ for the negative force per particle charge (see (1.2.22)). If, in addition, the field E is a gradient field, then, using (1.2.23), we set up the semi-classical electron ensemble Hamiltonian

$$H(x, p, t) = \sum_{i=1}^{M} \varepsilon\left(\frac{p_i}{\hbar}\right) - qV(x, t), \tag{1.2.44}$$

where we set $p = (p_1, \dots, p_M)$. Here

$$p_i = \hbar k_i \tag{1.2.45}$$

denotes the crystal momentum vector of the i-th electron (see [1.4]). The semi-classical equations of motion (1.2.42), (1.2.43) are then equivalent to the Hamiltonian equations (1.2.27), (1.2.28) corresponding to the semi-classical Hamiltonian function (1.2.44).

The semi-classical electron-ensemble Liouville equation reads:

$$\partial_t f + \sum_{i=1}^{M} v(k_i) \cdot \operatorname{grad}_{x_i} f + \frac{1}{\hbar} \mathscr{F} \cdot \operatorname{grad}_k f = 0, \qquad t > 0, \tag{1.2.46}$$

where $x \in \mathbb{R}_x^{3M}$, $k_i \in B$ for $i = 1, \dots, M$. We impose periodic boundary conditions for k_i, $i = 1, \dots, M$:

$$f(x, k_1, \dots, k_i, \dots, k_M, t) = f(x, k_1, \dots, -k_i, \dots, k_M, t), \qquad k_i \in \partial B. \tag{1.2.47}$$

The definitions of the electron ensemble position density and of the electron ensemble current density have to be modified:

$$n_{\text{class}, B}(x, t) := \int_{B^M} f(x, k, t)\, dk \tag{1.2.48}$$

$$J_{\text{class}, B}(x, t) := -q \int_{B^M} v(k) f(x, k, t)\, dk \tag{1.2.49}$$

where we set $v(k) := (v(k_1), \dots, v(k_M))$.

The periodicity of f in k_i and the point-symmetry of the Brillouin zone B imply that the conservation property (1.2.18) and the conservation law

(1.2.19) also hold for the semi-classical Liouville equation, if the driving force $\mathscr{F}$ is divergence-free with respect to k, i.e. if $\operatorname{div}_k \mathscr{F} \equiv 0$ holds.
When the parabolic energy-wave vector relationship

$$\varepsilon(k) = \frac{\hbar^2 |k|^2}{2m}, \qquad k \in \mathbb{R}_k^{3M} \tag{1.2.50}$$

for electrons in a vacuum is used, then we obtain $v = p/m = \hbar k/m$ and the semi-classical Liouville equation reduces to its classical counterpart (1.2.12).

Magnetic Fields

A further extension of the Liouville equation is concerned with the inclusion of magnetic field effects. Consider a single classical electron, which moves under the action of an electric field E and of a magnetic field with induction vector B_{ind}. The magnetic field generates a contribution to the driving force $\mathscr{F}$, which is now given by (see [1.4]):

$$\mathscr{F} = -q(E + v \times B_{\text{ind}}). \tag{1.2.51}$$

The corresponding classical single-electron Liouville equation reads:

$$\partial_t f + v \cdot \operatorname{grad}_x f - \frac{q}{m}(E + v \times B_{\text{ind}}) \cdot \operatorname{grad}_v f, \qquad t > 0, \tag{1.2.52}$$

where $f = f(x, v, t)$, $x \in \mathbb{R}_x^3$, $v \in \mathbb{R}_v^3$. In the semi-classical case we obtain

$$\partial_t f + v(k) \cdot \operatorname{grad}_x f - \frac{q}{\hbar}(E + v(k) \times B_{\text{ind}}) \cdot \operatorname{grad}_k f = 0, \qquad t > 0 \tag{1.2.53}$$

with $f = f(x, k, t)$, $x \in \mathbb{R}_x^3$, $k \in B$, subject to periodic boundary conditions on ∂B.

1.3 The Boltzmann Equation

For an ensemble of many interacting particles there are two fundamental difficulties connected with the Liouville equation:

- Models for the driving force, which comprise short range and long range interactions, are not readily available,
- The dimension of the M-particle ensemble phase space is $6M$, which is very large in practical applications.

Even disregarding the problem of constructing appropriate driving forces, the high dimensionality of the Liouville equation, when employed as a semiconductor transport model, is prohibitive for numerical simulations. Consider a VLSI-device with 10^4 conduction electrons in the active region. Then the Liouville equation for the electron ensemble is posed on the 6×10^4-dimensonal phase space!

The main goal of the subsequent analysis will be a reduction of the dimension of the Liouville equation. This will be accomplished as follows. At first we shall derive a system of equations for the position-velocity densities of subensembles consisting of d electrons with d ranging from 1 to M. This system, called the BBGKY-hierarchy (from the names Bogoliubov [1.9], Born and Green [1.10], Kirkwood [1.30] and Yvon [1.64]), is obtained by assuming a certain structure of the interaction field (weak two-particle interactions) and by integrating the Liouville equation with respect to the position and velocity coordinates of $M - d$ particles for $d = 1, \ldots, M$. Then the formal limit $M \to \infty$ is carried out and a particular solution of the hierarchy, determined by a single function of three position, three velocity coordinates and time, is constructed. This particular solution, based on the assumption that the particles of a small subensemble move independently of each other, represents the electron number density in the physical phase space $\mathbb{R}_x^3 \times \mathbb{R}_v^3$. It is the solution of the so-called Vlasov equation, which can be considered as an 'aggregated' one-particle Liouville equation supplemented by a self-consistent (mean) field relation. The Vlasov equation is a macroscopic equation describing the motion of a weakly interacting large particle ensemble. Usually, it is employed to model the Coulomb interaction caused by a typical (weak) long range force.

However, when charge transport in a semiconductor is considered on a sufficiently large time scale, then the motion of the particles is decisively influenced by strong short range forces, so-called scatterings, or in a fully classical picture, collisions of particles. For the accurate description of charge transport in semiconductors the short range interactions of the particles with their environment (crystal lattice) are usually more important than short range forces between particles, which only play a significant role when the particle density is very large. In order to account for these effects, we shall extend the Vlasov equation and obtain the Boltzmann equation for semiconductors.

The Boltzmann equation was derived by L. Boltzmann in 1872 as a model for the kinetics of gases. Its most distinguished feature is the appearance of a nonlinear and nonlocal 'collision operator', which is responsible for formidable mathematical difficulties in the analytical and numerical treatment. We shall discuss a modification of the collision operator, which allows a proper description of the motion of charged particles in semiconductors.

The Vlasov Equation

We consider an ensemble of M electrons with equal mass, denote—as in the previous Section—the position vector of the ensemble by $x = (x_1, \ldots, x_M)$ and the velocity vector by $v = (v_1, \ldots, v_M)$, where $x_i \in \mathbb{R}_x^3$, $v_i \in \mathbb{R}_v^3$ are the position and velocity coordinates, resp., of the i-th electron.

We make the following assumptions:

(i) the electrons move in a vacuum, or equivalently, the impact of the semiconductor crystal lattice on the motion is neglected,
(ii) the force field $\mathcal{F}$ acting on the ensemble is independent of the velocity vector, in particular magnetic field effects are ignored,
(iii) the motion is governed by an external electric field and by two-particle interaction forces.

The first two assumptions are for simplicity's sake only, they will be discarded of later on. The third is crucial for the derivation of the Vlasov equation.

We denote the force field per electron charge by $E = E(x, t)$ (see (1.2.22)), $E = (E_1, \ldots, E_M)$, where $E_i \in \mathbb{R}^3$ is the field exerted on the i-th electron (per unit charge) and set:

$$E_i(x, t) = E_{\mathrm{ext}}(x_i, t) + \sum_{\substack{j=1 \\ j \neq i}}^{M} E_{\mathrm{int}}(x_i, x_j). \tag{1.3.1}$$

E_{ext} denotes the external electric field and E_{int} the two-particle interaction field. The ansatz (1.3.1) means that the force exerted on the i-th electron is the sum of the electric field acting on the i-th electron and of the sum of the $M - 1$ two-body forces exerted on the i-th electron by the other electrons of the ensemble. Moreover, we suppose that the electrons are indistinguishable in the sense that the interaction force $E_{\mathrm{int}} = E_{\mathrm{int}}(x, y)$ is independent of the electron indices. Also, by the action-reaction law, the force exerted by the i-th electron on the j-th electron is equal to the negative force exerted by the j-th electron on the i-th electron:

$$E_{\mathrm{int}}(x_i, x_j) = -E_{\mathrm{int}}(x_j, x_i), \qquad x_i, x_j \in \mathbb{R}^3_x. \tag{1.3.2}$$

For notational convenience we set $E_{\mathrm{int}}(x, x) = 0$ (we shall later on consider forces $E_{\mathrm{int}}(x, y)$ with singularities at $x = y$).

The Liouville equation for the joint position-velocity density $f = f(x_1, \ldots, x_M, v_1, \ldots, v_M, t)$ of the ensemble then reads:

$$\begin{aligned} \partial_t f &+ \sum_{i=1}^{M} v_i \cdot \operatorname{grad}_{x_i} f - \frac{q}{m} \sum_{i=1}^{M} E_{\mathrm{ext}}(x_i, t) \cdot \operatorname{grad}_{v_i} f \\ &- \frac{q}{m} \sum_{i=1}^{M} \sum_{j=1}^{M} E_{\mathrm{int}}(x_i, x_j) \cdot \operatorname{grad}_{v_i} f = 0. \end{aligned} \tag{1.3.3}$$

Note that the assumption (1.3.2) implies that the density f is independent of the numbering of the particles for all times if it is initially, i.e.

$$f(x_1, \ldots, x_M, v_1, \ldots, v_M, t) = f(x_{\pi(1)}, \ldots, x_{\pi(M)}, v_{\pi(1)}, \ldots, v_{\pi(M)}, t), \qquad x_i \in \mathbb{R}^3_x, \quad v_i \in \mathbb{R}^3_v \tag{1.3.4}$$

holds for all permutations π of $\{1, \ldots, M\}$ and for all times t, if it holds for the initial datum $f_I = f(t = 0)$, which we shall assume henceforth.

We now set up the joint position-velocity density $f^{(d)}$ of a subensemble consisting of d electrons:

$$f^{(d)}(x_1, \ldots, x_d, v_1, \ldots, v_d, t) = \int_{\mathbb{R}_x^{3(M-d)}} \int_{\mathbb{R}_v^{3(M-d)}} f(x_1, \ldots, x_M, v_1, \ldots, v_M, t) \times dx_{d+1} \ldots dx_M \, dv_{d+1} \ldots dv_M, \tag{1.3.5}$$

with $1 \leqslant d \leqslant M - 1$. An equation for $f^{(d)}$ is obtained by integrating the Liouville equation (1.3.3) with respect to 3 $(M - d)$ position and velocity coordinates and by assuming that f decays to zero sufficiently fast as $|x_i| \to \infty$, $|v_i| \to \infty$:

$$\begin{aligned} &\partial_t f^{(d)} + \sum_{i=1}^{d} v_i \cdot \operatorname{grad}_{x_i} f^{(d)} - \frac{q}{m} \sum_{i=1}^{d} E_{\text{ext}}(x_i, t) \cdot \operatorname{grad}_{v_i} f^{(d)} \\ &- \frac{q}{m} \sum_{i=1}^{d} \sum_{j=1}^{d} E_{\text{int}}(x_i, x_j) \cdot \operatorname{grad}_{v_i} f^{(d)} - \frac{q}{m}(M - d) \\ &\times \sum_{i=1}^{d} \operatorname{div}_{v_i} \left(\int_{\mathbb{R}^3_{v_*}} \int_{\mathbb{R}^3_{x_*}} E_{\text{int}}(x_i, x_*) f_*^{(d+1)} \, dx_* \, dv_* \right) \\ &= 0, \end{aligned} \tag{1.3.6}$$

where we denoted $f_*^{(d+1)} = f^{(d+1)}(x_1, \ldots, x_d, x_*, v_1, \ldots, v_d, v_*, t)$. In order to derive (1.3.6) observe that the terms with index $i \geqslant d + 1$ in the sum involving the outer field E_{ext} vanish by the divergence theorem. The same holds true for the terms with $i \geqslant d + 1$ in the sum involving the spatial derivatives and for the terms with $i \geqslant d + 1$ in the double sum involving the interaction field E_{int}. By (1.3.4) each term with $1 \leqslant i \leqslant d$ gives an identical contribution for each $j \geqslant d$ represented by the last sum in (1.3.6).

The equations (1.3.6) for $1 \leqslant d \leqslant M - 1$ constitute the so-called Bogoliubov-Born-Green-Kirkwood-Yvon (BBGKY) hierarchy for the classical Liouville equation (see [1.13]). In general this system of equations cannot be solved explicitly, however it is accessible to an asymptotic analysis for M large compared to d, i.e. in the case of small subensembles of a large particle ensemble. This is particularly interesting for us since in semiconductor physics one is usually concerned with extremely large charge carrier ensembles. In order to be able to carry out the limit $M \to \infty$ at least formally, we assume that $|E_{\text{int}}|$ is of the order of magnitude $1/M$ for M large, which very reasonably implies that the total field strength $|E_i|$ exerted on each electron remains finite as $M \to \infty$.

For a fixed subensemble size d the equation (1.3.6) then becomes in the limit $M \to \infty$:

$$\begin{aligned} &\partial_t f^{(d)} + \sum_{i=1}^{d} v_i \cdot \operatorname{grad}_{x_i} f^{(d)} - \frac{q}{m} \sum_{i=1}^{d} E_{\text{ext}}(x_i, t) \cdot \operatorname{grad}_{v_i} f^{(d)} \\ &- \frac{q}{m} \sum_{i=1}^{d} \operatorname{div}_{v_i} \left(\int_{\mathbb{R}^3_{v_*}} \int_{\mathbb{R}^3_{x_*}} M f_*^{(d+1)} E_{\text{int}}(x_i, x_*) \, dx_* \, dv_* \right) \\ &= 0. \end{aligned} \tag{1.3.7}$$

Intuitively, it is reasonable to assume that the electrons of a subensemble, which is small compared to the total number of electrons, move independently of each other. In terms of the subensemble probability density $f^{(d)}$ this implies the ansatz:

$$f^{(d)}(x_1, \ldots, x_d, v_1, \ldots, v_d, t) = \prod_{i=1}^{d} P(x_i, v_i, t). \tag{1.3.8}$$

The one-particle density $P = f^{(1)}$ is obtained from (1.3.7) with $d = 1$ by using (1.3.8) with $d = 2$:

$$\partial_t P + v \cdot \operatorname{grad}_x P - \frac{q}{m} E_{\text{eff}}(x, t) \cdot \operatorname{grad}_v P = 0,$$

$$t > 0, \qquad x \in \mathbb{R}_x^3, \qquad v \in \mathbb{R}_v^3, \tag{1.3.9}$$

with

$$E_{\text{eff}}(x, t) = E_{\text{ext}}(x, t) + \int_{\mathbb{R}_{x_*}^3} \int_{\mathbb{R}_{v_*}^3} MP(x_*, v_*, t) E_{\text{int}}(x, x_*) \, dv_* \, dx_*,$$

$$t > 0, \qquad x \in \mathbb{R}_x^3. \tag{1.3.10}$$

A simple calculation shows that (1.3.8) is a particular solution of (1.3.7) for arbitrary $d \in \mathbb{N}$ if P solves (1.3.9), (1.3.10). Equivalently, the solution $f^{(d)}$, $d \in \mathbb{N}$, of the BBGKY-hierarchy (1.3.7) can be factored according to (1.3.8), if the initial data $f^{(d)}(t = 0)$, $d \in \mathbb{N}$, admit such a factorization.
We define:

$$F(x, v, t) = MP(x, v, t) \tag{1.3.11}$$

$$n(x, t) = \int_{\mathbb{R}_v^3} F(x, v, t) \, dv. \tag{1.3.12}$$

The quantities F and n represent the expected electron *number densities* in phase space and in position space resp., i.e. $F(x, v, t)$ is the number of electrons per unit volume in an infinitesimal neighbourhood of (x, v) at time t and $n(x, t)$ is the number of electrons per unit volume in an infinitesimal neighbourhood of x at time t.
We multiply (1.3.9) by M and obtain the so-called Vlasov equation [1.13], [1.33]:

$$\partial_t F + v \cdot \operatorname{grad}_x F - \frac{q}{m} E_{\text{eff}} \cdot \operatorname{grad}_v F = 0,$$

$$x \in \mathbb{R}_x^3, \qquad v \in \mathbb{R}_v^3, \qquad t > 0, \tag{1.3.13}$$

$$E_{\text{eff}}(x, t) = E_{\text{ext}}(x, t) + \int_{\mathbb{R}_{x_*}^3} n(x_*, t) E_{\text{int}}(x, x_*) \, dx_*,$$

$$x \in \mathbb{R}_x^3, \qquad t > 0. \tag{1.3.14}$$

The macroscopic electron current density is given by

$$J = -q \int_{\mathbb{R}_v^3} vF \, dv . \tag{1.3.15}$$

The equation (1.3.13) has the form of a single particle Liouville equation. Many-body physics enters only through the effective field E_{eff}, which in turn depends on the number density n (self-consistent field modeling). The characteristics

$$\dot{x} = v, \qquad \dot{v} = -\frac{q}{m} E_{\text{eff}}(x, t) \tag{1.3.16}$$

are the trajectories of electrons moving in the field E_{eff}. They are the limiting (x_i, v_i)-trajectories of the Liouville equation (1.3.3) as $M \to \infty$.
The Pauli principle of solid state physics states that two electrons cannot occupy the same state (x, v) at the same time t (see, e.g., [1.34], [1.31]). Since $F(x, v, t)$ can also be interpreted as the existence probability of a particle at the state (x, v) at time t, we expect

$$0 \leqslant F(x, v, t) \leqslant 1, \qquad x \in \mathbb{R}_x^3, \qquad v \in \mathbb{R}_v^3, \qquad t > 0 \tag{1.3.17}$$

to hold. It is easy to show by using the characteristics (1.3.16) that upper and lower bounds of solutions of the Vlasov equation are conserved in time. Thus, if we require

$$0 \leqslant F(x, v, t = 0) \leqslant 1, \qquad x \in \mathbb{R}_x^3, \qquad v \in \mathbb{R}_v^3, \tag{1.3.18}$$

then (1.3.17) holds and the Vlasov equation satisfies the Pauli principle (an existence probability $F(x, v, t)$ larger than 1 can formally be interpreted as a multiple occupancy of the state (x, v) at the time t).
The Vlasov equation is nonlinear with a nonlocal nonlinearity of quadratic type. It provides a macroscopic description of the motion of many-body systems under the assumption of a weak interaction caused by a long range force (see [1.13], [1.31], [1.4] for various applications). In particular, it does not account for scatterings of particles generated by strong short range forces and, thus, it only represents a useful model on a time scale much shorter than the mean time between two consecutive scattering events.

The Poisson Equation

The most important long range force acting between two electrons is the Coulomb force modeling the mass-action law (see, e.g. [1.4]). It is represented by the interaction field

$$E_{\text{int}}(x, y) = -\frac{q}{4\pi\varepsilon_s} \frac{x - y}{|x - y|^3}; \qquad x, y \in \mathbb{R}^3, \qquad x \neq y. \tag{1.3.19}$$

The permittivity ε_s accounts for the dispersive effect of the considered host material.

We obtain the corresponding effective field from (1.3.14)

$$E_{\text{eff}}(x, t) = E_{\text{ext}}(x, t) - \frac{q}{4\pi\varepsilon_s} \int_{\mathbb{R}^3_{x_*}} n(x_*, t) \frac{x - x_*}{|x - x_*|^3} \, dx_* . \tag{1.3.20}$$

A simple computation shows

$$\operatorname{div} E_{\text{eff}} = \operatorname{div} E_{\text{ext}} - \frac{q}{\varepsilon_s} n \tag{1.3.21}$$

and

$$\operatorname{rot} E_{\text{eff}} = \operatorname{rot} E_{\text{ext}} . \tag{1.3.22}$$

If the exterior field is vortex-free

$$\operatorname{rot} E_{\text{ext}} = 0, \tag{1.3.23}$$

then the effective field is vortex-free, too, and there are potential functions V_{eff}, V_{ext} such that

$$E_{\text{eff}} = -\operatorname{grad}_x V_{\text{eff}} \tag{1.3.24}$$

$$E_{\text{ext}} = -\operatorname{grad}_x V_{\text{ext}} \tag{1.3.25}$$

hold. Then (1.3.21) can be rewritten as

$$-\Delta V_{\text{eff}} = -\Delta V_{\text{ext}} - \frac{q}{\varepsilon_s} n . \tag{1.3.26}$$

The effective potential satisfies a Poisson equation with a right hand side which depends linearly on the electron number density n.

Assume now that the external field is generated by ions of charge $+q$, which are present in the material. Then, again by Coulomb's law we have

$$E_{\text{ext}}(x, t) = \frac{q}{4\pi\varepsilon_s} \int_{\mathbb{R}^3_{x_*}} C(x_*, t) \frac{x - x_*}{|x - x_*|^3} \, dx_* , \tag{1.3.27}$$

where $C(x, t)$ is the number density of the background ions (in position space at the time t). We calculate

$$\operatorname{div} E_{\text{ext}} = \frac{q}{\varepsilon_s} C \tag{1.3.28}$$

and

$$\Delta V_{\text{ext}} = -\frac{q}{\varepsilon_s} C \tag{1.3.29}$$

follows. By inserting into (1.3.26) we obtain the well-known form of the Poisson equation

$$-\varepsilon_s \Delta V_{\text{eff}} = \rho \tag{1.3.30}$$

where

$$\rho = q(C - n) \tag{1.3.31}$$

is the charge density of the system consisting of conduction electrons and positively charged background ions.
The Vlasov equation with the Coulomb interaction field is usually referred to as Vlasov-Poisson equation.

The Whole Space Vlasov Problem

We consider the Vlasov equation (1.3.12), (1.3.13), (1.3.14) on the whole position-velocity space $\mathbb{R}_x^3 \times \mathbb{R}_v^3$ supplemented by the initial condition

$$F(x, v, t = 0) = F_I(x, v), \qquad x \in \mathbb{R}_x^3, \qquad v \in \mathbb{R}_v^3. \tag{1.3.32}$$

By integrating (1.3.13) over $\mathbb{R}_v^3$ we obtain the macroscopic conservation law

$$q\partial_t n - \operatorname{div} J = 0 \tag{1.3.33}$$

assuming that F decays sufficiently fast to zero as $|v| \to \infty$. Integration over $\mathbb{R}_x^3$ gives the conservation of the total number of particles

$$\int_{\mathbb{R}_x^3} n(x, t)\, dx = \int_{\mathbb{R}_x^3} n_I(x)\, dx, \qquad t > 0 \tag{1.3.34}$$

if $J \to 0$ as $|x| \to \infty$. We denoted

$$n_I(x) := \int_{\mathbb{R}_v^3} F_I(x, v)\, dv.$$

Note that the macroscopic electron continuity equation (1.3.33) and the Poisson equation (1.3.30), (1.3.31) do not constitute a 'closed' system of partial differential equations, since a relation for the current density J in terms of the potential V_{eff} and the number density n is not available yet. The derivation of such an equation, which describes the current flow in semiconductors, is the subject of Chapter 2.
A rigorous mathematical analysis (existence, uniqueness and regularity of solutions) of the Vlasov equation is beyond the scope of this book. We refer the interested reader to the references [1.21], [1.14], [1.62], [1.5]. Here we only mention the basic decoupling approach to the construction of a solution:

(i) Given $F^{(0)} = F^{(0)}(x, v, t)$, compute the number density $n^{(0)} = n^{(0)}(x, t)$ from (1.3.12) and the effective field $E_{\text{eff}}^{(0)} = E_{\text{eff}}^{(0)}(x, t)$ from (1.3.14).
(ii) Set $E_{\text{eff}} = E_{\text{eff}}^{(0)}$ in (1.3.13) and solve the so obtained linear hyperbolic equation subject to the initial condition (1.3.32). This is done by using that the solution $F^{(1)} = F^{(1)}(x, v, t)$ is constant along the characteristics

$$\dot{x} = v$$

$$\dot{v} = -\frac{q}{m} E_{\text{eff}}^{(0)}(x, t).$$

For appropriate initial data the sequence $F^{(l)}$, obtained by iterating (i), (ii), can be shown to converge to a limit function, which is a solution of the initial value problem for the Vlasov equation. The mathematical sophistication goes into proving bounds for (derivatives of) $F^{(l)}$ and $E_{\text{eff}}^{(l)}$ which are uniform in l and which allow the passage to the limit as $l \to \infty$. Details can be found in [1.14], [1.62], [1.5].

Bounded Position Domains

In semiconductor simulation transport equations are usually solved on bounded position domains, which represent the device geometry. We now consider the Vlasov-Poisson problem (1.3.12), (1.3.13), (1.3.30), (1.3.31) for $x \in \Omega$, where $\Omega \subseteq \mathbb{R}_x^3$ is a bounded convex domain. The velocity variable v is still assumed to vary in all $\mathbb{R}_v^3$.

Obviously, a boundary condition for (1.3.13) has to be imposed on those subsets of $\partial\Omega \times \mathbb{R}_v^3$ at which the x-characteristics point into Ω. These so-called inflow boundaries are given by

$$\Gamma_- := \{(x, v) | x \in \partial\Omega,\ v \in \mathbb{R}_v^3,\ \nu(x) \cdot v < 0\}, \tag{1.3.35}$$

where $\nu(x)$ denotes the unit outward normal vector to $\partial\Omega$ at $x \in \partial\Omega$. The outflow segments are

$$\Gamma_+ := \{(x, v) | x \in \partial\Omega,\ v \in \mathbb{R}_v^3,\ \nu(x) \cdot v > 0\}. \tag{1.3.36}$$

Most simply, Dirichlet boundary conditions for the Vlasov equation are imposed on the inflow segments:

$$F(x, v, t) = F_D(x, v, t), \qquad (x, v) \in \Gamma_-, \qquad t > 0 \tag{1.3.37}$$

with $0 \leqslant F_D \leqslant 1$ given.

Clearly, the solution of the one-particle Liouville equation (1.3.13) then still satisfies the bounds (1.3.17), i.e. the Pauli principle also holds for the initial boundary value problem for the Vlasov equation. Moreover, the electron continuity equation (1.3.33) is still valid, however, instead of (1.3.34) we now obtain by employing the divergence theorem

$$\frac{d}{dt} \int_\Omega n(x, t)\, dx = \int_{\Gamma_-} F_D |\nu(x) \cdot v|\, ds(x)\, dv - \int_{\Gamma_+} F |\nu(x) \cdot v|\, ds(x)\, dv, \tag{1.3.38}$$

where $s(x)$ denotes the surface measure on $\partial\Omega$. Thus, the total number of particles is generally not conserved, its rate of change is the difference of the incoming and the outgoing fluxes. By integrating (1.3.38) with respect to t and using $F \geqslant 0$ we obtain the estimate:

$$\int_\Omega n(x, t)\, dx \leqslant \int_\Omega n_I(x)\, dx + \int_0^t \int_{\Gamma_-} F_D |\nu(x) \cdot v|\, ds(x)\, dv\, dt. \tag{1.3.39}$$

The right-hand side represents a bound for the growth of the total number of electrons caused by the inflow which is determined by the boundary condition (1.3.37).

Also, boundary conditions for the Poisson equation (1.3.30) are required. Usually, mixed Neumann-Dirichlet conditions are imposed. Therefore, we split the boundary $\partial\Omega$ into Dirichlet segments $\partial\Omega_D$ and Neumann segments $\partial\Omega_N$ with $\partial\Omega_D \cup \partial\Omega_N = \partial\Omega$, $\partial\Omega_D \cap \partial\Omega_N = \phi$. We impose

$$E_{\text{eff}} \cdot \nu = E_b \cdot \nu \quad \text{on } \partial\Omega_N \tag{1.3.40}$$

$$V_{\text{eff}} = V_b \quad \text{on } \partial\Omega_D, \tag{1.3.41}$$

where E_b is a given vector field on $\partial\Omega_N$ and V_b a given real-valued function on $\partial\Omega_D$.

Neumann segments model insulating device boundaries, artificial boundary segments introduced in order to separate the considered device from neighboring devices in a VLSI-chip and semiconductor-oxide interfaces in MOS-devices (see [1.37], [1.51] and Chapter 4 for details). Dirichlet boundaries represent contact segments on which a bias is applied to the device.

We remark that the inflow Dirichlet boundary condition (1.3.37) is—from a physical point of view—not fully compatible with the mixed Neumann-Dirichlet conditions (1.3.40), (1.3.41) for the Poisson equation. Actually, Dirichlet inflow conditions should only be prescribed at the 'Dirichlet inflow segments' $\Gamma_- \cap (\partial\Omega_D \times \mathbb{R}_v^3)$ and reflecting boundary conditions on the 'Neumann inflow segments' $\Gamma_- \cap (\partial\Omega_N \times \mathbb{R}_v^3)$:

$$F(x, v, t) = F(x, v - 2\nu(x)(\nu(x) \cdot v), t), \qquad (x, v) \in \Gamma_- \cap (\partial\Omega_N \times \mathbb{R}_v^3)$$

We refer the interested reader to [1.13] for details. An L^p-semigroup analysis of transport operators on bounded position domains can be found in [1.15].

The Semi-Classical Vlasov Equation

Instead of taking the classical Liouville equation as basis for the derivation of the Vlasov equation we can also start out from the semi-classical formulation (1.2.46). When the above assumptions on the external and internal fields are made, we obtain by proceeding as in the classical case:

$$\partial_t F + v(k) \cdot \text{grad}_x F - \frac{q}{\hbar} E_{\text{eff}} \cdot \text{grad}_k F = 0,$$

$$x \in \mathbb{R}_x^3, \qquad k \in B, \qquad t > 0, \tag{1.3.42}$$

$$E_{\text{eff}}(x, t) = E_{\text{ext}}(x, t) + \int_{\mathbb{R}_{x_*}^3} n(x_*, t) E_{\text{int}}(x, x_*) \, dx_*,$$

$$x \in \mathbb{R}_x^3, \qquad t > 0 \tag{1.3.43}$$

with the electron number density

$$n(x, t) = \int_B F(x, k, t)\, dk. \tag{1.3.44}$$

As before, B denotes the Brillouin zone of the semiconductor crystal and $v(k)$ the electron velocity corresponding to a specific energy band.
The position-wave vector number density F is assumed to satisfy periodic boundary conditions in k:

$$F(x, k, t) = F(x, -k, t), \qquad x \in \mathbb{R}_x^3, \qquad k \in \partial B, \qquad t > 0. \tag{1.3.45}$$

The distinguished feature of the semi-classical Vlasov equation is that it takes into account the (quantum) effects of the lattice periodic potential generated by the ions of the semiconductor crystal lattice.
The (semi-classical) electron current density is defined by

$$J(x, t) = -q \int_B v(k) F(x, k, t)\, dk. \tag{1.3.46}$$

Clearly, the semi-classical Vlasov-Poisson equation can also be posed on a bounded position domain Ω. The Dirichlet boundary condition on the inflow segments then reads:

$$F(x, k, t) = F_D(x, k, t), \qquad (x, k) \in \Gamma_-, \qquad t > 0 \tag{1.3.47}$$

with F_D given and

$$\Gamma_- := \{(x, k) | x \in \partial\Omega,\ k \in B,\ \nu(x) \cdot v(k) < 0\}. \tag{1.3.48}$$

The current continuity equation and the conservation of the total number of particles (see the case $\Omega = \mathbb{R}_x^3$) holds as in the classical case. Also, the Pauli principle is satisfied for all times, if it is satisfied initially and on the inflow boundaries.

Magnetic Fields—The Maxwell Equations

So far we neglected the effects of magnetic fields in the derivation of the classical and semi-classical Vlasov equations. Assume now that a magnetic field with effective induction vector $B_{\text{eff}} = B_{\text{eff}}(x, t) \in \mathbb{R}^3$ exerts influence on the motion of the electron ensemble, too. Then, by setting $B_{\text{ind}} = B_{\text{eff}}$ in the classical Liouville equation (1.2.53) we obtain the Vlasov equation

$$\partial_t F + v \cdot \text{grad}_x F - \frac{q}{m}(E_{\text{eff}} + v \times B_{\text{eff}}) \cdot \text{grad}_v F = 0, \tag{1.3.49}$$

supplemented by the effective field equation (1.3.14).
The electric and magnetic field quantities are not independent, their relationship is governed by the Maxwell equations (see, [1.26], [1.51]), which for an arbitrary medium read:

$$\operatorname{div} D = \rho \tag{1.3.50}$$

$$\operatorname{div} B_{\text{ind}} = 0 \tag{1.3.51}$$

$$\operatorname{rot} E = -\partial_t B \tag{1.3.52}$$

$$\operatorname{rot} H = \partial_t D + J_{\text{tot}} \tag{1.3.53}$$

supplemented by the material equations

$$D = \varepsilon_s E \tag{1.3.54}$$

$$B_{\text{ind}} = \mu H\,. \tag{1.3.55}$$

The three-dimensional field quantities have the following meaning:

D: electric displacement vector

B_{ind}: magnetic induction vector

E: electric field vector

H: magnetic field vector

J_{tot}: total current density vector

μ denotes the permeability of the medium and, as before, ε_s the permittivity. We assume that the medium is isotropic and homogeneous and, thus, μ and ε_s are constant positive scalars.
We insert (1.3.54) into (1.3.50) and set $E = E_{\text{eff}}$

$$\varepsilon_s \operatorname{div} E_{\text{eff}} = \rho\,. \tag{1.3.56}$$

Setting $B_{\text{ind}} = B_{\text{eff}}$ gives

$$\operatorname{div} B_{\text{eff}} = 0 \tag{1.3.57}$$

$$\operatorname{rot} E_{\text{eff}} = -\partial_t B_{\text{eff}} \tag{1.3.58}$$

and by inserting (1.3.55), (1.3.54) into (1.3.53) we obtain

$$\frac{1}{\mu} \operatorname{rot} B_{\text{eff}} = \varepsilon_s \partial_t E_{\text{eff}} + J_{\text{tot}}\,. \tag{1.3.59}$$

Assume now that the external field is generated by positively charged ions. Then the equations (1.3.49), (1.3.56)–(1.3.59) constitute a 'closed' system of partial differential equations when supplemented by (1.3.31), (1.3.12) and by

$$J_{\text{tot}} = J_{\text{ion}} + J\,, \tag{1.3.60}$$

where J is given by (1.3.15) and J_{ion} denotes the current density caused by the flux of the positively charged ions with the number density $C = C(x, t)$. A mathematical analysis of the so-called Vlasov-Maxwell system can be found in [1.22].
We remark that a semi-classical Vlasov-Maxwell system can easily be derived by combining the 'ansätze' of this and of the previous paragraph.

Collisions—The Boltzmann Equation

The Vlasov equation accounts for long range interactions of particles. Short range interactions, which occur in the form of 'collisions' of the particles with other particles of the ensemble and with their environment, were neglected so far. These collisions have the effect that the particles are instantaneously scattered from one state to another in such a way that their velocity vector and, consequently, their momentum and wave vectors, change extremely fast, while the change of the position vector takes place slowly.

The goal of the following considerations is to derive an extension of the Vlasov equation, which includes a description of the long range interactions and a statistical account of the scattering events. The starting point for a phenomenological derivation of this equation, formulated first by L. Boltzmann in 1872 for the description of nonequilibrium phenomena in dilute gases, is the observation that the rate of change of the number density $F = F(x, v, t)$ of the ensemble due to the convection caused by the effective field E_{eff} vanishes along the characteristics (1.3.16) when collisions are neglected:

$$\left(\frac{dF}{dt}\right)_{\text{conv}} = 0. \tag{1.3.61}$$

It is reasonable to postulate that the rate of change of F due to convection and the rate of change of F due to collisions balance:

$$\left(\frac{dF}{dt}\right)_{\text{conv}} = \left(\frac{dF}{dt}\right)_{\text{coll}}. \tag{1.3.62}$$

Explicitly, (1.3.62) reads:

$$\partial_t F + v \cdot \text{grad}_x F - \frac{q}{m} E_{\text{eff}} \cdot \text{grad}_v F = \left(\frac{dF}{dt}\right)_{\text{coll}}, \tag{1.3.63}$$

where the effective field E_{eff} satisfies (1.3.14).

The rate $P(x, v' \to v, t)$ of a particle with position vector x, at the time t, to change its velocity vector v' into v due to a scattering event is assumed to be proportional to the occupation probability $F(x, v', t)$ of the state (x, v') at time t. Also, in order to account for the Pauli principle, it is assumed to be proportional to $1 - F(x, v, t)$, which is the probability that the state (x, v) is not occupied at the time t. We thus set

$$P(x, v' \to v, t) = s(x, v', v) F(x, v', t)(1 - F(x, v, t)), \tag{1.3.64}$$

where s is the so-called scattering rate. More precisely speaking, $s(x, v', v)\, dv'$ is the transition rate for an electron with position vector x to change its velocity vector v' belonging to the volume element dv' (around v') to v. Clearly, the scattering rate is determined by the physics of the considered scattering mechanism. Those scattering mechanisms, which are important in semiconductor physics, will be discussed below.

The rate of change of the number density F due to collisions at (x, v, t) is given by the 'sum' of the rates of particles being scattered from all possible states (x, v') into the state (x, v) at the time t minus the sum of the rates of the particles being scattered from the state (x, v) into any possible state (x, v') at the time t:

$$\left(\frac{dF}{dt}\right)_{\text{coll}}(x, v, t) = \int_{\mathbb{R}^3_{v'}} [P(x, v' \to v, t) - P(x, v \to v', t)]\, dv'. \tag{1.3.65}$$

We insert (1.3.64) into (1.3.65), set

$$Q(F) := \left(\frac{dF}{dt}\right)_{\text{coll}} \tag{1.3.66}$$

and obtain

$$Q(F)(x, v, t) = \int_{\mathbb{R}^3_{v'}} [s(x, v', v)F'(1 - F) - s(x, v, v')F(1 - F')]\, dv', \tag{1.3.67}$$

where we denoted:

$$F = F(x, v, t), \qquad F' = F(x, v', t). \tag{1.3.68}$$

Q is called collision operator and $Q(F)$ collision integral.
The Boltzmann equation (1.3.63) then can be written in the form

$$\partial_t F + v \cdot \operatorname{grad}_x F - \frac{q}{m} E_{\text{eff}} \cdot \operatorname{grad}_v F = Q(F),$$
$$x \in \mathbb{R}^3_x, \qquad v \in \mathbb{R}^3_v, \qquad t > 0 \tag{1.3.69}$$

supplemented by the effective field equation (1.3.14). When the Coulomb force is used to model the long range interaction, then (1.3.69), (1.3.14), (1.3.12) is often referred to as Boltzmann-Poisson problem.
We remark that the presented derivation of the Boltzmann equation is purely phenomenological. A more rigorous approach for gas-dynamics can be found in [1.13].
In addition to the nonlinearity caused by the self-consistent field, the collision integral $Q(F)$ introduces another quadratic nonlinearity, which is nonlocal in the velocity direction. A rigorous mathematical analysis of the Boltzmann equation—even with a given electric field—is extremely complicated and by far beyond the scope of this book. Below, we shall only sketch an existence proof and discuss some qualitative properties, in particular those which are important for the derivation of the drift diffusion approximation (see Chapter 2). We refer the reader to the book [1.13] for a wealth of information on the (gas-dynamical) Boltzmann equation and for a huge collection of references. Also, for the mathematically oriented reader, we mention the recent paper [1.23], where existence, globally in time, of solutions of the (gas-dynamical) Boltzmann equation in the field-free case $E_{\text{eff}} \equiv 0$ is shown.

The Semi-Classical Boltzmann Equation

For the modeling of transport in semiconductors the Boltzmann equation in the semi-classical formulation is usually employed in order to incorporate the quantum effects of the semiconductor crystal lattice as discussed in Section 1.2. Therefore, we start out with the semi-classical Vlasov equation (1.3.42), (1.3.43) and include the collision effects as above. We obtain:

$$\partial_t F + v(k)\cdot \operatorname{grad}_x F - \frac{q}{\hbar} E_{\text{eff}} \cdot \operatorname{grad}_k F = Q(F), \qquad x \in \mathbb{R}^3_x, \quad k \in B, \quad t > 0, \tag{1.3.70}$$

where the collision integral $Q(F)$ is given by

$$Q(F)(x,k,t) = \int_B [s(x,k',k)F'(1-F) - s(x,k,k')F(1-F')]\,dk'. \tag{1.3.71}$$

We denoted

$$F = F(x,k,t), \qquad F' = F(x,k',t). \tag{1.3.72}$$

Clearly, the periodic boundary condition (1.3.45) has to be imposed.
Also, the (semi-)classical Boltzmann-Poisson problem can be posed on a bounded position domain $\Omega \subseteq \mathbb{R}^3_x$. The boundary conditions on $\partial\Omega$ are the same as for the Vlasov-Poisson equation.
We impose the initial condition

$$F(x,k,t=0) = F_I(x,k), \tag{1.3.73}$$

which is assumed to obey the Pauli principle, i.e.

$$0 \leqslant F_I(x,k) \leqslant 1 \tag{1.3.74}$$

holds. In the bounded position domain case the inflow datum F_D also has to be between 0 and 1.
We shall now sketch the existence and uniqueness proof for the Boltzmann-Poisson problem in the whole space case as presented in [1.44]. The proof proceeds by a decoupling iterative approach similar to the existence proof for the Vlasov-Poisson problem. Only the collision integral has to be taken care of accordingly.
We set $F^{(0)} = 0$ and construct a sequence of approximations $\{F^{(l)}\}_{l \in \mathbb{N}_0}$ as follows. Given $F^{(l)}$, we compute the number density

$$n^{(l)} = \int_B F^{(l)}\,dk$$

and the effective field $E^{(l)}_{\text{eff}}$ by inserting $n^{(l)}$ into (1.3.14) using the Poission kernel (1.3.19). Then we have

$$\partial_t F^{(l+1)} + v(k)\cdot \operatorname{grad}_x F^{(l+1)} - \frac{q}{\hbar} E^{(l)}_{\text{eff}} \cdot \operatorname{grad}_k F^{(l+1)}$$
$$= Q_{\text{lin}}(F^{(l+1)}, F^{(l)}) \tag{1.3.75}$$

subject to the initial condition (1.3.73) and the periodic boundary condition on ∂B, where

$$Q_{\text{lin}}(F^{(l+1)}, F^{(l)})$$
$$= \int_B [s(x, k', k) F^{(l)'}(1 - F^{(l+1)}) - s(x, k, k') F^{(l+1)}(1 - F^{(l)'})]\, dk'. \tag{1.3.76}$$

We used the obvious notation

$$F^{(l)'} = F^{(l)}(x, k', t).$$

Clearly, Q_{lin} can be written as

$$Q_{\text{lin}}(F^{(l+1)}, F^{(l)}) = A^{(l)}(1 - F^{(l+1)}) - B^{(l)} F^{(l+1)} \tag{1.3.77}$$

where

$$A^{(l)} = \int_B s(x, k', k) F^{(l)'}\, dk',$$
$$B^{(l)} = \int_B s(x, k, k')(1 - F^{(l)'})\, dk'. \tag{1.3.78}$$

The problem (1.3.75) is a linear transport equation, which can be solved by the method of characteristics or by semigroup theory.
The characteristic form of (1.3.75) reads

$$\frac{dF^{(l+1)}}{dt} + (A^{(l)} + B^{(l)}) F^{(l+1)} = A^{(l)} \tag{1.3.79}$$

along the characteristics determined by $E^{(l)}_{\text{eff}}$:

$$\dot{x} = v(k), \qquad \hbar\dot{k} = -qE^{(l)}_{\text{eff}}. \tag{1.3.80}$$

Since $s > 0$ we conclude $A^{(l)} \geqslant 0$ from (1.3.78) if $F^{(l)} \geqslant 0$. Integration of (1.3.79) gives $F^{(l+1)} \geqslant 0$. For $G^{(l+1)} := 1 - F^{(l+1)}$ we obtain from (1.3.79)

$$\frac{dG^{(l+1)}}{dt} + (A^{(l)} + B^{(l)}) G^{(l+1)} = B^{(l)}. \tag{1.3.81}$$

If $F^{(l)} \leqslant 1$ we obtain $B^{(l)} \geqslant 0$ and $G^{(l+1)} \geqslant 0$ follows by integrating (1.3.81). Thus, if the initial datum F_I satisfies the Pauli principle, all iterates $F^{(l)}$ satisfy the Pauli principle for all times $t > 0$ and, by passing to the limit $l \to \infty$, we conclude that the Boltzmann equation conserves the upper bound 1 and the lower bound 0, i.e. the solution F satisfies

$$0 \leqslant F(x, k, t) \leqslant 1, \qquad t \geqslant 0. \tag{1.3.82}$$

It was actually shown in [1.44] that the sequence $\{F^{(l)}\}_{l \in \mathbb{N}_0}$ converges to the unique solution of the Boltzmann-Poisson problem if the transition rate s is sufficiently regular and positive.

Conservation and Relaxation

We integrate the collision integral (1.3.71) with respect to the wave-vector k and obtain

$$\int_B Q(F)\,dk = \int_B \int_B [s(x,k',k)F'(1-F) - s(x,k,k')F(1-F')]\,dk'\,dk = 0. \qquad (1.3.83)$$

This implies the conservation law (1.3.33) and—for the whole space problem—the conservation of the total number of electrons (1.3.34). Collision processes neither destroy nor generate particles.
Another important property of the collision operator is related to the relaxation of the state of the ensemble towards local thermodynamical equilibrium. The so-called principle of detailed balance asserts that the local scattering probabilities vanish for all states (x, k), (x, k') in thermal equilibrium (see [1.4], [1.8]), i.e.

$$s(x,k',k)F_e'(1-F_e) = s(x,k,k')F_e(1-F_e') \qquad (1.3.84)$$

holds, where F_e denotes the equilibrium number density. It follows from standard statistical mechanics that F_e is given by the Fermi-Dirac statistics

$$F_e(k) = F_D\left(\frac{\varepsilon(k)-\varepsilon_F}{k_B T}\right), \qquad (1.3.85)$$

where $\varepsilon(k)$ is the considered energy band of the semiconductor, ε_F denotes the Fermi-energy, k_B the Boltzmann constant, T the lattice temperature and

$$F_D(u) = \frac{1}{1+e^u} \qquad (1.3.86)$$

(see [1.31]).
From (1.3.84) we obtain the following property of the scattering rate s by a simple calculation:

$$s(x,k,k') = \exp\left(\frac{\varepsilon(k')-\varepsilon(k)}{k_B T}\right) s(x,k',k). \qquad (1.3.87)$$

It was shown in [1.44] that the condition (1.3.87) on s is sufficient and necessary to guarantee that the null manifold of the collision operator Q consists of Fermi-Dirac distributions, i.e.

$$Q(F) = 0, \qquad 0 \leqslant F \leqslant 1 \qquad (1.3.88)$$

implies that F is of the form (1.3.85) for some Fermi-energy $-\infty \leqslant \varepsilon_F = \varepsilon_F(x,t) \leqslant \infty$ if (1.3.87) holds.

The quantity

$$\lambda(x, k) := \int_B s(x, k, k')\, dk' \tag{1.3.89}$$

is called collision frequency. It measures the strength of the interaction at the state (x, k) corresponding to the transition rate s. Its reciprocal

$$\tau(x, k) := \frac{1}{\lambda(x, k)} \tag{1.3.90}$$

is the relaxation time describing the average time between two consecutive collisions at (x, k). We shall see below that τ represents the time scale on which the density F relaxes towards an equilibrium state (1.3.85).

Low Density Approximation

In many semiconductor device applications the particle density F is small, i.e.

$$0 \leqslant F(x, k, t) \ll 1 \tag{1.3.91}$$

holds. Very often the quadratic terms in the collision operator are ignored (set to zero) in these cases. Then the so obtained simplified linear collision operator is given by

$$Q_L(F)(x, k, t) = \int_B [s(x, k', k)F' - s(x, k, k')F]\, dk'. \tag{1.3.92}$$

Obviously, Q_L also satisfies the conservation property

$$\int_B Q_L(F)\, dk = 0. \tag{1.3.93}$$

The principle of detailed balance now gives

$$s(x, k', k)F_e' = s(x, k, k')F_e. \tag{1.3.94}$$

In the context of the low density approximation the Fermi-Dirac distribution is usually approximated by the Maxwellian distribution, which reads

$$M(k) = N^* \exp\left(-\frac{\varepsilon(k)}{k_B T}\right), \qquad N^* := \left(\int_B \exp\left(-\frac{\varepsilon(k)}{k_B T}\right) dk\right)^{-1}. \tag{1.3.95}$$

With $F_e = M(k)$ we obtain from (1.3.94)

$$s(x, k', k)M(k') = s(x, k, k')M(k), \tag{1.3.96}$$

which is equivalent to the condition (1.3.87) obtained for the nonlinear collision operator under the assumption of the Fermi-Dirac equilibrium distribution.

We easily conclude that the scattering rate s can now be written as

$$s(x, k, k') = \phi(x, k', k)M(k'), \tag{1.3.97}$$

where ϕ is symmetric with respect to k and k':

$$\phi(x, k, k') = \phi(x, k', k). \tag{1.3.98}$$

The function ϕ is called collision cross-section.
It is now easy to show that the null-space of the low-density collision operator Q_L is spanned by the Maxwellian:

$$Q_L(F) = 0 \leftrightarrow F(x, k, t) = n(x, t)M(k). \tag{1.3.99}$$

By using (1.3.97) and (1.3.98) we can now write Q_L in the form

$$Q_L(F)(x, k, t) = \int_B \phi(x, k', k)[M(k)F' - M(k')F]\,dk'. \tag{1.3.100}$$

The Relaxation Time Approximation

It is particularly interesting to investigate whether the solutions of the Boltzmann equation converge to an equilibrium state as $t \to \infty$. Since this analysis is very complicated for the nonlinear collision operator Q as well as for the low density approximation Q_L we shall carry out another simplification of the collision integral. When the initial datum F_I is close to a multiple of the Maxwellian it is natural (and mathematically convenient) to approximate F' in (1.3.100) by $n(x, t)M(k')$. By using the definition of the relaxation time τ given by (1.3.90) we obtain

$$Q_R(F)(x, k, t) = -\frac{1}{\tau(x, k)}(F(x, k, t) - M(k)n(x, t)). \tag{1.3.101}$$

Note that the so-called relaxation time approximation collision operator Q_R is linear and local in the wave vector k. When the characteristics $x = x(t)$, $k = k(t)$, which satisfy

$$\dot{x} = v(k), \qquad x(t = 0) = x_0$$

$$\hbar\dot{k} = -qE_{\text{eff}}, \qquad k(t = 0) = k_0$$

are introduced, then the Boltzmann equation with the collision operator Q_R takes the form

$$\frac{d}{dt}F = -\frac{1}{\tau}(F - Mn) \tag{1.3.102}$$

along $(x(t), k(t))$. By straightforward integration we obtain

$$F(x(t), k(t), t) = e^{-t/\tau}\left(F_I(x_0, k_0) + \frac{1}{\tau}\int_0^t n(x(s), s)M(k(s))e^{s/\tau}\,ds\right), \tag{1.3.103}$$

where—for the sake of simplicity—we assumed the relaxation time τ to be constant. Also, the electron density $n = \int_B F\,dk$ is assumed to be known in (1.3.103). It is now an easy exercise to show that

$$F(x(t), k(t), t) - n(x(t), t)M(k(t)) \sim e^{-t/\tau}, \qquad t \to \infty \tag{1.3.104}$$

holds, i.e. the relaxation time τ is the scale on which F returns to the equilibrium density from the perturbed state F_I along the characteristics. In the collisionless case (Vlasov equation) there is no mechanism, which forces the state of the particle ensemble to relax towards thermodynamical equilibrium in the large time limit. This is also expressed by the fact that the Vlasov equation is time reversible for static exterior fields while the Boltzmann equation is not. The relaxation behaviour of the solutions of the Boltzmann equation is precisely caused by the effect of collisions. This is represented mathematically by the famous H-Theorem (see [1.13] for the gas-dynamics case and [1.44] for the semiconductor case).

Polar Optical Scattering

Typically, the transition rates, determined by the physics of the considered scattering process in the semiconductor crystal are highly nonsmooth functions of k, k' (more precisely speaking they are, in general, distributions). As a typical example we present the transition rate for the polar optical scattering process modeling collisions of electrons with phonons, which are quantized vibrations of the semiconductor crystal lattice (see [1.31] for details on the physics). It is of the following form (see [1.43]):

$$s_{po}(k, k') = \frac{s_{\text{opt}}}{|k - k'|^2}((N_0 + 1)\delta(\varepsilon(k') - \varepsilon(k) + \hbar\omega_0) + N_0\delta(\varepsilon(k') - \varepsilon(k) - \hbar\omega_0)), \tag{1.3.105}$$

where $\hbar\omega_0$ is the (constant) energy of a polar optical phonon, N_0 is the phonon occupation number given by the Bose-Einstein statistics

$$N_0 = \left(\exp\left(\frac{\hbar\omega_0}{k_B T}\right) - 1\right)^{-1} \tag{1.3.106}$$

and

$$s_{\text{opt}} = \frac{q^2\hbar\omega_0}{8\pi^2\hbar\varepsilon_0}\left(\frac{1}{\varepsilon_\infty} - \frac{1}{\varepsilon_r}\right). \tag{1.3.107}$$

Here ε_0 denotes the vacuum permittivity, ε_∞ the high frequency relative permittivity and ε_r the low frequency relative permittivity of the semiconductor. δ stands for the Dirac distribution.

Particle-Particle Interaction

The transition rate s, which models particle-particle scatterings depends on the density F itself (see [1.28]) and, consequently, this scattering mechanism introduces another nonlinearity into the Boltzmann equation. The corresponding model is of the form

$$s_{pp}[F](x,k,k') = \int_B \int_B H(|k-k'|,|k_0-k_0'|)F_0'(1-F_0)\delta_{k_0}\delta_\varepsilon \, dk_0' \, dk_0, \quad (1.3.108)$$

where H is nonnegative, $F_0' = F(x,k_0',t)$, $F_0 = F(x,k_0,t)$,

$$\delta_{k_0} = \delta(k_0 + k' - k_0' - k), \qquad \delta_\varepsilon = \delta(\varepsilon(k_0) + \varepsilon(k') - \varepsilon(k_0') - \varepsilon(k)).$$

When $s_{pp}[F]$ is inserted into the collision integral, it becomes apparent that the corresponding collision operator is nonlinear of fourth order. Particle-particle scattering is only relevant for very high local densities. It is neglected in most practical situations.

We remark that there are various other scattering mechanisms which play a role in semiconductor physics (see [1.48]). For a particular practical application the relevant mechanisms have to be identified, the corresponding scattering rates then have to be added to obtain the total scattering rate modeling all the considered interactions.

1.4 The Quantum Liouville Equation

In ultra-integrated semiconductor devices the characteristic length of the active region is usually under 1 μm. Very often these devices are operated at large applied voltages, which leads to extremely high local electric field strengths. With today's technology electric field peaks of 10^6 V/cm are usually reached. It is well-known that potential variations of this order of magnitude lead to quantum effects, which cannot be properly described by the so far presented classical or semi-classical transport models. On the other hand, there is a group of semiconductor devices, whose performance explicitly relies on a quantum mechanical phenomenon, namely the tunneling effect (e.g. the so-called tunnel diode, see [1.57]).

For these reasons and, in particular, since tomorrow's semiconductor technology promises an even higher degree of miniaturization and integration, it is of great importance to devise transport models capable of describing quantum phenomena, which are still sufficiently simple to allow for reasonably efficient numerical simulation. In this Section we shall consider a quantum transport model based on Wigner functions. Introduced by E. Wigner in 1932 as quantum equivalent of classical particle distribution functions, Wigner functions were closely scrutinized by theoretical physicists but only recently their value for semiconductor simulation was discovered.

We start the presentation with the basic quantum mechanical equation of motion.

The Schrödinger Equation

In a quantum mechanical set-up the motion of an electron is described by the Schrödinger equation:

$$i\hbar\partial_t\psi = H\psi, \tag{1.4.1}$$

where the quantum Hamiltonian operator H for a single particle in a potential field

$$E(x, t) = -\operatorname{grad}_x V(x, t) \tag{1.4.2}$$

is given by

$$H = -\frac{\hbar^2}{2m}\Delta_x - qV(x, t) \tag{1.4.3}$$

(see, e.g. [1.34], [1.25]).
Note that the quantum Hamiltonian is obtained from the classical Hamiltonian function (1.2.26) by inserting the momentum operator

$$p = -i\hbar \operatorname{grad}_x. \tag{1.4.4}$$

A solution $\psi = \psi(x, t)$ of the Schrödinger equation is called a wave function. The square of its modulus

$$n_{\text{quan}} := |\psi|^2 \tag{1.4.5}$$

represents the quantum mechanical probability density for the position of the electron, i.e. the number

$$\int_A |\psi(x, t)|^2\, dx \tag{1.4.6}$$

is the probability of finding the electron in the subset A of the position space $\mathbb{R}_x^3$ at the time t.
Note that $\int_{\mathbb{R}_x^3} |\psi|^2\, dx$ is conserved by the motion. Multiplying (1.4.1) by the complex conjugate $\bar{\psi}$ of the wave function ψ, integrating over $\mathbb{R}_x^3$ and taking imaginary parts gives:

$$\frac{d}{dt}\int_{\mathbb{R}_x^3} |\psi|^2\, dx = 0 \tag{1.4.7}$$

assuming that the potential V is real valued (as we shall do henceforth) and that ψ decays sufficiently fast to zero as $|x| \to \infty$. Thus,

$$\int_{\mathbb{R}_x^3} n_{\text{quan}}(x, t)\, dx = \int_{\mathbb{R}_x^3} n_{\text{quan},I}(x)\, dx, \qquad n_{\text{quan},I} = |\psi_I|^2 \tag{1.4.8}$$

follows, where ψ_I is the initial datum for the Schrödinger equation

$$\psi(x, t = 0) = \psi_I(x). \tag{1.4.9}$$

Clearly, (1.4.8) is the quantum equivalent of (1.2.18) in the single electron case.

We calculate the rate of change of the probability (1.4.6):

$$\begin{aligned}\frac{d}{dt}\int_A |\psi|^2\,dx &= \int_A (\psi\partial_t\bar{\psi} + \bar{\psi}\partial_t\psi)\,dx\\ &= \frac{i}{\hbar}\int_A (\psi H\bar{\psi} - \bar{\psi}H\psi)\,dx\\ &= -\frac{i\hbar}{2m}\int_A (\psi\Delta\bar{\psi} - \bar{\psi}\Delta\psi)\,dx\end{aligned}$$

(bars denote complex conjugation).

Since $\psi\Delta\bar{\psi} - \bar{\psi}\Delta\psi = \operatorname{div}_x(\psi\operatorname{grad}_x\bar{\psi} - \bar{\psi}\operatorname{grad}_x\psi)$ we obtain

$$\frac{d}{dt}\int_A |\psi|^2\,dx = \frac{1}{q}\int_A \operatorname{div} J_{\text{quan}}\,dx \tag{1.4.10}$$

with

$$J_{\text{quan}} = \frac{i\hbar q}{2m}(\bar{\psi}\operatorname{grad}_x\psi - \psi\operatorname{grad}_x\bar{\psi}). \tag{1.4.11}$$

By the divergence theorem

$$\frac{d}{dt}\int_A |\psi|^2\,dx = \frac{1}{q}\int_{\partial A} J_{\text{quan}}\cdot\nu_A\,ds \tag{1.4.12}$$

holds, where ν_A is the outer unit normal to ∂A. Thus, the vector J_{quan} is called quantum mechanical electron current density. From (1.4.10) we obtain the conservation law for the one-electron case:

$$q\partial_t n_{\text{quan}} - \operatorname{div} J_{\text{quan}} = 0, \tag{1.4.13}$$

which is analogous to the classical equation (1.2.19).

The eigenvalue problem for the Schrödinger equation

$$-\frac{\hbar^2}{2m}\Delta\psi - qV(x)\psi = \varepsilon\psi \tag{1.4.14}$$

is obtained by looking for time-periodic solutions of the form $\exp[-(i/\hbar)\varepsilon t]\,\psi(x)$. The spectral values ε of (1.4.14) are the possible energies of the electron (cf. Section 1.2).

Tunneling

We shall now present the maybe most simple, explicitly solvable model for the tunneling of a particle through a potential barrier. Therefore we consider

the one-dimensional steady state Schrödinger equation (1.4.14) with $V(x) = -(m/q)\delta(x)$, i.e. we assume that the potential barrier is infinitely high and infinitely thin. In Section 1.2 we analyzed the motion of classical electrons in the same potential field and showed that they are reflected at the barrier $x = 0$ no matter how large their velocity. As we shall see now their quantum mechanical behaviour is entirely different.
We remark that we proceed analogously to [1.6] in this paragraph.
The equation of motion for electrons with the energy ε now reads:

$$\frac{\mu^2}{2}\psi_{xx} - \delta(x)\psi = -\varepsilon\psi, \qquad -\infty < x < \infty \tag{1.4.15}$$

with $\mu = \frac{\hbar}{m}$, $\varepsilon > 0$. It can be shown that (1.4.15) is equivalent to

$$\frac{\mu^2}{2}\psi_{xx} + \varepsilon\psi = 0, \qquad x \neq 0 \tag{1.4.16}$$

$$\psi(0-) = \psi(0+) \tag{1.4.17}$$

$$\frac{\mu^2}{2}(\psi_x(0+) - \psi_x(0-)) = \psi(0) \tag{1.4.18}$$

(see Problem 1.20).
By solving (1.4.16) we obtain:

$$\psi(x) = \begin{cases} a \exp\left(-i\frac{\sqrt{2\varepsilon}}{\mu}x\right) + b \exp\left(i\frac{\sqrt{2\varepsilon}}{\mu}x\right), & x < 0 \\ c \exp\left(-i\frac{\sqrt{2\varepsilon}}{\mu}x\right) + d \exp\left(i\frac{\sqrt{2\varepsilon}}{\mu}x\right), & x > 0 \end{cases}. \tag{1.4.19}$$

We now assume that a monoenergetic beam of particles, represented by a right moving wave of the form $\exp(-i\sqrt{2\varepsilon}\,x/\mu)$, is aimed at $x = 0$ from $x = -\infty$. We then expect a reflected left-moving wave for $x < 0$ and a transmitted right-moving wave for $x > 0$. This gives $a = 1$ and $d = 0$:

$$\psi(x) = \begin{cases} \exp\left(-i\frac{\sqrt{2\varepsilon}}{\mu}x\right) + b \exp\left(i\frac{\sqrt{2\varepsilon}}{\mu}x\right), & x < 0 \\ c \exp\left(-i\frac{\sqrt{2\varepsilon}}{\mu}x\right), & x > 0 \end{cases}. \tag{1.4.20}$$

The interface conditions (1.4.17), (1.4.18) give

$$b = \frac{\mu\sqrt{2\varepsilon}\,i - 1}{2\mu^2\varepsilon + 1}, \qquad c = \frac{\mu\sqrt{2\varepsilon}(\mu\sqrt{2\varepsilon} + i)}{2\mu^2\varepsilon + 1}. \tag{1.4.21}$$

$R := |b|^2$ is the probability that a particle of energy ε is reflected at the barrier, it is therefore called reflection coefficent. $T = |c|^2$ is the probability that the particle is transmitted through the barrier, it is called transmission coefficient. We calculate

$$R = \frac{1}{2\mu^2\varepsilon + 1}, \qquad T = \frac{2\mu^2\varepsilon}{2\mu^2\varepsilon + 1}. \tag{1.4.22}$$

Obviously $R + T = 1$ holds. Also $R = 1$, $T = 0$ for $\varepsilon = 0$ and $R = 0$, $T = 1$ for $\varepsilon = \infty$. Thus, a particle with zero energy (corresponding to zero velocity) is reflected and a particle with infinite energy (infinite velocity) is transmitted. These are the only energy values for which the classical and the quantum cases agree. For all energy values $0 < \varepsilon < \infty$ there is a nonzero reflection probability and a nonzero transmission probability. Also, note that in the classical limit $\hbar \to 0$, which implies $\mu \to 0$, we obtain $R = 1$ and $T = 0$.

Particle Ensembles and Density Matrices

The motion of a particle ensemble consisting of M electrons is described by the Schrödinger equation (1.4.1) with the M-body Hamiltonian

$$H = -\frac{\hbar^2}{2m}\sum_{l=1}^{M} \Delta_{x_l} - qV(x_1, \ldots, x_M, t), \tag{1.4.23}$$

where $x_i \in \mathbb{R}_x^3$ denotes the position vector of the i-th electron.

As in the classical case we shall in the sequel denote the position vector of the ensemble by $x = (x_1, \ldots, x_M) \in \mathbb{R}_x^{3M}$. Then the quantum probability ensemble position density n_{quan} and the quantum ensemble current density J_{quan} are defined as in (1.4.5) and (1.4.11) resp. The conservation law (1.4.13) also holds true for the electron ensemble. In the conservation property (1.4.8) the integration has to be stretched over $\mathbb{R}_x^{3M}$.

For future reference we introduce the density matrix ρ, corresponding to the wave function ψ of the M-electron ensemble, which is defined by

$$\rho(r, s, t) = \overline{\psi(r, t)}\psi(s, t), \qquad r, s \in \mathbb{R}_x^{3M}. \tag{1.4.24}$$

The diagonal elements represent the ensemble position density

$$\rho(x, x, t) = n_{\text{quan}}(x, t) \tag{1.4.25}$$

and the ensemble current density is given by

$$J_{\text{quan}}(x, t) = \frac{i\hbar q}{2m}(\text{grad}_s - \text{grad}_r)\rho(\cdot, \cdot, t)|_{r=s=x}. \tag{1.4.26}$$

Differentiating (1.4.24) with respect to t and using the Schrödinger equation (1.4.1) gives the evolution equation for the density matrix ρ:

$$i\hbar\partial_t\rho = (H_s - H_r)\rho, \tag{1.4.27}$$

where H_s, H_r stand for the Hamiltonian acting on the s and, resp., r variable. With the M-body Hamiltonian (1.4.23) the equation (1.4.27) reads explicitly

$$i\hbar\partial_t\rho = -\frac{\hbar^2}{2m}(\Delta_s\rho - \Delta_r\rho) - q(V(s,t) - V(r,t))\rho. \tag{1.4.28}$$

This equation is the Heisenberg equation of motion.
We refer those readers, who have a deeper interest in quantum mechanics to the textbooks [1.25], [1.34]. For the mathematically oriented reader interested in analytical results on the Schrödinger equation (which has been the subject for an intensive mathematical scrutiny), we recommend [1.46].

Wigner Functions

We shall now reformulate the quantum equations of motion in kinetic form. Therefore we introduce the change of coordinates

$$r = x + \frac{\hbar}{2m}\eta, \qquad s = x - \frac{\hbar}{2m}\eta \tag{1.4.29}$$

in the density matrix ρ and set

$$u(x,\eta,t) = \rho\left(x + \frac{\hbar}{2m}\eta, x - \frac{\hbar}{2m}\eta, t\right). \tag{1.4.30}$$

In the sequel we shall often consider Fourier transforms of functions which depend on η. Since $(\hbar/2m)\eta$ has the dimension of length, we conclude that η has the dimension of inverse velocity, and, thus, the dual variable of η has the dimension of velocity. Therefore, we denote it by v and define the Fourier transform

$$\mathscr{F}g(\eta) := \int_{\mathbb{R}_v^{3M}} g(v)e^{-iv\cdot\eta}\,dv \tag{1.4.31}$$

of a function $g = g(v)$, $g\colon \mathbb{R}_v^{3M} \to \mathbb{C}$. The inverse Fourier transform of a function $h = h(\eta)$, $h\colon \mathbb{R}_\eta^{3M} \to \mathbb{C}$ reads

$$\mathscr{F}^{-1}h(v) := \frac{1}{(2\pi)^{3M}}\int_{\mathbb{R}_\eta^{3M}} h(\eta)e^{iv\cdot\eta}\,d\eta. \tag{1.4.32}$$

The Wigner function w, which corresponds to the wave function ψ (or, equivalently, to the density matrix ρ given by (1.4.24)) is defined as the inverse Fourier transform of u with respect to η:

$$w := \mathscr{F}^{-1}u \tag{1.4.33}$$

or, explicitly:

$$w(x,v,t) = \frac{1}{(2\pi)^{3M}}\int_{\mathbb{R}_\eta^{3M}} \rho\left(x + \frac{\hbar}{2m}\eta, x - \frac{\hbar}{2m}\eta, t\right)e^{i\eta\cdot v}\,d\eta. \tag{1.4.34}$$

It was introduced by E. Wigner in 1932 (see [1.63]) and, as we shall see below, its construction constitutes a major break-through in the quest for a kinetic formulation of quantum transport.

Using (1.4.25) we immediately derive that the mean value of the Wigner function w with respect to the velocity v is the quantum electron ensemble position density

$$n_{\text{quan}}(x, t) = \int_{\mathbb{R}_v^{3M}} w(x, v, t)\, dv, \tag{1.4.35}$$

since $u(x, \eta = 0, t) = \mathscr{F} w(x, \eta = 0, t) = \rho(x, x, t)$ holds. Also, we obtain from (1.4.26), (1.4.29):

$$J_{\text{quan}} = -q \operatorname{grad}_\eta \rho|_{\eta=0}\,. \tag{1.4.36}$$

By taking the gradient of (1.4.31) with respect on η we conclude $\operatorname{grad}_\eta \mathscr{F} g(\eta) = -i\mathscr{F}(vg)(\eta)$. Thus,

$$J_{\text{quan}}(x, t) = -q \int_{\mathbb{R}_v^{3M}} vw(x, v, t)\, dv \tag{1.4.37}$$

follows. The first order moment of the Wigner function w with respect to the velocity v, multiplied by $-q$, is the quantum current density of the electron ensemble. Thus, as far as the zeroth and first order moments are concerned, the Wigner function behaves as the classical particle distribution. However, as will be demonstrated later on, the Wigner function does not necessarily stay nonnegative in its evolution process. Unlike in the classical case, it can therefore not be interpreted as a probability density. In the literature it is often referred to as 'quasi-distribution' of particles. For precisely this reason Wigner functions were not employed for practical simulations until recently, when they were rediscovered as the maybe only quantum transport model for semiconductors which is accessible to numerical simulations. On a theoretical physics level, however, Wigner functions have been intensively scrutinized (see [1.58], [1.12], [1.20]).

The Quantum Transport Equation

The evolution equation for the Wigner functions is obtained by transforming the Heisenberg equation (1.4.28) for the density matrix ρ to the (x, η)-coordinates given by (1.4.29):

$$\partial_t u + i \operatorname{div}_\eta(\operatorname{grad}_x u) + iq \frac{V\left(x + \dfrac{\hbar}{2m}\eta, t\right) - V\left(x - \dfrac{\hbar}{2m}\eta, t\right)}{\hbar} u = 0 \tag{1.4.38}$$

and by Fourier transformation

$$\partial_t w + v \cdot \operatorname{grad}_x w + \frac{q}{m}\theta_\hbar[V]w = 0,$$

$$x \in \mathbb{R}_x^{3M}, \qquad v \in \mathbb{R}_v^{3M}, \qquad t > 0. \tag{1.4.39}$$

The operator $\theta_\hbar[V]$ is defined by

$$(\mathscr{F}\theta_\hbar[V]w)(x,\eta,t) = im\frac{V\left(x+\frac{\hbar}{2m}\eta,t\right)-V\left(x-\frac{\hbar}{2m}\eta,t\right)}{\hbar}(\mathscr{F}w)(x,\eta,t) \qquad (1.4.40)$$

or, explicitly:

$$(\theta_\hbar[V]w)(x,v,t) = \frac{im}{(2\pi)^{3M}}\int_{\mathbb{R}^{3M}_\eta}\int_{\mathbb{R}^{3M}_{v'}}\frac{V\left(x+\frac{\hbar}{2m}\eta,t\right)-V\left(x-\frac{\hbar}{2m}\eta,t\right)}{\hbar}w(x,v',t) \times e^{i(v-v')\cdot\eta}\,dv'\,d\eta. \qquad (1.4.41)$$

An operator, whose Fourier transform acts as a multiplication operator on the Fourier transform of the function, is called a linear pseudo-differential operator and the multiplicator is called the symbol of the pseudo-differential operator. For the mathematical analysis of this type of operators we refer the reader to [1.52], [1.59], [1.60], [1.61].
Thus, $\theta_\hbar[V]$ is a pseudo-differential operator with the symbol

$$(\delta V)_\hbar(x,\eta,t) := im\frac{V\left(x+\frac{\hbar}{2m}\eta,t\right)-V\left(x-\frac{\hbar}{2m}\eta,t\right)}{\hbar} \qquad (1.4.42)$$

and the quantum Liouville equation (1.4.39) is a linear pseudo-differential equation.
The local term $\partial_t w + v\cdot\operatorname{grad}_x w$ describes the motion of free electrons just as in the classical case, the nonlocal term $(q/m)\,\theta_\hbar[V]w$, which generally couples all velocities and frequencies, models the acceleration by the field

$$E(x,t) = -\operatorname{grad}_x V(x,t). \qquad (1.4.43)$$

It is the quantum analogue of the term $q/m \operatorname{grad}_x V\cdot\operatorname{grad}_v f$, which appears in the classical Liouville equation (1.2.12). Formally, the symbol satisfies

$$(\delta V)_\hbar \overset{\hbar\to 0}{\longrightarrow} i\operatorname{grad}_x V\cdot\eta \qquad (1.4.44)$$

and in the formal limit $\hbar\to 0$ the equation (1.4.38) reduces to

$$\partial_t u + i\operatorname{div}_\eta(\operatorname{grad}_x u) + i\frac{q}{m}\operatorname{grad}_x V\cdot\eta u = 0, \qquad (1.4.45)$$

which is the Fourier transformed Liouville equation

$$\partial_t w + v\cdot\operatorname{grad}_x w + \frac{q}{m}\operatorname{grad}_x V\cdot\operatorname{grad}_v w = 0 \qquad (1.4.46)$$

since $\mathscr{F}^{-1}(i\eta u) = \operatorname{grad}_v(\mathscr{F}^{-1}u)$.

Thus, the quantum Liouville equation becomes the classical Liouville equation when the so-called classical limit $\hbar \to 0$ is carried out formally. Later on we shall make this statement mathematically precise.
By the above derivation the quantum Liouville equation follows directly from the Schrödinger equation. Many-body physics enters through the number of coordinates ($3M$ position and $3M$ velocity coordinates for an M-electron ensemble) and through the form of the many-body potential V. Thus, the quantum Liouville equation is by no means simpler than the many-body Schrödinger equation, actually, the number of coordinates doubled. As we shall see in the next Section, its advantage is the kinetic form, which is accessible to a one-body approximation in which many-body physics only enters through an averaged potential. Also, the kinetic equation allows a formulation on bounded position domains, subject to (more or less) physically reasonable boundary conditions. This is of particular importance for the numerical simulation of semiconductor devices.
Very often, the following generic notation for pseudo-differential operators is used: For the operator

$$(Ag)(v) = \frac{1}{(2\pi)^{3M}} \int_{\mathbb{R}^{3M}_\eta} \int_{\mathbb{R}^{3M}_{v'}} a(\eta) g(v') e^{i(v-v')\cdot\eta} \, dv' \, d\eta \tag{1.4.47}$$

with the symbol $a = a(\eta)$, one writes

$$a\left(\frac{1}{i}\operatorname{grad}_v\right) g := Ag. \tag{1.4.48}$$

Then, the convection operator $\theta_\hbar[V]$ can be expressed as

$$\begin{aligned} \theta_\hbar[V] &= im \frac{V\left(x + \frac{\hbar \operatorname{grad}_v}{2mi}, t\right) - V\left(x - \frac{\hbar \operatorname{grad}_v}{2mi}, t\right)}{\hbar} \\ &= (\delta V)_\hbar\left(x, \frac{1}{i}\operatorname{grad}_v, t\right) \end{aligned} \tag{1.4.49}$$

and the quantum Liouville equation (1.4.39) takes the form

$$\partial_t w + v \cdot \operatorname{grad}_x w + iq \frac{V\left(x + \frac{\hbar \operatorname{grad}_v}{2mi}, t\right) - V\left(x - \frac{\hbar \operatorname{grad}_v}{2mi}, t\right)}{\hbar} w = 0. \tag{1.4.50}$$

Pure and Mixed States

We consider the whole space problem for the $3M$-dimensional quantum Liouville equation (1.4.50) subject to the initial condition

$$w(x, v, t = 0) = w_I(x, v), \qquad x \in \mathbb{R}^{3M}_x, \qquad v \in \mathbb{R}^{3M}_v. \tag{1.4.51}$$

The solution of this initial value problem is the Wigner function

$$w(x, v, t) = \frac{1}{(2\pi)^{3M}} \int_{\mathbb{R}_\eta^{3M}} \overline{\psi\left(x + \frac{\hbar}{2m}\eta, t\right)} \psi\left(x - \frac{\hbar}{2m}\eta, t\right) e^{iv\cdot\eta}\, d\eta \tag{1.4.52}$$

for all times $t \geqslant 0$ *if and only if* the initial datum w_I satisfies

$$w_I(x, v) = \frac{1}{(2\pi)^{3M}} \int_{\mathbb{R}_\eta^{3M}} \overline{\psi_I\left(x + \frac{\hbar}{2m}\eta\right)} \psi_I\left(x - \frac{\hbar}{2m}\eta\right) e^{iv\cdot\eta}\, d\eta \tag{1.4.53}$$

for some function $\psi_I = \psi_I(x)$, which is the initial wave function of the state ψ. (1.4.53) is equivalent to the following conditions on the initial density matrix:

$$\text{(a)} \quad \frac{\partial^2}{\partial r \partial s} \ln \rho_I(r, s) = 0, \qquad \text{(b)} \quad \rho_I(r, s) = \overline{\rho_I(s, r)}, \tag{1.4.54}$$

where we set

$$\rho_I(r, s) := \mathscr{F} w_I(x, \eta) \tag{1.4.55}$$

evoking the coordinate transformation (1.4.29). Note that (1.4.54) (b) is equivalent to w_I being real valued.

If (1.4.52) holds, then the quantum state of the electron is fully described by the single wave function $\psi = \psi(x, t)$. In quantum physics this is referred to as a pure quantum state.

Clearly, for initial data w_I, which do not satisfy (1.4.54), the solution w of the quantum Liouville equation is not of the pure state form (1.4.52), and, thus, a more general solution representation has to be sought.

Let $\psi^{(1)} = \psi^{(1)}(x, t)$, $\psi^{(2)} = \psi^{(2)}(x, t)$ be two solutions of the Schrödinger equation. By a simple computation it is immediately verified that the product $\overline{\psi^{(1)}(r, t)}\psi^{(2)}(s, t)$ solves the Heisenberg equation (1.4.27) and by linearity we conclude that all linear combinations of such products of the form

$$\rho(r, s, t) := \sum_{l,j} \rho_{lj} \overline{\psi^{(l)}(r, t)} \psi^{(j)}(s, t) \tag{1.4.56}$$

are solutions of (1.4.27), too. A solution of the quantum Liouville equation is then obtained by setting $r = x + (\hbar/2m)\eta$, $s = x - (\hbar/2m)\eta$ and by inverse Fourier transformation. To solve the initial value problem (1.4.50), (1.4.51), the coefficients ρ_{lj} and the wave functions $\psi^{(m)}$ at $t = 0$ have to be adapted to the initial function w_I. We must require

$$\rho_I(r, s) = \sum_{l,j} \rho_{l,j} \overline{\psi_I^{(l)}(r)} \psi_I^{(j)}(s). \tag{1.4.57}$$

This gives a clear indication on how an L^2-theory for the quantum Liouville equation should be set up. For the following we assume

$$w_I \in L^2(\mathbb{R}_x^{3M} \times \mathbb{R}_v^{3M}) \tag{1.4.58}$$

and choose a complete orthonormed system of $L^2(\mathbb{R}^{3M})$-functions $\{\psi_I^{(l)}\}_{l\in\mathbb{N}}$. Then $\{\psi_I^{(l)}(r)\psi_I^{(j)}(s)\}_{l,j\in\mathbb{N}}$ is a complete orthonormed system in $L^2(\mathbb{R}_r^{3M}\times\mathbb{R}_s^{3M})$ (see, e.g., [1.46]). We compute the initial density matrix ρ_I from the initial datum w_I by using (1.4.55) and expand ρ_I into the Fourier series (1.4.57). We obtain the Fourier coefficients ρ_{lj}:

$$\rho_{lj} = \int_{\mathbb{R}_r^{3M}\times\mathbb{R}_s^{3M}} \rho_I(r,s)\psi_I^{(l)}(r)\overline{\psi_I^{(j)}(s)}\,dr\,ds. \tag{1.4.59}$$

The next step is to solve the Schrödinger equations

$$i\hbar\partial_t\psi = -\frac{\hbar^2}{2m}\Delta\psi - qV(x,t)\psi, \qquad x\in\mathbb{R}^{3M}, \qquad t>0 \tag{1.4.60}$$

$$\psi(x,t=0) = \psi_I^{(l)}(x), \qquad x\in\mathbb{R}^{3M} \tag{1.4.61}$$

for $\psi = \psi^{(l)}(x,t)$ and $l\in\mathbb{N}$. Then the solution of the initial value problem for the quantum Liouville equation is obtained by employing the coordinate transformation (1.4.29) and by Fourier transformation:

$$w(x,v,t) = \frac{1}{(2\pi)^{3M}}\sum_{l,j\in\mathbb{N}}\rho_{lj}\int_{\mathbb{R}_\eta^{3M}}\overline{\psi^{(l)}\left(x+\frac{\hbar}{2m}\eta,t\right)}\,\psi^{(j)}\left(x-\frac{\hbar}{2m}\eta,t\right) \times e^{iv\cdot\eta}\,d\eta. \tag{1.4.62}$$

This solution w exists for all $t\geqslant 0$ (as convergent series in $L^2(\mathbb{R}_x^{3M}\times\mathbb{R}_v^{3M})$) if the Schrödinger equation (1.4.60) has a solution globally in time for all initial data in $L^2(\mathbb{R}^{3M})$. Conditions on the potential V, which guarantee the global existence of L^2 wave-functions for L^2 initial data can be found in [1.46], [1.29].

It is easy to show that the solution w remains real valued for all $t>0$, if it is real valued initially (which we shall assume henceforth). For such initial data a more convenient solution representation can be obtained. By a simple functional analytic argument (see [1.38]) we conclude the existence of a complete orthonormal system of $L^2(\mathbb{R}^{3M})$ functions $\{\phi^{(k)}\}_{k\in\mathbb{N}}$ such that

$$\rho_I(r,s) = \sum_{k\in\mathbb{N}}\lambda_k\overline{\phi^{(k)}(r)}\phi^{(k)}(s) \tag{1.4.63}$$

holds with

$$\lambda_k = \int_{\mathbb{R}_r^{3M}\times\mathbb{R}_s^{3M}}\rho_I(r,s)\phi^{(k)}(r)\overline{\phi^{(k)}(s)}\,dr\,ds. \tag{1.4.64}$$

By proceeding as above we obtain the diagonal representation of the density matrix

$$\rho(r,s,t) = \sum_{k\in\mathbb{N}}\lambda_k\overline{\phi^{(k)}(r,t)}\phi^{(k)}(s,t) \tag{1.4.65}$$

and of the solution of the quantum Liouville equation

$$w(x, v, t) = \frac{1}{(2\pi)^{3M}} \sum_{k \in \mathbb{N}} \lambda_k \int_{\mathbb{R}_\eta^{3M}} \overline{\phi^{(k)}\left(x + \frac{\hbar}{2m}\eta, t\right)} \times \phi^{(k)}\left(x - \frac{\hbar}{2m}\eta, t\right) e^{iv\cdot\eta} d\eta, \quad (1.4.66)$$

where $\phi^{(k)} = \phi^{(k)}(x, t)$ denotes the solution of the Schrödinger equation (1.4.60) with initial datum $\phi^{(k)} = \phi^{(k)}(x)$.

We conclude from (1.4.66) that the general L^2-solution of the initial value problem for the quantum Liouville equation can be written as an infinite sum of Wigner-functions. Thus, the quantum Liouville equation is capable of describing arbitrary *mixed quantum states*, which cannot be represented by a single wave function.

It is an easy exercise to show that initially orthonormal wave functions remain orthonormal for all times. Since the functions $\phi^{(k)}(x)$ are orthonormal in $L^2(\mathbb{R}^{3M})$, the wave functions $\phi^{(k)}(x, t)$ are orthonormal for $t > 0$ and we obtain from Parseval's inequality

$$\|\rho(\cdot, \cdot, t)\|_{L^2(\mathbb{R}_r^{3M} \times \mathbb{R}_s^{3M})} = \|\rho_I\|_{L^2(\mathbb{R}_r^{3M} \times \mathbb{R}_s^{3M})}, \qquad t > 0. \quad (1.4.67)$$

Since for every function $g \in L^2(\mathbb{R}_v^{3M})$

$$\|\mathscr{F} g\|_{L^2(\mathbb{R}_\eta^{3M})} = (2\pi)^{3M/2} \|g\|_{L^2(\mathbb{R}_v^{3M})} \quad (1.4.68)$$

holds (see [1.46]), we conclude from (1.4.66), (1.4.67):

$$\|w(\cdot, \cdot, t)\|_{L^2(\mathbb{R}_x^{3M} \times \mathbb{R}_v^{3M})} = \|w_I\|_{L^2(\mathbb{R}_x^{3M} \times \mathbb{R}_v^{3M})}, \qquad t > 0. \quad (1.4.69)$$

The L^2-norm of the solution of the quantum Liouville equation is time-invariant.

An analysis of the steady states of the quantum Liouville equation, also based on the representation (1.4.62), can be found in [1.18].

We now formally integrate (1.4.66) term by term over $\mathbb{R}_v^{3M}$, use

$$\int_{\mathbb{R}_v^{3M}} g(v)\, dv = (\mathscr{F} g)(\eta = 0)$$

and obtain

$$n_{\text{quan}}(x, t) = \int_{\mathbb{R}_v^{3M}} w(x, v, t)\, dv = \sum_{k \in \mathbb{N}} \lambda_k |\phi^{(k)}(x, t)|^2. \quad (1.4.70)$$

Another formal term by term integration, now over $\mathbb{R}_x^{3M}$, gives

$$\int_{\mathbb{R}_x^{3M}} n_{\text{quan}}(x, t)\, dx = \sum_{k \in \mathbb{N}} \lambda_k = \int_{\mathbb{R}_x^{3M}} n_{\text{quan},I}(x)\, dx. \quad (1.4.71)$$

Here we used $\int_{\mathbb{R}_x^{3M}} |\phi^{(k)}(x, t)|^2\, dx = 1$. (1.4.71) establishes the conservation of the integral of the quantum ensemble position density for mixed states.

We remark that the main ingredient for the existence of n_{quan} and for the mathematical justification of the performed term by term integrations is the

assumption $\lambda_k \geqslant 0$, $\forall\, k \in \mathbb{N}$, which by (1.4.70) guarantees the nonnegativity of the position density n_{quan} (see also [1.38]). We shall come back to this point later on.
If the potential V is continuous in x and if w decays sufficiently fast as $|x| \to \infty$, $|v| \to \infty$, we obtain from (1.4.42):

$$\int_{\mathbb{R}_v^{3M}} \theta_\hbar[V]w\, dv = (\delta V)_\hbar(\mathscr{F}w)(x, \eta = 0, t) = 0. \tag{1.4.72}$$

Thus, integrating the quantum Liouville equation with respect to v gives the conservation law

$$q\partial_t n_{\text{quan}} - \operatorname{div} J_{\text{quan}} = 0 \tag{1.4.73}$$

for the quantum current density.
We refer the mathematically oriented reader to the reference [1.38] for a rigorous presentation of the results of this paragraph based on a functional analytic framework for the Schrödinger and the quantum Liouville equations. A different approach for bounded potentials can be found in [1.39].

The Classical Limit

We consider a quadratic potential of the form

$$V(x, t) = \tfrac{1}{2}x^T A(t)x + b(t)\cdot x + c(t), \tag{1.4.74}$$

where $A(t)$ is a realvalued symmetric $3M \times 3M$-matrix, $b(t)$ a real $3M$-vector and $c(t) \in \mathbb{R}$. The superscript 'T' denotes transposition. Evaluation of the symbol $(\delta V)_\hbar$ defined in (1.4.42) gives

$$(\delta V)_\hbar(x, \eta, t) = i(A(t)x + b(t))\cdot\eta \tag{1.4.75}$$

and the quantum Liouville equation becomes

$$\partial_t w + v\cdot\operatorname{grad}_x w + \frac{q}{m}(A(t)x + b(t))\cdot\operatorname{grad}_v w = 0. \tag{1.4.76}$$

Thus, in the case of a quadratic potential (linear field) the quantum Liouville equation and the classical Liouville equation are identical. Clearly, this does not hold true for more general potentials.
However, since (1.4.44) holds for sufficiently smooth potentials, we are led to the conjecture that limits as $\hbar \to 0$ of solutions of the quantum Liouville equation are solutions of the classical Liouville equation, if the potential V is sufficiently smooth. Results of this type were proven in [1.38], [1.39] by employing functional analytic methods. Here we shall present a more basic technique based on asymptotic expansions. We shall proceed somewhat formally, but a justification of the expansion method is possible even without much mathematical sophistication.
For the sake of simplicity we consider the one-dimensional motion of an electron in a static potential $V = V(x)$, which is assumed to be infinitely

differentiable. Then, by formal power series expansion, we obtain

$$(\delta V)_\hbar(x,\eta) \sim i \sum_{k=0}^{\infty} \frac{\eta^{2k+1}}{4^k(2k+1)!} \frac{d^{2k+1}V(x)}{dx^{2k+1}} \mu^{2k}, \tag{1.4.77}$$

where we set $\mu = \hbar/m$. We are therefore motivated to expand the solution $w = w^\hbar$ in powers of μ^2, too. We make the ansatz

$$u^\hbar(x,\eta,t) \sim \sum_{k=0}^{\infty} u_{2k}(x,\eta,t)\mu^{2k} \tag{1.4.78}$$

for the Fourier transform $u^\hbar = \mathscr{F}w^\hbar$. The coefficients u_{2k} are as yet unknown. In order to keep matters as simple as possible we assume that the initial datum w_I and, consequently, $u_I = \mathscr{F}w_I$ are independent of $\hbar$.
We insert the expansions (1.4.77), (1.4.78) into the Fourier transformed quantum Liouville equation (1.4.38) and obtain by equating coefficients of equal powers of μ^2:

$$\begin{aligned} &\partial_t u_0 + i\frac{\partial u_0}{\partial x\, \partial \eta} + i\frac{q}{m}\frac{dV}{dx}(x)\eta u_0 = 0, \\ &u_0(x,\eta,t=0) = u_I(x,\eta) \end{aligned} \tag{1.4.79}$$

for $k = 0$ and

$$\begin{aligned} &\partial_t u_{2k} + i\frac{\partial u_{2k}}{\partial x\, \partial \eta} + i\frac{q}{m}\frac{dV}{dx}(x)\eta u_{2k} \\ &\quad = \frac{q}{m}\sum_{l=1}^{k} \frac{\eta^{2l+1}}{4^l(2l+1)!}\frac{d^{2l+1}V(x)}{dx^{2l+1}} u_{2k-2l}, \\ &u_{2k}(x,\eta,t=0) = 0 \end{aligned} \tag{1.4.80}$$

for $k > 0$. We set $w_{2k} = \mathscr{F}^{-1}u_{2k}$ and obtain equations for the coefficients w_{2k} of the expansion

$$w^\hbar(x,v,t) \sim \sum_{k=0}^{\infty} w_{2k}(x,v,t)\mu^{2k} \tag{1.4.81}$$

by inverse Fourier transformation of (1.4.79), (1.4.80). The leading term w_0 satisfies the classical Liouville equation

$$\begin{aligned} &\partial_t w_0 + v\partial_x w_0 + \frac{q}{m}\frac{dV(x)}{dx}\partial_v w_0 = 0, \\ &w_0(x,v,t=0) = w_I(x,v) \end{aligned} \tag{1.4.82}$$

and the higher order coefficients solve inhomogeneous versions of the classical equation:

$$\begin{aligned} &\partial_t w_{2k} + v\partial_x w_{2k} + \frac{q}{m}\frac{dV(x)}{dx}\partial_v w_{2k} = R_{2k}, \\ &w_{2k}(x,v,t=0) = 0, \end{aligned} \tag{1.4.83}$$

where the right-hand side R_{2k} depends on v-derivatives of $w_0, \ldots, w_{2k-2}$ and on x-derivatives of V.

The presented expansion procedure is only formal. It was shown in [1.53] that approximations of arbitrary high order, say $O(\mu^{2r})$ for $r \in \mathbb{N}$, are obtained by cutting the expansion (1.4.81) at the index $k = r - 1$, if the potential and the initial datum are sufficiently smooth. This approximation result can easily be extended to more dimensions. For weaker convergence results (under less stringent regularity assumptions) we refer to [1.38], [1.39].

So far, the theory for the classical limit of the quantum Liouville equation in the case of nonsmooth potentials is not well developed. However, an asymptotic analysis for a highly irregular potential, namely the one-dimensional barrier $V(x) = -(m/q)\delta(x)$ discussed in the Paragraph on tunneling, was presented in [1.53]. It is shown that the solutions of the corresponding quantum Liouville equation tend to the classical limit (total reflection, see Section 1.2). The quantum corrections, e.g. the tunneling current, are of order $\hbar^2$.

To get a feeling for the 'actual size' of $\hbar$, the quantum Liouville equation has to be scaled appropriately (see Problem 1.29 and Section 1.6). Then it becomes apparent that the 'scaled Planck constant' is indirectly proportional to the square of the characteristic device length. Quantum effects become more pronounced as the active device length decreases.

Nonnegativity of Wigner Functions

At first we consider a pure quantum state with wave function $\psi(\cdot, t) \in L^2(\mathbb{R}_x^{3M})$ and Wigner function $w = w_\psi$ given by

$$w_\psi(x, v, t) = \frac{1}{(2\pi)^{3M}} \int_{\mathbb{R}_\eta^{3M}} \overline{\psi\left(x + \frac{\hbar}{2m}\eta, t\right)} \psi\left(x - \frac{\hbar}{2m}\eta, t\right) e^{iv\cdot\eta}\, d\eta. \tag{1.4.84}$$

A well-known result (see [1.27]) asserts that w_ψ is nonnegative everywhere if and only if either $\psi \equiv 0$ or if ψ is the exponential of a quadratic in x, i.e.

$$\psi(x, t) = \exp\left(-\frac{1}{2}x^T\Lambda(t)x - \alpha(t)\cdot x + \beta(t)\right), \quad x \in \mathbb{R}^{3M}, \quad t \geqslant 0, \tag{1.4.85}$$

where $\Lambda(t)$ is a complex $3M \times 3M$-matrix with a positive definite symmetric real part, $\alpha(t)$ is an arbitrary complex $3M$-vector and $\beta(t) \in \mathbb{C}$. Since the proof of this result is instructive we shall present it here for the one-dimensional case.

Theorem 1.4.1: *Let*

$$w_\psi(x, v) = \frac{1}{2\pi} \int_{\mathbb{R}_\eta} \overline{\psi\left(x + \frac{\hbar}{2m}\eta\right)} \psi\left(x - \frac{\hbar}{2m}\eta\right) e^{i\eta v}\, d\eta, \quad x \in \mathbb{R}_x, \quad v \in \mathbb{R}_v \tag{1.4.86}$$

be the Wigner function of the state $\psi \in L^2(\mathbb{R})$. *Then*

$$w_\psi(x, v) \geqslant 0, \qquad x \in \mathbb{R}_x, \qquad v \in \mathbb{R}_v \tag{1.4.87}$$

holds if and only if either there are complex coefficients λ, α, γ *with* $\operatorname{Re} \lambda > 0$ *such that* ψ *is given by*

$$\text{(a)} \quad \psi(x) = \exp\left(-\frac{\lambda}{2}x^2 - \alpha x + \gamma\right), \qquad x \in \mathbb{R} \tag{1.4.88}$$

or

$$\text{(b)} \quad \psi \equiv 0. \tag{1.4.88}$$

Proof: At first note that $\psi \equiv 0$ if and only if $w_\psi \equiv 0$. Now let $\psi = \psi_{\lambda,\alpha,\gamma}$ be given by (1.4.88) (a). Using the well-known formula

$$\int_{\mathbb{R}} e^{-(\lambda/2)x^2 + zx}\, dx = \sqrt{\frac{2\pi}{\lambda}}\, e^{z^2/2\lambda}, \quad \operatorname{Re} \lambda > 0, \quad \operatorname{Re} \sqrt{\frac{2\pi}{\lambda}} > 0 \tag{1.4.89}$$

for $z \in \mathbb{C}$, we can easily compute the Wigner function $w_\psi =: w_{\lambda,\alpha,\gamma}$ from (1.4.86). It is of the form

$$w_{\lambda,\alpha,\gamma}(x, v) = e^{p_2(x,v)} > 0, \tag{1.4.90}$$

where p_2 is a real polynomial of degree two in x and v.
To establish the necessity of (1.4.88) we assume $w_\psi \geqslant 0$, $\psi \neq 0$. A simple argument shows that

$$\left| \int_{\mathbb{R}_x} \int_{\mathbb{R}_v} w_\psi(x, v) w_{1,z,0}(x, v)\, dx\, dv \right| = \frac{1}{2\pi} \left| \int_{\mathbb{R}} \overline{\psi(x)} \psi_{1,z,0}(x)\, dx \right| \tag{1.4.91}$$

holds for $z \in \mathbb{C}$. Since $w_\psi \geqslant 0$, $w_\psi \not\equiv 0$ and since $w_{1,z,0}$ is of the form (1.4.90), the left hand side of (1.4.91) is positive. Thus, the right hand side is nonzero for $z \in \mathbb{C}$ and, consequently, the entire function

$$F(z) = \int_{\mathbb{R}} \bar{\psi}(x) e^{-(x^2/2) - zx}\, dx \tag{1.4.92}$$

has no zeros in $\mathbb{C}$. We estimate (1.4.92) using (1.4.89):

$$|F(z)|^2 \leqslant \|\psi\|^2_{L^2(\mathbb{R})} \sqrt{\pi}\, e^{(\operatorname{Re} z)^2} \tag{1.4.93}$$

and conclude from Hadamard's Theorem (see, e.g., [1.50]) that F is of the form

$$F(z) = e^{az^2 + bz + c}. \tag{1.4.94}$$

We set $z = -iy$ and obtain from (1.4.92) that $F(iy) = e^{-ay^2 - iby + c}$ is the Fourier transform of $\bar{\psi}(x) e^{-x^2/2}$. Since the only class of functions whose Fourier transforms are exponentials of quadratic polynomials are of that type themselves (see, e.g. [1.46]), we conclude (1.4.88) (a). □

The class of potentials V, which generate wave functions of the type (1.4.85) can easily be determined by inserting (1.4.85) into the Schrödinger equation. A simple computation shows that $V(x, t)$ is quadratic in x, i.e. it is of the form (1.4.74). As we already know, the quantum Liouville equation reduces to the classical Liouville equation for such potentials and the preservation of the nonnegativity for arbitrary initial data is immediate. The more interesting—and quite striking—part of the result, however, is the necessity of (1.4.88) and, consequently, of the class of quadratic potentials for the nonnegativity of the Wigner function of a pure state.

The situation for mixed quantum states, i.e. for arbitrary initial data $w_I \in L^2(\mathbb{R}_x^{3M} \times \mathbb{R}_v^{3M})$ for the quantum Liouville equation is more complicated and a necessary condition for the nonnegativity of the solution w has not been obtained yet.

A sufficient condition for the nonnegativity of the electron position density n_{quan}, however, can easily be obtained from the representations (1.4.66) and (1.4.70). At first we remark that the coefficients λ_k are the eigenvalues of the operator R_I: $L^2(\mathbb{R}^{3M}) \to L^2(\mathbb{R}^{3M})$ defined by

$$(R_I f)(r) := \int_{\mathbb{R}^{3M}} \rho_I(r, s) f(s)\, ds \tag{1.4.95}$$

(see [1.38]). The function ρ_I, which is the initial density matrix of the mixed state, is called positive semi-definite, if the integral operator R_I is positive semi-definite, i.e. if $\lambda_k \geqslant 0$, $\forall\, k \in \mathbb{N}$.

By (1.4.70) the positive semi-definiteness of ρ_I is sufficient to guarantee the nonnegativity of the electron density n_{quan} for $x \in \mathbb{R}_x^{3M}$, $t \geqslant 0$.

For potentials more general than the quadratic (1.4.74) we cannot expect an initially nonnegative solution of the quantum Liouville equation to remain nonnegative for all time, $t \geqslant 0$. In spite of the nonnegativity of the electron position density for positive semi-definite initial density matrices a fully probabilistic interpretation of the quantum Liouville equation is therefore not possible. This fact was quite a deterrent for the practical use of Wigner functions. Only recently they were employed for semiconductor device simulations out of sheer need of a quantum transport model, which is simple enough to allow a reasonably efficient numerical solution. The results obtained are very convincing (see [1.49], [1.32]) and further research on the quantum Liouville equation is ongoing.

An Energy-Band Version of the Quantum Liouville Equation

So far the presented quantum transport model does not account for the effect of the crystal lattice on the motion of the electrons. For semiconductor simulation it is desirable to employ a transport model, which is capable of describing quantum effects like tunneling and which also contains a description of the crystal lattice structure like the semi-classical Liouville equation of Section 1.2. The model presented below was introduced in [1.1].

We consider a single electron in the (fixed) energy band $\varepsilon = \varepsilon(k)$, defined for k in the first Brillouin zone B of the semiconductor (and extended periodically to $\mathbb{R}_k^3$). Then the semi-classical Hamiltonian is given by

$$H(x, k, t) = \varepsilon(k) - qV(x, t). \tag{1.4.96}$$

Note that the position vector x and the crystal momentum vector $p = \hbar k$ are canonically conjugate.
By the correspondence principle of quantum mechanics (see, e.g., [1.40]) the quantum mechanical Hamiltonian operator in wave vector formulation is obtained by substituting the position operator $i\,\text{grad}_k$ for the position variable x in the Hamiltonian function (1.4.96). The corresponding Schrödinger equation (in wave vector formulation) then reads formally:

$$i\hbar\, \partial_t \hat{\psi} = [\varepsilon(k) - qV(i\,\text{grad}_k)]\hat{\psi}, \qquad \hat{\psi} = \hat{\psi}(k, t). \tag{1.4.97}$$

The pseudo-differential operator $V(i\,\text{grad}_k)$ will be defined below. We assume that the wave function $\hat{\psi}$ is periodic in k with the periodicity of the reciprocal crystal lattice $\hat{L}$. Clearly, the reason for this assumption is the periodicity of the energy band $\varepsilon = \varepsilon(k)$.
Therefore, we can expand $\hat{\psi}$ into a Fourier series and, hence, its Fourier transform is a discretely defined function on the direct lattice L:

$$\hat{\psi}(k, t) = \sum_{x \in L} \psi(x, t) e^{-ik\cdot x}, \qquad k \in B \tag{1.4.98}$$

$$\psi(x, t) = \frac{1}{|B|} \int_B \hat{\psi}(k, t) e^{ik\cdot x}\, dk, \qquad x \in L. \tag{1.4.99}$$

Here, $|B|$ denotes the volume of the Brillouin zone.
We now define the potential operator $V(i\,\text{grad}_k)$ by

$$(V(i\,\text{grad}_k)\hat{f})(k) = \frac{1}{|B|} \int_B \sum_{x \in L} V(x) \hat{f}(k') e^{ix\cdot(k-k')}\, dk' \tag{1.4.100}$$

for $\hat{f} = \hat{f}(k)$, $k \in B$.
The Schrödinger equation in space representation is obtained by Fourier transforming (1.4.97) using (1.4.98), (1.4.99):

$$ih\, \partial_t \psi = [\varepsilon(-i\,\text{grad}_x) - qV(x, t)]\psi, \qquad \psi = \psi(x, t), \tag{1.4.101}$$

where we denoted

$$(\varepsilon(-i\,\text{grad}_x) f)(x) = \frac{1}{|B|} \int_B \sum_{x \in L} \varepsilon(k) f(x') e^{ik(x-x')}\, dk \tag{1.4.102}$$

for $f = f(x)$, $x \in L$.
In order to define the corresponding Wigner function we set up the density matrix

$$\rho(r, s, t) = \overline{\psi(r, t)}\psi(s, t), \qquad (r, s) \in L \times L, \tag{1.4.103}$$

which satisfies the Heisenberg equation

$$i\hbar\,\partial_t\rho = [\varepsilon(-i\,\text{grad}_s) - \varepsilon(-i\,\text{grad}_r) - qV(s,t) + qV(r,t)]\rho, \qquad (r,s) \in L \times L. \tag{1.4.104}$$

Similarly to the vacuum case we introduce the coordinate transformation

$$r = x + \frac{v}{2}, \qquad s = x - \frac{v}{2} \tag{1.4.105}$$

and define the function

$$u(x, v, t) := \rho\left(x + \frac{v}{2}, x - \frac{v}{2}, t\right) \tag{1.4.106}$$

for $x \in \frac{1}{2}L$ and for all v which can be represented as a difference of two points in L. Since the grid, on which u lives, is not rectangular in the (x, v)-space, it is convenient to introduce additional gridpoints such that $(x, v) \in \frac{1}{2}L \times L$ and to set $u = 0$ on these (artificially introduced) points initially. We remark that this only simplifies the notation. It has no effect on the solution since u remains zero on the additional gridpoints for all times if it is zero there initially.

The Wigner function is then defined as the Fourier transform of u with respect to v:

$$w(x, k, t) = \sum_{v \in L} u(x, v)e^{-ik\cdot v}, \qquad x \in \frac{1}{2}L, \qquad k \in B. \tag{1.4.107}$$

It satisfies the following pseudo-differential equation

$$\partial_t w + \frac{i}{\hbar}\left[\varepsilon\left(k + \frac{1}{2i}\,\text{grad}_x\right) - \varepsilon\left(k - \frac{1}{2i}\,\text{grad}_x\right) + qV\left(x + \frac{1}{2i}\,\text{grad}_k\right) - qV\left(x - \frac{1}{2i}\,\text{grad}_k\right)\right]w = 0, \quad x \in \frac{1}{2}L, \quad k \in B, \tag{1.4.108}$$

which, in explicit notation, reads:

$$\partial_t w + \frac{1}{i\hbar}\left[\frac{1}{2|B|}\int_{2B}\sum_{x' \in L/2}\left(\varepsilon\left(k + \frac{\mu}{2}\right) - \varepsilon\left(k - \frac{\mu}{2}\right)\right) \times e^{i\mu(x-x')}w(x', k)\,d\mu + \frac{q}{|B|}\sum_{v \in L}\int_B\left(V\left(x + \frac{v}{2}\right) - V\left(x - \frac{v}{2}\right)\right) \times e^{iv\cdot(k-k')}w(x, k')\,dk'\right] = 0. \tag{1.4.109}$$

The solution w of (1.4.109) subject to an initial condition $w(t = 0) = w_I$ is the quasi-distribution of the electron in the phase space $\frac{1}{2}L \times B$. As usual, the electron position density is given by

$$n(x, t) = \int_B w(x, k, t)\,dk, \qquad x \in \frac{1}{2}L \tag{1.4.110}$$

and the electron current density

$$J(x, t) = -q \int_B v(k) w(x, k, t)\, dk. \tag{1.4.111}$$

The band-diagram quantum Liouville equation (1.4.108) has two properties, which are different from the vacuum quantum Liouville equation (1.4.50). Firstly, the admissible values of the position vector x are discrete (more precisely, they are restricted to $\frac{1}{2}$ times the crystal lattice L) and the k-vector is restricted to the first Brillouin zone B of the crystal. Secondly, the equation (1.4.108) is nonlocal in the position vector x as well as in the wave vector k. The nonlocality in x is introduced by the band-diagram $\varepsilon = \varepsilon(k)$.
We remark that the derivation of the band diagram quantum Liouville equation is based on the Hamiltonian operator

$$H_B = \varepsilon(-i \operatorname{grad}_x) - qV(x, t),$$

obtained formally from the semi-classical Hamiltonian function $\varepsilon(k) - qV(x, t)$ by employing the correspondence principle. It can be shown that H_B is an approximation for the 'full' quantum Hamiltonian $H_L - qV$, where H_L, given by (1.2.35), represents the quantum effects of the periodic lattice potential. The approximation quality deteriorates if the motion of the electron is not confined to the given band ε ([1.56]).
Obviously, it does not make sense to carry out the classical limit $\hbar \to 0$ in (1.4.108). In order to analyze the relationship of the full quantum band diagram transport model (1.4.108) to the semi-classical single electron Liouville equation

$$\partial_t f + \frac{1}{\hbar} \operatorname{grad}_k \varepsilon(k) \cdot \operatorname{grad}_x f + \frac{q}{\hbar} \operatorname{grad}_x V \cdot \operatorname{grad}_k f = 0, \qquad x \in \mathbb{R}_x^3, \quad k \in B, \tag{1.4.112}$$

which is subject to periodic boundary conditions on ∂B, both equations have to be appropriately scaled first. Following [1.1] we introduce the scaling

$$x_s = \frac{x}{x_0}, \qquad k_s = lk, \qquad t_s = \frac{t}{t_0}, \qquad L_s = \frac{1}{x_0} L, \qquad B_s = lB$$
$$\varepsilon_s(k_s) = \frac{1}{\varepsilon_0} \varepsilon(k), \qquad V_s(x_s) = \frac{1}{V_0} V(x) \tag{1.4.113}$$

where we denoted scaled quantities by the index 's'. x_0 is the length of the active region of the considered semiconductor device and l is a characteristic lattice spacing of L (for the sake of simplicity we assume that the grid-point distances of L in all three directions are of equal order of magnitude, otherwise we would have to scale the three components of k separately). Note that the volume of the Brillouin zone is of the order of magnitude $1/l^3$. The scaling factors t_0, V_0, ε_0 stand for a characteristic observation time, characteristic potential value and maximal value of ε on B resp.

With this scaling the equation (1.4.108) reads (after dropping the index 's'):

$$w_t + \frac{i}{\alpha}\left[a\varepsilon\left(k + \frac{\alpha}{2i}\,\mathrm{grad}_x\right) - a\varepsilon\left(k - \frac{\alpha}{2i}\,\mathrm{grad}_x\right)\right. \tag{1.4.114}$$

$$\left. + bV\left(x + \frac{\alpha}{2i}\,\mathrm{grad}_k\right) - bV\left(x - \frac{\alpha}{2i}\,\mathrm{grad}_x\right)\right]w = 0,$$

where the dimensionless constants are given by

$$\alpha = \frac{l}{x_0}, \qquad a = \frac{\varepsilon_0 t_0 l}{\hbar x_0}, \qquad b = \frac{qV_0 t_0 l}{\hbar x_0}. \tag{1.4.115}$$

For comparison, the scaled semi-classical Liouville equation (1.4.112) is of the form

$$f_t + a\,\mathrm{grad}_k\,\varepsilon(k)\cdot\mathrm{grad}_x\,f + b\,\mathrm{grad}_x\,V\cdot\mathrm{grad}_k\,f = 0. \tag{1.4.116}$$

For a typical tunneling diode the following scaling factors may be chosen

$$x_0 = 10^{-8}\,\mathrm{m}, \qquad l = 10^{-10}\,\mathrm{m}, \qquad t_0 = 10^{-14}\,\mathrm{s}, \tag{1.4.117}$$
$$qV_0 = 10^{-18}\,\mathrm{J}, \qquad \varepsilon_0 = 10^{-18}\,\mathrm{J}.$$

Then the constants α, a, b are

$$\alpha \approx 0.01, \qquad a \approx 1, \qquad b \approx 0.1. \tag{1.4.118}$$

Usually, α is small while a and b are constants of the order of magnitude 1. Physically, a small value of α means that the characteristic device length is large compared to the crystal lattice spacing. Thus it makes sense to consider the limit of the scaled equation (1.4.114) as $\alpha \to 0$ while the constants a, b are kept fixed. We remember that the equation (1.4.114) is posed on the phase space

$$x \in L = \alpha L_0, \qquad k \in B \tag{1.4.119}$$

where the scaled Brillouin zone B has a volume which is of the order of magnitude 1 and L_0 is the crystal lattice scaled by $2l$. Thus, the lattice spacing of L_0 is of the order of magnitude 1.

It is apparent now that three different limits have to be carried out simultaneously in (1.4.114) in order to obtain the scaled semiclassical transport equation (1.4.116):

(i) $\alpha \to 0$ 'in the lattice'. The lattice αL_0 becomes finer as $\alpha \to 0$ and we expect the discretely defined Wigner functions w to converge to a function defined on $\mathbb{R}_x^3 \times B$.

(ii) $\alpha \to 0$ 'in the band operator'

$$\frac{ai}{\alpha}\left[\varepsilon\left(k + \frac{\alpha}{2i}\,\mathrm{grad}_x\right) - \varepsilon\left(k - \frac{\alpha}{2i}\,\mathrm{grad}_x\right)\right],$$

which tends formally to the differential operator $a\,\mathrm{grad}_k\,\varepsilon(k)\cdot\mathrm{grad}_x$.

(iii) $\alpha \to 0$ in the potential operator

$$\frac{bi}{\alpha}\left[V\left(x + \frac{\alpha}{2i}\operatorname{grad}_k\right) - V\left(x - \frac{\alpha}{2i}\operatorname{grad}_k\right)\right],$$

which tends formally to the differential operator $b \operatorname{grad}_x V(x) \cdot \operatorname{grad}_k$.

For sufficiently smooth energy bands and potentials we then expect the solutions $w = w^\alpha$ of (1.4.114) to converge to the solution of (1.4.116). A rigorous mathematical treatment of the one-dimensional case can be found in [1.54].

For numerical simulations it is desirable to derive an energy band quantum transport model, which is simpler than (1.4.114) but still capable of modeling quantum effects like tunneling. The maybe most intriguing way to achieve this is to perform the limits (i) and (ii) but to leave the potential energy term unchanged. The transport equation obtained in this way then reads

$$w_t + a \operatorname{grad}_k \varepsilon(k) \cdot \operatorname{grad}_x w + \frac{bi}{\alpha}\left[V\left(x + \frac{\alpha}{2i}\operatorname{grad}_k\right) - V\left(x - \frac{\alpha}{2i}\operatorname{grad}_k\right)\right] w = 0, \qquad x \in \mathbb{R}_x^3, \qquad k \in B. \tag{1.4.120}$$

We remark that the equation (1.4.120) is—even for pure quantum states—not equivalent to the Schrödinger equation.

For a mathematical analysis of (1.4.120) (coupled with a self-consistent potential model, see Sections 1.3 and 1.5) we refer to [1.16], [1.17].

A many-body version of the energy band quantum transport model can easily be derived by starting out from the many-body Hamiltonian (1.2.44). Since the derivation does not give new insights we do not present the details here.

We conclude this Section with the remark that—similarly to the classical case—magnetic field effects (and spin effects) can also be taken into account in setting up the quantum Liouville equation. Since the models are highly complicated (particularly when the spin is included) and since they are not employed in practical semiconductor device simulations we only refer to [1.3] for the derivation and mathematical analysis of the electromagnetic quantum Liouville equation for electrons with spin.

1.5 The Quantum Boltzmann Equation

The application of the quantum Liouville equation, which is the quantum analogue of the classical Liouville equation, to the modeling of many-body systems involves the same problems as in the classical case. Firstly, realistic models for the many-body potential, which comprise long range and short range interactions, are generally not available. Secondly, the dimension of the phase space on which the M-particle quantum Liouville equation is

posed, equals $6M$, which is by far too large for numerical simulations in practically relevant applications.
In this Section we shall derive single particle approximations of the quantum Liouville equation, which contain a self-consistent potential equation to account for the many-body effects. Just as in the classical case presented in Section 1.3 we shall at first consider long range forces only and derive the corresponding quantum Vlasov equation. It has the form of a single particle quantum Liouville equation supplemented by a Poisson equation for the effective potential, when the particle interaction is modeled by the Coulomb force. Then we shall discuss short range interactions (scattering events), which lead to the quantum Boltzmann equation.
For the derivation of the quantum Vlasov equation we shall proceed similarly to the classical case. We shall set up the quantum analogue of the BBGKY-hierarchy and use the Hartree approximation to obtain an equation for the one-body density matrix whose Fourier transform with respect to the dual velocity variable is the quantum Vlasov equation. Short range interactions will be incorporated as in the classical case, namely by a collision integral operator, which appears on the right-hand side of the quantum Boltzmann equation.

Subensemble Density Matrices

We consider an ensemble of M electrons of equal mass m, whose motion is governed by the Schrödinger equation (1.4.1) with the M-body Hamiltonian (1.4.23). The ensemble density matrix ρ is defined by

$$\begin{aligned}\rho(r_1, \ldots, r_M, s_1, \ldots, s_M, t) \\ = \overline{\psi(r_1, \ldots, r_M, t)}\psi(s_1, \ldots, s_M, t), \qquad r_i, s_i \in \mathbb{R}^3,\end{aligned} \tag{1.5.1}$$

where ψ is the wave function of the ensemble. We recall that ρ satisfies the Heisenberg equation (1.4.27). For the following we shall assume that the electrons of the ensemble are indistinguishable in the sense that the density matrix remains invariant under any permutation of the r- and s-arguments, i.e.

$$\rho(r_1, \ldots, r_M, s_1, \ldots, s_M, t) = \rho(r_{\pi(1)}, \ldots, r_{\pi(M)}, s_{\pi(1)}, \ldots, s_{\pi(M)}, t) \tag{1.5.2}$$

holds for any permutation π of the set $\{1, \ldots, M\}$ and for all $r_i, s_i \in \mathbb{R}^3, t \geqslant 0$. The condition (1.5.2) is satisfied if either the wave function ψ is antisymmetric

$$\psi(x_1, \ldots, x_M, t) = \operatorname{sgn}(\pi)\psi(x_{\pi(1)}, \ldots, x_{\pi(M)}, t), \qquad \forall\, \pi, \quad \forall\, x_i, \quad t \geqslant 0 \tag{1.5.3}$$

or if it is symmetric

$$\psi(x_1, \ldots, x_M, t) = \psi(x_{\pi(1)}, \ldots, x_{\pi(M)}, t), \qquad \forall\, \pi, \quad \forall\, x_i, \quad t \geqslant 0. \tag{1.5.4}$$

The property (1.5.3) holds for ensembles of Fermions and (1.5.4) for ensembles of Bosons (see [1.34]). The former represents the Pauli principle of quantum mechanics mentioned in Section 1.3, which prohibits the double occupancy of states, i.e. the wave functions of Fermions satisfy

$$\psi(x_1, \ldots, x_M, t) = 0 \quad \text{if} \quad x_i = x_j \quad \text{for} \quad i \neq j. \tag{1.5.5}$$

Since the particles considered in this book (electrons and holes) are Fermions, we shall assume (1.5.3) and, consequently, (1.5.2) to hold henceforth. Note that, by the Schrödinger equation, the potential V has to satisfy

$$V(x_1, \ldots, x_M, t) = V(x_{\pi(1)}, \ldots, x_{\pi(M)}, t), \qquad \forall \pi, \quad \forall x_i, \quad t \geq 0 \tag{1.5.6}$$

for an ensemble of Fermions. It is easy to show that the anti-symmetry of ψ and consequently (1.5.2) are conserved in the evolution process if (1.5.6) holds.
To model the motion of subensembles we shall now introduce subensemble density matrices.
The density matrix corresponding to a subensemble consisting of d particles is obtained by evaluating the ensemble density matrix ρ at $r_i = s_i$ for $i = d + 1, \ldots, M$ and by integrating with respect to these coordinates:

$$\begin{aligned} &\rho^{(d)}(r_1, \ldots, r_d, s_1, \ldots, s_d, t) \\ &\quad := \int_{\mathbb{R}^{3(M-d)}} \rho(r_1, \ldots, r_d, u_{d+1}, \ldots, u_M, s_1, \ldots, s_d, u_{d+1}, \ldots, u_M, t) \\ &\qquad \times du_{d+1}, \ldots, du_M . \end{aligned} \tag{1.5.7}$$

The trace of $\rho^{(d)}$ represents the quantum electron position density of the d-particle subensemble:

$$n_{\text{quan}}^{(d)}(x_1, \ldots, x_d, t) = \rho^{(d)}(x_1, \ldots, x_d, x_1, \ldots, x_d, t). \tag{1.5.8}$$

The subensemble quantum electron current density is given by

$$\begin{aligned} &J_{\text{quan}}^{(d)}(x_1, \ldots, x_d, t) \\ &\quad = \frac{i\hbar q}{2m}(\text{grad}_{s^{(d)}} - \text{grad}_{r^{(d)}})\rho^{(d)}(\ldots, \ldots, t)|_{r^{(d)}=s^{(d)}=x^{(d)}} \end{aligned} \tag{1.5.9}$$

where we set $r^{(d)} = (r_1, \ldots, r_d)$, $s^{(d)} = (s_1, \ldots, s_d)$ and $x^{(d)} = (x_1, \ldots, x_d)$. Clearly, the indistinguishability property (1.5.2) is inherited by the subensemble density matrices, i.e.

$$\rho^{(d)}(r_1, \ldots, r_d, s_1, \ldots, s_d, t) = \rho^{(d)}(r_{\sigma(1)}, \ldots, r_{\sigma(d)}, s_{\sigma(1)}, \ldots, s_{\sigma(d)}, t) \tag{1.5.10}$$

holds for all permutations σ of $\{1, \ldots, d\}$ and all $r_i, s_i \in \mathbb{R}^3$, $t > 0$.

The Quantum Vlasov Equation

As in the classical case we assume now that the potential V is the sum of an external potential and an internal potential stemming from two-particle interactions:

$$V(x_l, \ldots, x_M, t) = \sum_{l=1}^{M} V_{\text{ext}}(x_l, t) + \frac{1}{2}\sum_{l=1}^{M}\sum_{j=1}^{M} V_{\text{int}}(x_l, x_j), \tag{1.5.11}$$

where V_{int} is symmetric

$$V_{\text{int}}(x_l, x_j) = V_{\text{int}}(x_j, x_l), \qquad l, j = 1, \ldots, N. \tag{1.5.12}$$

The factor $\frac{1}{2}$ in (1.5.11) is necessary since each particle pair is counted twice in the sum representing the accumulated two-particle interactions.
The Heisenberg equation of motion for the ensemble density matrix ρ then reads:

$$\begin{aligned} i\hbar\, \partial_t \rho = &-\frac{\hbar^2}{2m}\sum_{l=1}^{M} (\Delta_{s_l}\rho - \Delta_{r_l}\rho) - q\sum_{l=1}^{M} (V_{\text{ext}}(s_l, t) - V_{\text{ext}}(r_l, t))\rho \\ &-\frac{q}{2}\sum_{l=1}^{M}\sum_{j=1}^{M} (V_{\text{int}}(s_l, s_j) - V_{\text{int}}(r_l, r_j))\rho. \end{aligned} \tag{1.5.13}$$

We remark that the ensemble is assumed to move in a vacuum and that magnetic field effects are ignored at this point.
We set $u_l = s_l = r_l$ for $l = d + 1, \ldots, M$ in the equation (1.5.13) and integrate over $\mathbb{R}^3_{u_{d+1}} \times \cdots \times \mathbb{R}^3_{u_M}$. Assuming that ρ decays to zero sufficiently fast as $|r_l| \to \infty$, $|s_l| \to \infty$, we obtain by using the definition of the subensemble density matrix $\rho^{(d)}$ given by (1.5.7) and the indistinguishability property (1.5.10):

$$\begin{aligned} i\hbar\, \partial_t \rho^{(d)} = &-\frac{\hbar^2}{2m}\sum_{l=1}^{d} (\Delta_{s_l}\rho^{(d)} - \Delta_{r_l}\rho^{(d)}) \\ &-\sum_{l=1}^{d} (V_{\text{ext}}(s_l, t) - V_{\text{ext}}(r_l, t))\rho^{(d)} \\ &-q(M-d)\sum_{l=1}^{d} \int_{\mathbb{R}^3} [V_{\text{int}}(s_l, u_*) \\ &\qquad - V_{\text{int}}(r_l, u_*)]\rho_*^{(d+1)}\, du_* \end{aligned} \tag{1.5.14}$$

for $1 \leqslant d \leqslant M - 1$, where we denoted

$$\rho_*^{(d+1)} = \rho^{(d+1)}(r_1, \ldots, r_d, u_*, s_1, \ldots, s_d, u_*, t). \tag{1.5.15}$$

The system of equations (1.5.14) constitutes the quantum equivalent of the BBGKY-hierarchy presented in Section 1.3. As in the classical case, it is not possible to solve the system exactly for finite M, therefore we shall again consider the limit $M \to \infty$ for small subensembles. Then, at least a particular solution can be obtained. Clearly, this limiting procedure is reasonable since

we are interested in a single particle type approximation ($d = 1$) of the quantum Liouville equation for large electron ensembles.
Analogously to the classical case, we assume that the two-body interaction potential V_{int} is of the order of magnitude $1/M$ as $M \to \infty$ which implies that the total potential generated by each particle

$$V_{l,\text{tot}}(x_1, \dots, x_M, t) = \sum_{j=1}^{M} V_{\text{int}}(x_l, x_j) + V_{\text{ext}}(x_l, t), \quad 1 \leqslant l \leqslant M$$

remains finite as $M \to \infty$.
For a fixed subensemble size d we obtain by going to the limit $M \to \infty$ in (1.5.14);

$$\begin{aligned} i\hbar\, \partial_t \rho^{(d)} = {} & -\frac{\hbar^2}{2m} \sum_{l=1}^{d} (\Delta_{s_l} \rho^{(d)} - \Delta_{r_l} \rho^{(d)}) - q \sum_{l=1}^{d} (V_{\text{ext}}(s_l, t) \\ & - V_{\text{ext}}(r_l, t))\rho^{(d)} \\ & - q \sum_{l=1}^{d} \int_{\mathbb{R}^3} (V_{\text{int}}(s_l, u_*) - V_{\text{int}}(r_l, u_*)) M \rho_*^{(d+1)}\, du_* \,. \end{aligned} \quad (1.5.16)$$

As in the classical case we now assume that the particles in the subensemble move independently from each other (which, again, is reasonable for small subensembles). This is reflected by the so-called Hartree ansatz (see [1.7]):

$$\rho^{(d)}(r_1, \dots, r_d, s_1, \dots, s_d, t) = \prod_{i=1}^{d} R(r_i, s_i, t). \quad (1.5.17)$$

We obtain an equation for the one-particle density matrix $R := \rho^{(1)}$ by setting $d = 1$ in (1.5.16) and by employing the ansatz (1.5.17) for $d = 2$:

$$i\hbar\, \partial_t R = -\frac{\hbar^2}{2m}(\Delta_s R - \Delta_r R) - q(V_{\text{eff}}(s, t) - V_{\text{eff}}(r, t))R, \qquad r, s \in \mathbb{R}^3, \quad t > 0 \quad (1.5.18)$$

with the effective potential relation

$$V_{\text{eff}}(x, t) = V_{\text{ext}}(x, t) + \int_{\mathbb{R}^3_{x_*}} M R(x_*, x_*, t) V_{\text{int}}(x, x_*)\, dx_* \,. \quad (1.5.19)$$

It is now an easy exercise to show that a particular solution of (1.5.16) for arbitrary d is given by (1.5.17), if R satisfies (1.5.18), (1.5.19).
We multiply (1.5.18) by the total number of particles M, introduce the coordinate transformation

$$r = x + \frac{\hbar}{2m}\eta, \qquad s = x - \frac{\hbar}{2m}\eta \quad (1.5.20)$$

and obtain

$$\partial_t U + i \operatorname{div}_\eta(\operatorname{grad}_x U) + iq \frac{V_{\text{eff}}\left(x + \frac{\hbar}{2m}\eta, t\right) - V_{\text{eff}}\left(x - \frac{\hbar}{2m}\eta, t\right)}{\hbar} U = 0,$$

$$x \in \mathbb{R}_x^3, \quad \eta \in \mathbb{R}_\eta^3, \quad t > 0, \quad (1.5.21)$$

where we set

$$U(x, \eta, t) = MR(r, s, t). \quad (1.5.22)$$

Inverse Fourier transformation of (1.5.21) with respect to η gives

$$\partial_t W + v \cdot \operatorname{grad}_x W + \frac{q}{m} \theta_\hbar[V_{\text{eff}}] W = 0, \quad x \in \mathbb{R}_x^3, \quad v \in \mathbb{R}_v^3, \quad t > 0, \quad (1.5.23)$$

where the velocity v is the dual variable of η and $W := \mathscr{F}^{-1} U$. The pseudo-differential operator $\theta_\hbar$ is defined in (1.4.41).
We have

$$MR(x, x, t) = U(x, \eta = 0, t) = \int_{\mathbb{R}_v^3} W(x, v, t)\, dv$$

and, thus, the effective potential equation (1.5.19) can be rewritten as

$$V_{\text{eff}}(x, t) = V_{\text{ext}}(x, t) + \int_{\mathbb{R}_{x_*}^3} n(x_*, t) V_{\text{int}}(x, x_*)\, dx_*,$$

$$x \in \mathbb{R}_x^3, \quad t > 0, \quad (1.5.24)$$

where

$$n(x, t) = \int_{\mathbb{R}_v^3} W(x, v, t)\, dv, \quad x \in \mathbb{R}_x^3, \quad t > 0 \quad (1.5.25)$$

denotes the quantum electron number density. The macroscopic quantum current density is given by

$$J(x, t) = -q \int_{\mathbb{R}_v^3} v W(x, v, t)\, dv, \quad x \in \mathbb{R}_x^3, \quad t > 0. \quad (1.5.26)$$

The equation (1.5.23) supplemented by the effective potential relation (1.5.24) is called quantum (or nuclear) Vlasov equation (see [1.41]). Analogously to the classical case it has the form of a single-particle quantum Liouville equation. Many-body effects only come in by the equation (1.5.24) for the effective potential, in which the electron number density n enters.
The quantum Vlasov equation is a nonlinear pseudo-differential equation. The symbol

$$(\delta V_{\text{eff}})_\hbar(x, \eta, t) = im \frac{V_{\text{eff}}\left(x + \frac{\hbar}{2m}\eta, t\right) - V_{\text{eff}}\left(x - \frac{\hbar}{2m}\eta, t\right)}{\hbar} \quad (1.5.27)$$

depends on n by (1.5.24) and, thus, on the solution W itself. The nonlinearity $\theta_\hbar[V_{\text{eff}}]W$ is of quadratic nonlocal nature.
Since (1.5.23) is a single-particle quantum Liouville equation, the analysis of Section 1.4 applies to the linear problem with V_{eff} given. In particular, contrary to the classical Vlasov equation, the quantum Vlasov equation does not preserve the non-negativity of the solution W. The number density n, however, remains non-negative for all times, if the initial single-particle density matrix $R(t = 0)$ is positive semi-definite (see Section 1.4). Also, the quantum Vlasov equation formally converges to the classical Vlasov equation with $E_{\text{eff}} = -\text{grad}_x V_{\text{eff}}$ as $\hbar \to 0$.
The quantum Vlasov equation models the quantum mechanical motion of a large particle ensemble moving in a vacuum under the influence of an exterior potential field taking into account weak, long range interactions of particles. Thus, as the classical Vlasov equation, it is a transport model useful on a time-scale much shorter than the mean time between two consecutive scattering events. Contrary to the classical Vlasov equation, the quantum Vlasov equation is capable of modeling the tunneling effect, which makes it particularly attractive as a tool for the numerical simulation of ultra-integrated semiconductor devices.

The Poisson Equation

To account for the Coulomb interaction we set

$$V_{\text{int}}(x, y) = -\frac{q}{4\pi\varepsilon_s}\frac{1}{|x-y|}, \qquad x, y \in \mathbb{R}^3, \qquad x \neq y. \tag{1.5.28}$$

Again, ε_s denotes the permittivity of the semiconductor. Obviously

$$\text{grad}_x V_{\text{int}} = -E_{\text{int}}, \qquad x \neq y \tag{1.5.29}$$

holds, where E_{int} is the Coulomb interaction field (1.3.19). V_{int} is up to the factor q/ε_s the normalized fundamental solution of the Laplace equation. The self-consistent potential equation (1.5.24) then reads

$$V_{\text{eff}}(x, t) = V_{\text{ext}}(x, t) - \frac{q}{4\pi\varepsilon_s}\int_{\mathbb{R}^3_x}\frac{n(x_*, t)}{|x - x_*|}\,dx_* \tag{1.5.30}$$

and the Poisson equation

$$-\varepsilon_s\,\Delta V_{\text{eff}} = -\varepsilon_s\,\Delta V_{\text{ext}} - qn, \qquad x \in \mathbb{R}^3_x, \qquad t > 0 \tag{1.5.31}$$

holds. If the external potential field V_{ext} is generated by ions of charge $+q$ present in the material, then (1.3.27) follows, where $C = C(x, t)$ denotes the number density of the background ions. In this case we obtain (1.3.29) and, consequently, (1.3.30), (1.3.31).
We remark that different effective potential equations, which also include local values of the number density n, exist in the literature, too. They can be seen as attempts to describe medium range particle interactions. Since these

models are not used in semiconductor simulations, we shall not discuss them here in detail. The interested reader is referred to [1.7], [1.41].

We supplement the Vlasov equation (1.5.23), (1.5.24), (1.5.25) by the initial condition

$$W(x, v, t = 0) = W_I(x, v), \qquad x \in \mathbb{R}^3_x, \qquad v \in \mathbb{R}^3_v. \tag{1.5.32}$$

The main difficulty in the mathematical analysis of the quantum Vlasov equation lies in the fact that—contrary to the classical case—an L^1-theory for the linear quantum Liouville equation, which would guarantee the existence of the number density n, does not exist yet for a sufficiently broad class of potentials. Clearly, the L^2-theory is not sufficient since $w(\cdot, \cdot, t) \in L^2(\mathbb{R}^3_x \times \mathbb{R}^3_v)$ does not imply that n is well-defined. Recently, the existence and uniqueness of a solution, globally defined in t, was proven in the one- and three-dimensional cases by reformulating the quantum Vlasov equation as system of countably many Schrödinger equations coupled to the self-consistent Newtonian potential relation (see [1.55] for the one-dimensional and [1.11] for the three-dimensional case). This reformulation is based on the equivalence of the linear quantum Liouville equation to a system of countably many Schrödinger equations presented in Section 1.4. The quantum Vlasov equation (1.5.23), (1.5.24), (1.5.25), (1.5.32) can be written as

$$\left.\begin{aligned} i\hbar\, \partial_t \phi^{(k)} &= -\frac{\hbar^2}{2m}\Delta\phi^{(k)} - qV_{\text{eff}}\phi^{(k)}, \quad x \in \mathbb{R}^3_x, \quad t > 0 \\ \phi^{(k)}(x, t = 0) &= \phi^{(k)}(x), \quad x \in \mathbb{R}^3_x \end{aligned}\right\} k \in \mathbb{N}$$

$$V_{\text{eff}} = V_{\text{ext}} - \frac{q}{4\pi\varepsilon_s}\int_{\mathbb{R}^3_{x_*}} \frac{n(x_*, t)}{|x - x_*|}\, dx_*$$

where n is given by (1.4.70):

$$n(x, t) = \sum_{k=1}^{\infty} \lambda_k |\phi^{(k)}(x, t)|^2 .$$

In the one-dimensional case the Green's function $-1/(4\pi|x - x_*|)$ has to be substituted by $-|x - x_*|$.

Note that the scalars λ_k and the initial data $\phi^{(k)}(x)$ only depend on the initial Wigner function $W_I(x, v)$ (see Section 1.4 for details).

This Schrödinger-Poisson system can be analyzed under reasonable assumptions on λ_k and $\phi^{(k)}(x)$ in a more straightforward way than the pseudo-differential equation form of the quantum Vlasov-Poisson system.

An analysis of the one-dimensional steady state Schrödinger-Poisson problem, posed on a bounded interval, was presented in [1.19].

An existence and uniqueness result for the one-dimensional quantum Vlasov-Poisson problem with periodic boundary conditions in x can be found in [1.2]. The spectral properties of the linearized equation are also discussed in that paper and the convergence of the solution to the solution of the corresponding classical problem as $\hbar \to 0$ was proven, too.

By integrating the quantum Vlasov equation with respect to the velocity v and by proceeding as for the quantum Liouville equation in Section 1.4 we obtain the conservation law:

$$q\partial_t n - \operatorname{div} J = 0. \tag{1.5.33}$$

Also, if W is sufficiently regular, then the conservation of the total number of particles follows by integrating (1.5.33) with respect to x:

$$\int_{\mathbb{R}_x^3} n(x, t)\, dx = \int_{\mathbb{R}_x^3} n_I(x)\, dx, \qquad t > 0, \tag{1.5.34}$$

where we set $n_I(x) = \int_{\mathbb{R}_v^3} W_I(x, v)\, dv$.

The Quantum Vlasov Equation on a Bounded Position Domain

As its classical analogue, the quantum Vlasov equation can also be posed on a bounded position domain which, in semiconductor device modeling, represents the device geometry. Given the bounded convex domain $\Omega \subseteq \mathbb{R}_x^3$, the inflow boundary condition

$$W(x, v, t) = W_D(x, v, t), \qquad (x, v) \in \Gamma_-, \qquad t > 0, \tag{1.5.35}$$

where the inflow segment Γ_- is defined in (1.3.35), can be imposed. Then the Poisson equation (1.5.31) is also posed on Ω and supplemented by Dirichlet or mixed Dirichlet-Neumann boundary conditions for V_{eff} on $\partial\Omega$.
However, due to the nonlocal character of the pseudo-differential operator $\theta_\hbar[V_{\text{eff}}]$ the potential V_{eff} still has to be defined on the whole position space $\mathbb{R}_x^3$. Thus, the solution of the Poisson equation has to be extended from Ω to $\mathbb{R}_x^3$ in order to be used as an input for the Vlasov equation. The problem of determining physically reasonable extensions has not been solved satisfactorily yet. In one-dimensional simulations a piecewise constant continuous extension is normally used.
A disadvantage of the inflow boundary conditions is that they do not exclude the reflection of incoming waves. Absorbing boundary conditions, which provide a better model for Ohmic contacts, were derived in [1.24] by means of the theory of pseudo-differential operators.
Analytical results for the linear quantum transport equation (with given bounded potential) subject to inflow boundary conditions can be found in [1.39].

The Energy-Band Version of the Quantum Vlasov Equation

The quantum Vlasov equation presented above does not take into account the impact of the semiconductor crystal lattice on the motion of the particles. In order to do this the (multi-particle version of the) energy-band quantum Liouville equation (1.4.108) has to be taken as starting point for the quantum

BBGKY-hierarchy. Since the involved calculations are along the lines of the vacuum problem presented above, we do not give them here, but merely state the result. The quantum Vlasov equation on the Brillouin zone B is just the single particle energy band quantum Liouville equation (1.4.108)

$$\partial_t W+\frac{i}{\hbar}\left[\varepsilon\left(k+\frac{1}{2i}\operatorname{grad}_x\right)-\varepsilon\left(k-\frac{1}{2i}\operatorname{grad}_x\right)+qV_{\text{eff}}\left(x+\frac{1}{2i}\operatorname{grad}_k\right)\right.$$
$$\left.-qV_{\text{eff}}\left(x-\frac{1}{2i}\operatorname{grad}_k\right)\right]W=0,\quad x\in\frac{1}{2}L,\quad k\in B,\quad t>0,\tag{1.5.36}$$

where L denotes the crystal lattice and $\varepsilon = \varepsilon(k)$ the considered energy band of the semiconductor. Note that the (quasi) distribution $W = W(x, k, t)$ is defined for $x \in \frac{1}{2}L$, $k \in B$ and $t > 0$. The equation for the effective potential is obtained by replacing the integration in (1.5.24) by a sum over $x \in \frac{1}{2}L$:

$$V_{\text{eff}}(x,t) = V_{\text{ext}}(x,t) + \sum_{\substack{x_*\in\frac{1}{2}L\\ x_*\neq x}} n(x_*,t)\,V_{\text{int}}(x,x_*),$$
$$x\in\tfrac{1}{2}L,\qquad t>0.\tag{1.5.37}$$

Clearly, the number density n is the integral of W over the Brillouin zone B

$$n(x,t)=\int_B W(x,k,t)\,dk,\qquad x\in\frac{1}{2}L,\qquad t>0\tag{1.5.38}$$

and the current density is given by

$$J(x,t)=-q\int_B v(k)W(x,k,t)\,dk,\qquad x\in\frac{1}{2}L,\qquad t>0\tag{1.5.39}$$

with the velocity

$$v(k)=\frac{1}{\hbar}\operatorname{grad}_k\varepsilon(k).\tag{1.5.40}$$

We now scale the equations (1.5.36), (1.5.37), (1.5.38) by using (1.4.113) and, additionaly

$$W(x,k,t)=\frac{l^3}{x_0^3}W_s\left(\frac{x}{x_0},lk,\frac{t}{t_0}\right),\qquad n(x,t)=\frac{1}{x_0^3}n_s\left(\frac{x}{x_0},\frac{t}{t_0}\right).\tag{1.5.41}$$

Then, after dropping the index 's', (1.5.36) reads

$$\partial_t W+\frac{i}{\alpha}\left[a\varepsilon\left(k+\frac{\alpha}{2i}\operatorname{grad}_x\right)-a\varepsilon\left(k-\frac{\alpha}{2i}\operatorname{grad}_x\right)\right.$$
$$\left.+bV_{\text{eff}}\left(x+\frac{\alpha}{2i}\operatorname{grad}_k\right)-bV_{\text{eff}}\left(x-\frac{\alpha}{2i}\operatorname{grad}_k\right)\right]W=0,$$
$$x\in\alpha L_0,\qquad k\in B,\qquad t>0.\tag{1.5.42}$$

The constants a, b and α are given in (1.4.115). When the Coulomb interaction potential (1.5.28) is taken, then the scaled discrete effective potential relation (1.5.37) takes the form:

$$V_{\text{eff}}(x, t) = V_{\text{ext}}(x, t) - c \sum_{\substack{x_* \in \alpha L_0 \\ x \neq x_0}} \alpha^3 n(x_*, t) \frac{1}{|x - x_*|}, \qquad x \in \alpha L_0, \quad t > 0, \tag{1.5.43}$$

with

$$c = \frac{q}{4\pi\varepsilon_s x_0 V_0}. \tag{1.5.44}$$

When the typical numerical values (1.4.117) are taken, then c is of the order of magnitude 1. The scaled number density n is obtained by integrating W over the scaled Brillouin zone B.

An existence and uniqueness result for the initial value problem (1.5.42), (1.5.43), (1.5.38) can be found in [1.16]. We remark that, due to the boundedness of the Brillouin zone B, an L^2-theory for (1.5.42) is sufficient to guarantee the existence of n, since $W(x, \cdot, t) \in L^2(B)$ implies that $n(x, t)$ is well-defined.

As discussed in Section 1.4 the formal limit of (1.5.42) as $\alpha \to 0$ is the scaled semi-classical Liouville equation (1.4.116) (with V substituted by V_{eff}). Obviously, in the limit $\alpha \to 0$ the sum in (1.5.43) has to be replaced by the integral and the discrete effective potential relation (1.5.43) becomes (the scaled version of) (1.5.30). A mathematical justification of this semi-classical limit in the one-dimensional case can be found in [1.54].

In many applications the exterior potential V_{ext} has locally large gradients or even jump-discontinuities. Then the tunneling effect becomes important and the semi-classical Vlasov equation does not give realistic results. Since, however, the energy band ε is a smooth function of k it is even in these cases reasonable to carry out the partial limits '$\alpha \to 0$ in the grid' and '$\alpha \to 0$ in the pseudo-differential operator involving ε' and to leave the potential energy pseudo-differential operator unchanged. Then the model equation (1.4.120) (with V replaced by V_{eff}) supplemented by the 'continuous' effective potential equation (1.5.24) is obtained. A mathematical analysis of this model (with a justification of the partial limit procedure) can be found in [1.16], [1.17]. We believe that this quantum transport model is highly appropriate for the simulation of ballistic phenomena in ultra-integrated semiconductor devices since it allows for a description of the band structure of the crystal and for the modeling of tunneling. Also, from the numerical point of view, it is significantly simpler than the 'discrete-x' problem (1.5.36), (1.5.37), (1.5.38).

Collisions

Just as its classical counterpart, the quantum Vlasov equation is time reversible (for static exterior fields), i.e. it does not contain a mechanism which forces the ensemble to relax towards thermodynamical equilibrium in the large time limit $t \to \infty$. In order to achieve this relaxation property we have to include the effects of short range interactions modeled by scattering events of particles. This, however, cannot be achieved by the purely phenomenological approach presented in Section 1.3 for the (semi-) classical case, since the notion of characteristics does not make sense for the quantum Vlasov equation. Principally, two different approaches are used for the derivation of the quantum Boltzmann equation. The first is based on Green's function techniques (see [1.28]) and the second on the Wigner formalism combined with a modification of the Hartree ansatz (see [1.12]). Since both approaches are highly complicated, we shall not present them here, but merely state the result, which is intuitive when one is familiar with the semi-classical Boltzmann equation.

The quantum Boltzmann equation has the form of an inhomogeneous quantum Vlasov equation, where the inhomogeneity represents the quantum collision integral

$$\partial_t W + v \cdot \operatorname{grad}_x W + \frac{q}{m} \theta_\hbar[V_{\text{eff}}] W = Q_\hbar(W),$$

$$x \in \mathbb{R}_x^3, \qquad v \in \mathbb{R}_v^3, \qquad t > 0. \quad (1.5.45)$$

The quantum collision operator $Q_\hbar$ is nonlocal in the velocity direction and, except when particle-particle interactions are considered, quadratically nonlinear. The particle-particle scattering quantum collision operator is nonlinear of fourth order in W (see [1.28]).

We remark that quantum scattering operators, which are nonlocal in the time direction, can also be found in the literature (see [1.36]).

The equation (1.5.45) is supplemented by an effective potential relation of the form (1.5.24) and by the initial condition (1.5.32).

As in the classical case the collision operator satisfies the conservation property

$$\int_{\mathbb{R}_v^3} Q_\hbar(W)\, dv = 0. \quad (1.5.46)$$

Its precise form depends on the considered scattering processes, however, to our knowledge, simulation of semiconductor devices with physically realistic quantum scattering operators have not been performed due to the enormous numerical complexity involved. For a numerical study of tunneling devices using the relaxation time approximation for the scattering operator we refer to [1.32].

1.6 Applications and Extensions

In this Section we shall discuss specific applications of kinetic transport equations to the modeling of semiconductors. In the course of this we shall

extend the transport models derived in the previous Sections to cover the particular requirements of semiconductor device physics.
At first we will present a multi-valley semi-classical transport model, which is of particular importance for the simulation of GaAs (Gallium-Arsenide) devices. Then we proceed to discuss bipolar semi-classical models, which constitute the basis for the derivation of the hydrodynamic and drift diffusion models in Chapter 2. Finally, we shall summarize the state-of-the-art of quantum modeling of ultra-integrated semiconductor devices.

Multi-Valley Models

It is well-known that the energy-wave vector function $\varepsilon = \varepsilon(k)$ has several minima for, e.g., the semiconductor GaAs. These minima, also termed energy-valleys, are separated by energy-shifts and, very often, the band diagram $\varepsilon(k)$ is approximated by a parabola in the neighbourhood of each valley. For GaAs three types of valleys have to be distinguished: the low energy Γ-valley and the higher energetic L- and X-valleys with energy shifts each of the order of magnitude 0.4 eV (see [1.48] for precise data). For the sake of simplicity we shall for the following neglect the X-valley and only consider a model, which comprises the Γ- and the L-valleys. This approximation is justified by the fact that the highest energy X-valley can only be occupied at very high electric field strengths.
Also we remark that, due to the symmetry properties of the Brillouin zone, several L- (and X-) valleys exist. In the sequel we shall treat them as equivalent.
A parabolic band approximation for the energy-wave vector relation in the Γ-valley reads

$$\varepsilon_\Gamma(k) = \frac{\hbar^2}{2}(k_1^2/m_1 + k_2^2/m_2 + k_3^2/m_3), \qquad k = (k_1, k_2, k_3)^T$$

where the origin of the k-space has been placed at the location of the band minimum and a suitable rotation has been performed. The parameters m_1, m_2, m_3 are called effective masses. This is motivated by a comparison of the velocity

$$v_\Gamma(k) = \frac{1}{\hbar}\,\mathrm{grad}_k\, \varepsilon_\Gamma(k) = \hbar(k_1/m_1, k_2/m_2, k_3/m_3)^T$$

with the velocity-wave vector relation $v = \hbar k/m$ for electrons in a vacuum. Formally, the parabolic band approximation can be obtained by a scaling of the wave vector which magnifies the vicinity of the band minimum. Accordingly, the boundary of the Brillouin zone is moved towards infinity and, as an approximation, B is replaced by $\mathbb{R}^3$. Apart from that, we shall use the common, although not rigorously justified, assumption that the effective masses in the different directions are equal:

$$m_1 = m_2 = m_3 = m_\Gamma$$

With analogous assumptions for the L-valleys we obtain

$$\varepsilon_\Gamma(k) = \frac{\hbar^2|k|^2}{2m_\Gamma}, \qquad \varepsilon_L(k) = \Delta + \frac{\hbar^2|k|^2}{2m_L}, \tag{1.6.1}$$

where m_L is the effective mass of an electron in the L-valley and Δ is the energy difference between the bottoms of the two valleys.
It is convenient to split the electron distribution function F into a part corresponding to the Γ-valley and a part corresponding to the L-valley since the electrons move within each valley even under low field strengths, while the transfer from the lower valley into a 'higher' one requires the presence of high electric fields. Taking into account the multiplicity of the L-valleys we set:

$$F(x, k, t) = F_\Gamma(x, k, t) + N_L F_L(x, k, t), \tag{1.6.2}$$

where N_L denotes the number of L-valleys.
Each of the distribution functions F_Γ, F_L is assumed to satisfy a Boltzmann equation:

$$\partial_t F_\Gamma + v_\Gamma(k)\cdot \operatorname{grad}_x F_\Gamma - \frac{q}{\hbar} E_{\mathrm{eff}}\cdot \operatorname{grad}_k F_\Gamma = Q_\Gamma(F_\Gamma) + Q_{\Gamma,L}(F_\Gamma, F_L), \tag{1.6.3}$$

$$\partial_t F_L + v_L(k)\cdot \operatorname{grad}_x F_L - \frac{q}{\hbar} E_{\mathrm{eff}}\cdot \operatorname{grad}_k F_L = Q_L(F_L) + Q_{L,\Gamma}(F_\Gamma, F_L), \tag{1.6.4}$$

where $v_\Gamma(k) = 1/\hbar \operatorname{grad}_k \varepsilon_\Gamma(k)$, $v_L(k) = 1/\hbar \operatorname{grad}_k \varepsilon_L(k)$ denote the velocities of electrons in the Γ- and L-valleys, resp. Q_Γ and Q_L are the intravalley collision operators. They are both of the form (1.3.67) (with appropriate Γ- and L-valley collision rates s_Γ and s_L, resp.). The Boltzmann equations (1.6.3), (1.6.4) are coupled by the intervalley collision integrals $Q_{L,\Gamma}(F_\Gamma, F_L)$ and $Q_{\Gamma,L}(F_\Gamma, F_L)$, which, in the low density approximation, are given by:

$$Q_{\Gamma,L}(F_\Gamma, F_L) = \int_{\mathbb{R}^3} (s_{L,\Gamma}(x, k', k)F_L' - s_{\Gamma,L}(x, k, k')F_\Gamma)N_L\, dk', \tag{1.6.5}$$

$$Q_{L,\Gamma}(F_\Gamma, F_L) = \int_{\mathbb{R}^3} (s_{\Gamma,L}(x, k', k)F_\Gamma' - s_{L,\Gamma}(x, k, k')F_L)N_L\, dk', \tag{1.6.6}$$

where $s_{\Gamma,L}(x, k, k')$ and $s_{L,\Gamma}(x, k, k')$ denote the transition rates from the state (x, k) of the Γ-valley into a state (x, k') of one of the L-valleys and, resp., vice versa.
The effective field E_{eff} is related to the electron number density

$$n = n_\Gamma + N_L n_L, \qquad n_\Gamma = \int_{\mathbb{R}^3} F_\Gamma\, dk, \qquad n_L = \int_{\mathbb{R}^3} F_L\, dk \tag{1.6.7}$$

by the equation (1.3.14).

The intervalley transition rates $s_{\Gamma,L}$, $s_{L,\Gamma}$ can be expressed in terms of the Maxwellian

$$M_\Gamma(k) = N_\Gamma^* \exp\left(-\frac{\varepsilon_\Gamma(k)}{k_B T}\right), \quad M_L(k) = N_L^* \exp\left(-\frac{\varepsilon_L(k)}{k_B T}\right), \tag{1.6.8a}$$

$$N_\Gamma^* = \left(\int_{\mathbb{R}^3} \exp\left(-\frac{\varepsilon_\Gamma(k)}{k_B T}\right) dk\right)^{-1}, \quad N_L^* = \left(\int_{\mathbb{R}^3} \exp\left(-\frac{\varepsilon_L(k)}{k_B T}\right) dk\right)^{-1}, \tag{1.6.8b}$$

by introducing the inter-valley cross-sections $\sigma_{\Gamma,L}$ and $\sigma_{L,\Gamma}$:

$$s_{\Gamma,L}(x, k, k') = \sigma_{\Gamma,L}(x, k, k') M_L(k'), \tag{1.6.9}$$

$$s_{L,\Gamma}(x, k, k') = \sigma_{L,\Gamma}(x, k, k') M_\Gamma(k'). \tag{1.6.10}$$

Clearly, $\sigma_{\Gamma,L}$ and $\sigma_{L,\Gamma}$ are nonnegative. If, in addition, they satisfy

$$\sigma_{L,\Gamma}(x, k', k) = \sigma_{\Gamma,L}(x, k, k') \quad \forall\, x, k, k', \tag{1.6.11}$$

then the property

$$Q_{\Gamma,L}(F_\Gamma, F_L) = Q_{L,\Gamma}(F_\Gamma, F_L) = 0 \leftrightarrow (F_\Gamma, F_L) = \frac{n}{1 + N_L}(M_\Gamma, M_L) \tag{1.6.12}$$

holds, i.e. the kernel of the inter-valley collision operator is spanned by the Maxwellians and we are led to expect the pair of distributions (F_Γ, F_L) to relax towards an element of this kernel in the large time limit $t \to \infty$.
For the precise structure of the cross-sections $\sigma_{\Gamma,L}$ and $\sigma_{L,\Gamma}$ and for numerical results of the two-valley model for GaAs we refer to [1.43].

Bipolar Model

In a typical semiconductor the conduction band is rather scarcely populated. For the technologically most relevant semiconductor silicon the intrinsic carrier concentration n_i at room-temperature is of the order of magnitude $10^{11}/\text{cm}^3$. Most of the electrons are valence electrons, i.e. they are responsible for the chemical compound of the semiconductor crystal. When the crystal is electrically neutral, then to each conduction electron there corresponds a 'hole' in the valence band, to which the positive charge $+q$ can be assigned. Since the gap between the valence and the conduction band (usually referred to as the bandgap) is significantly large for semiconductors, quite a lot of energy is necessary to transfer electrons from the valence band to the conduction band. This process is called generation of electron-hole pairs, i.e. an electron is generated in the conduction band and a hole in the valence band. The inverse process, that is the transfer of a conduction electron into the lower energetic valence band, is termed recombination of electron-hole pairs. Obviously, the somewhat artificial introduction of

positively charged holes in semiconductor physics gives a simple way of accounting for the valence electrons, whose motion renders a contribution to the current flow in the crystal (hole current). For more information on the basic physical properties of semiconductors we refer to [1.51], [1.57].
In the sequel we shall equip quantities, which correspond to electrons, with the index n and quantities which correspond to holes, with the index p. For example, F_n now stands for the electron distribution and F_p for the hole distribution. We denote the number densities by

$$n = \int_B F_n \, dk, \qquad p = \int_B F_p \, dk \tag{1.6.14}$$

and the current densities

$$J_n = -q \int_B v_n(k) F_n \, dk, \qquad J_p = q \int_B v_p(k) F_p \, dk, \tag{1.6.15}$$

where v_n and v_p denote the electron and hole velocities resp. related to the electron and hole band diagrams by $v_n = (1/\hbar)\nabla_k \varepsilon_n$, $v_p = -(1/\hbar)\nabla_k \varepsilon_p$.
The temporal evolution of the distribution functions F_n and F_p is—in the semi-classical framework—governed by the system of Boltzmann equations:

$$\partial_t F_n + v_n(k) \cdot \operatorname{grad}_x F_n - \frac{q}{\hbar} E_{\text{eff}} \cdot \operatorname{grad}_k F_n = Q_n(F_n) + I_n(F_n, F_p), \tag{1.6.16}$$

$$\partial_t F_p + v_p(k) \cdot \operatorname{grad}_x F_p + \frac{q}{\hbar} E_{\text{eff}} \cdot \operatorname{grad}_k F_p = Q_p(F_p) + I_p(F_n, F_p). \tag{1.6.17}$$

Q_n and Q_p stand for the electron and, resp., hole collision operators. They are supposed to model the short range interactions of the corresponding type of particles with their environment, i.e. with crystal impurities, phonons etc. Mathematically, they are of the form (1.3.67) with transition rates s_n and s_p resp., determined by the physics of the considered collision processes. Most importantly, they satisfy

$$s_n(x, k, k') = \exp\left(\frac{\varepsilon_n(k') - \varepsilon_n(k)}{k_B T}\right) s_n(x, k', k) \tag{1.6.18}$$

$$s_p(x, k, k') = \exp\left(\frac{\varepsilon_p(k') - \varepsilon_p(k)}{k_B T}\right) s_p(x, k', k), \tag{1.6.19}$$

which leads to the relaxation properties of Q_n, Q_p.
The operators I_n, I_p model a recombination and generation process of electron-hole pairs. They are given by:

$$I_n(F_n, F_p) = \int_B [g(x, k', k)(1 - F_n)(1 - F_p') - r(x, k, k') F_n F_p'] \, dk' \tag{1.6.20}$$

$$I_p(F_n, F_p) = \int_B [g(x, k, k')(1 - F_n')(1 - F_p) - r(x, k', k)F_n'F_p] \, dk' , \tag{1.6.21}$$

where $g(x, k, k')$ represents the rate of generation of an electron at the state (x, k) and of a hole at the state (x, k'). $r(x, k, k')$ is the analogous local recombination rate. The expressions (1.6.20), (1.6.21) are derived by accounting procedures similar to the derivation of the single particle scattering integral presented in Section 1.3.

The nonnegative functions g and r are related by:

$$r(x, k, k') = \exp\left(\frac{\varepsilon_n(k) - \varepsilon_p(k')}{k_B T}\right) g(x, k', k). \tag{1.6.22}$$

This equation guarantees that the null-manifold of I_n, I_p consists solely of pairs of Fermi-Dirac distributions with the same Fermi level (see [1.44]). Recombination and generation of carriers balance in thermal equilibrium.

The effective field E_{eff} enters in both Boltzmann equations (1.6.16), (1.6.17). The sign of E_{eff} in the hole transport equation (1.6.17) is reversed due to the opposite flow direction of the positively charged holes in the electric field E_{eff}.

Obviously, both electrons and holes contribute to the space charge density ρ. Also, for practically all semiconductor devices, ionized impurities, which mainly determine the performance of the device under consideration are present in the semiconductor crystal. These impurities are implanted into the semiconductor crystal in the fabrication of the device by a technologically highly complicated process (see [1.51]).

We shall denote the so-called impurity (or doping) profile by C. It is given by the difference of the number densities of positively charged donor ions and negatively charged acceptor ions. For the following we shall exclude mobile impurities, i.e. we shall assume that C is a function of the position variable x only, i.e. $C = C(x)$.

By simply adding up the charges, we obtain the total charge density

$$\rho = -q(n - p - C). \tag{1.6.23}$$

When the Coulomb interaction is accounted for, we have the following effective field equation:

$$E_{\mathrm{eff}} = E_{\mathrm{ext}} + \frac{1}{4\pi\varepsilon_s} \int_{\mathbb{R}^3_{x_*}} \rho(x_*, t) \frac{x - x_*}{|x - x_*|^3} \, dx_* , \tag{1.6.24}$$

where E_{ext} represents an exterior electric field acting on the semiconductor device.

A mathematical analysis, which gives a global (in $t > 0$) existence and uniqueness result for the electron-hole Boltzmann-Poisson system (1.6.16), (1.6.17), (1.6.24) subject to initial conditions and periodic boundary conditions on ∂B, can be found in [1.44]. It is based on an iteration method, which is in spirit similar to the single particle Boltzmann equation method presented in Section 1.3. We remark that the bipolar problem still preserves the upper bound 1 and the lower bound 0 for F_n and F_p, i.e. $0 \leqslant F_n \leqslant 1$, $0 \leqslant F_p \leqslant 1$ holds for all times $t > 0$ if it holds initially (Pauli principle).

For real-life simulations of semiconductor devices the bipolar Boltzmann-Poisson problem has to be formulated on a bounded position domain $\Omega \subseteq \mathbb{R}^3$ representing the geometry of the semiconductor device. Then, boundary conditions for the distribution functions F_n, F_p have to be prescribed on the inflow segments as discussed in the previous Sections and the effective field equation (1.6.24) is replaced by the Poisson equation

$$-\varepsilon_s \,\Delta V_{\text{eff}} = \rho\,, \qquad x \in \Omega \tag{1.6.25}$$

subject to Neumann-Dirichlet boundary conditions on $\partial\Omega$. The exterior field then originates from the Dirichlet boundary condition for V_{eff}, which represents voltages externally applied to the device.

The occurance of recombination-generation of carriers modifies the conservation laws for the current and for the number of carriers. By integrating the Boltzmann equations (1.6.16), (1.6.17) over the Brillouin zone B we obtain

$$q\partial_t n - \operatorname{div} J_n = -qR \tag{1.6.26}$$

$$q\partial_t p + \operatorname{div} J_p = -qR\,, \tag{1.6.27}$$

where R is the recombination-generation rate, which, expressed in terms of the distribution functions F_n, F_p, reads:

$$R = -\int_B I_p(F_n, F_p)\,dk = -\int_B I_n(F_n, F_p)\,dk\,. \tag{1.6.28}$$

Note that the conservation laws (1.6.26), (1.6.27) are nonlinearly coupled due to recombination-generation processes.

The total number of each type of particles is not conserved anymore. Subtracting (1.6.26) from (1.6.27) and using the definition of the charge density (1.6.23) gives the conservation law for the total current density J defined as the sum of the electron and hole current densities:

$$J = J_n + J_p\,. \tag{1.6.29a}$$

This conservation law reads:

$$\partial_t \rho + \operatorname{div} J = 0\,. \tag{1.6.29b}$$

Note that we used the assumption $\partial_t C \equiv 0$.

For the whole space case we obtain the conservation of the charge by integrating (1.6.29b) over $\mathbb{R}^3_x$:

$$\int_{\mathbb{R}^3_x} \rho(x,t)\,dx = \int_{\mathbb{R}^3_x} \rho(x, t=0)\,dx\,, \qquad \forall\, t > 0\,. \tag{1.6.30}$$

The modification of (1.6.30) for the bounded x-domain case is obvious.

A low density approximation of the electron-hole Boltzmann system can be obtained by assuming

$$0 \leqslant F_n \ll 1\,, \qquad 0 \leqslant F_p \ll 1 \tag{1.6.31}$$

and by setting all quadratic terms in F_n, F_p, which appear in the collision and recombination-generation integrals, equal to zero. Also, in many applications, the generation and recombination relaxation times

$$\tau_G(x, k) = \left(\int_B g(x, k, k')\, dk'\right)^{-1},$$
$$\tau_R(x, k) = \left(\int_B r(x, k, k')\, dk'\right)^{-1} \tag{1.6.32}$$

are large compared to the collision relaxation time. In these cases the recombination-generation integrals I_n, I_p are usually neglected, which leads to a significantly weaker coupling of the two Boltzmann equations. This approach is physically meaningful in close-to-thermal-equilibrium conditions.

Tunneling Devices

Charge transport in semiconductors is collision dominated when the observation time period is significantly larger than the collision relaxation time. Thus, the simulation of low and medium frequency devices (like MOSFETs, bipolar transistors and thyristors) must be based on mathematical models, which stem from the Boltzmann equation. For extremely high frequency devices, however, the interesting time-scale for simulations is usually short and, consequently, the charge transport is mainly ballistic, i.e. collisionsless models can be used. Such devices are very small (the width of the active region may be of the order of magnitude 20 nm) and they operate under high electric field strengths (very thin potential barriers with height 0.3 eV often occur). Therefore, the device operation is driven by quantum effects and simulations based on a (semi-) classical Vlasov model give totally unrealistic results. For such situations the quantum Vlasov equation, which models the collisionless quantum transport of electrons, is well suited.

As a typical example we consider the resonant tunneling diode depicted in Fig. 1.6.1. The device has two AlGaAs (aluminium-gallium arsenide) quan-

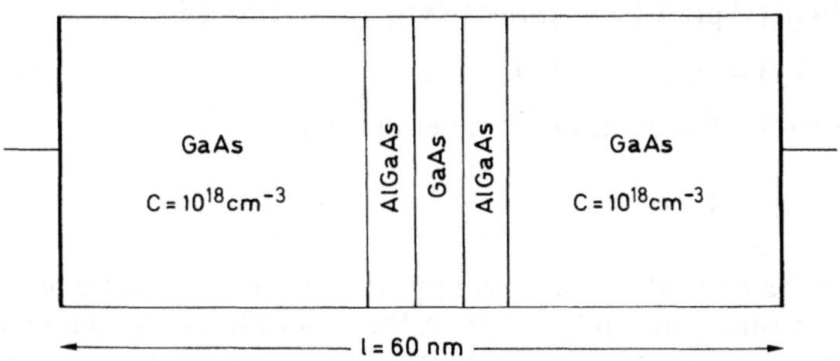

Fig. 1.6.1

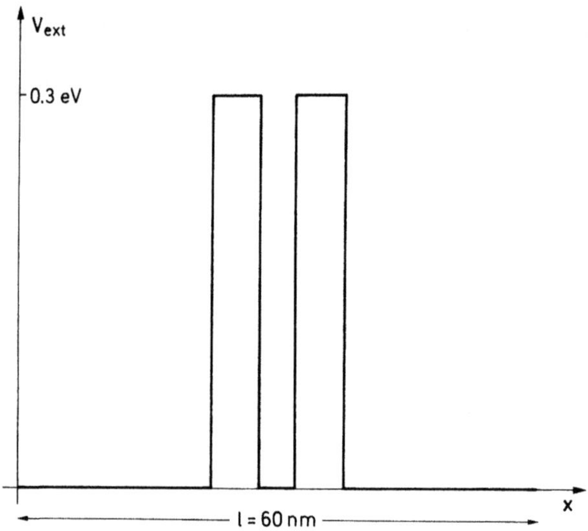

Fig. 1.6.2

tum barriers of thickness 5 nm separated by a GaAs quantum well, which has the thickness 5 nm. The barrier and the well are undoped, while the bulk regions outside the barriers are doped with donors of concentration 10^{18} cm^{-3}, i.e. the doping profile $C = 10^{18}$ cm^{-3} in the bulk regions and $C \equiv 0$ in the barriers and in the well. The quantum barriers are 0.3 eV high (see Fig. 1.6.2) and both bulk regions are contacted.

The (already somewhat simplified) device geometry suggests a one-dimensional quantum Vlasov-Poisson model (see Section 1.5):

$$\partial_t W + v \cdot \operatorname{grad}_x W + \frac{q}{m} \theta_\hbar [V_{\text{eff}}] W = 0, \quad 0 < x < l, \quad v \in \mathbb{R}, \quad t > 0 \tag{1.6.33}$$

$$W(x, v, t = 0) = W_I(x, v), \qquad 0 < x < l, \qquad v \in \mathbb{R}, \tag{1.6.34}$$

supplemented by

$$\operatorname{div}(\varepsilon_s(x) \operatorname{grad} V) = q(n - C(x)), \qquad 0 < x < l \tag{1.6.35}$$

$$V(0, t) = V_0, \qquad V(l, t) = V_1, \tag{1.6.36}$$

where n denotes the quantum number density

$$n = \int_{\mathbb{R}_v} W \, dv. \tag{1.6.37}$$

Note that the permittivity ε_s now appears 'inside' the divergence operator since it is position dependent due to the two different materials of which the device is made up. The boundary data V_0, V_1 determine the biasing condition.

The effective potential V_{eff} is obtained by adding the material potential V_{ext} of Fig. 1.6.2 to the solution V of the Poisson problem (1.6.35), (1.6.36) and by extending to $\mathbb{R}_x$:

$$V_{\text{eff}}(x,t) = \begin{cases} V_0, & x \leqslant 0 \\ V(x,t) + V_{\text{ext}}(x), & 0 < x < l. \\ V_1, & x > l \end{cases}$$

Also, we have to define boundary conditions for the quantum Vlasov equation at $x = 0$ and $x = l$. As discussed in Section 1.5 the simplest choice is to take inflow Dirichlet data

$$W(0, v, t) = W_0(v, t), \qquad v > 0, \qquad t > 0 \tag{1.6.39}$$

$$W(l, v, t) = W_1(v, t), \qquad v < 0, \qquad t > 0. \tag{1.6.40}$$

For the simulation of ideal contacts it is more appropriate to choose non-reflecting boundary conditions (see [1.24] and the discussion in Section 1.5). For appropriate choices of the initial and boundary data and for numerical results of simulations of the resonant tunneling diode we refer to [1.32]. These results clearly demonstrate the power of the quantum Vlasov-Poisson problem in modeling ultra-integrated semiconductor devices on sufficiently short time-scales. Due to the ongoing miniaturization of VLSI structures we expect this research area to become extremely important in the near future. In particular the inclusion of physically realistic quantum scattering models, which will allow simulations on significantly larger time scales, and the band diagram Wigner-Poisson model presented in Section 1.5, which incorporates the crystal structure of the semiconductor, are going to play a major role soon.

Problems

1.1 Solve the initial value problem (1.2.9), (1.2.13) for a constant electric field E. Draw the (x, v)-phase portrait of the characteristics in the one-dimensional case.

1.2 Show that the L^p-norms, $1 \leqslant p \leqslant \infty$ of non-negative solutions of the Liouville equation (1.2.9), (1.2.13) are conserved in the evolution process.
Remark: The L^p-norm of f is defined by:

$$\|f\|_{L^p} := \left(\int_{\mathbb{R}_v^3} \int_{\mathbb{R}_x^3} |f|^p \, dx \, dv \right)^{1/p}, \qquad 1 \leqslant p < \infty$$

$$\|f\|_{L^\infty} := \sup_{x,v} |f(x, v)| .$$

Hint: Multiply (1.2.9) by f^{p-1} and integrate over $\mathbb{R}_v^3 \times \mathbb{R}_x^3$ assuming that the solution decays sufficiently fast as $|x| \to \infty$, $|v| \to \infty$.

1.3 Draw the phase portrait of the characteristic equations (1.2.3), (1.2.4) for the one-dimensional potential barrier given by

$$V(x) = \frac{m}{q} \begin{cases} 0, & x \leqslant 0 \\ \frac{V_0}{\varepsilon} x, & 0 < x < \varepsilon, \quad \varepsilon > 0, \quad V_0 \in \mathbb{R}. \\ V_0, & x \geqslant \varepsilon \end{cases}$$

Discuss the cases $V_0 > 0$, $V_0 < 0$ and consider the limits $\varepsilon \to 0$, $V_0 \to -\infty$, $V_0 \to \infty$.

1.4 Let $H = H(x, p, t)$ be a general Hamiltonian function. Derive the Liouville equation corresponding to the equations of motion (1.2.27), (1.2.28). Show that the L^2-norm of the solution of the corresponding initial value problem is preserved in time.

1.5 Solve the three-dimensional Liouville equation (1.2.52) (subject to an initial condition) for constant electric and magnetic fields.

1.6 a) Are the conservation laws (1.2.18), (1.2.19) valid for the 'magnetic field problem' (1.2.52)?
b) Is nonnegativity preserved by (1.2.52)?

1.7 Definition: A map $w: \mathbb{R}^m \to \mathbb{R}^m$ is called volume preserving, if $\mathrm{vol}(A) = \mathrm{vol}(w(A))$ for all measurable subsets $A \subseteq \mathbb{R}^m$.
Prove that the characteristic map $w(t, \cdot, \cdot): \mathbb{R}^{6M} \to \mathbb{R}^{6M}$ defined by (1.2.20), (1.2.10), (1.2.11) is volume preserving for all $t \geqslant 0$ if (1.2.14) holds. Prove the analogous result for the case of a non-vanishing magnetic field.
The flow associated with a volume-preserving map is called imcompressible. Conclude the conservation of the L^p-norms of the solutions of the Liouville equation from the incompressibility of the flow defined by the characteristic map.

1.8 Let $f_I(x, v) = \delta(x - x_0, v - v_0)$, $x_0 \in \mathbb{R}_x^{3M}$, $v_0 \in \mathbb{R}_v^{3M}$ be the initial datum for the Liouville equation. Show that the solution of the initial value problem is given by $f(x, v, t) = \delta(x - x(t; x_0, v_0), v - v(t; x_0, v_0))$.

1.9 Show that the Liouville equation is time-reversible for static force-fields, i.e. if a solution exists for $t > 0$, then it also exists for $t < 0$ and the solution for $t < 0$ can be constructed from its values for $t > 0$.

1.10 Verify that the function $f^{(d)}$ given by (1.3.8) with P satisfying (1.3.9), (1.3.10) is a solution of (1.3.7).

1.11 Linearize the Vlasov equation (1.3.13), (1.3.14) at the equilibrium solution $F = F_e(v)$, $E_{\mathrm{eff}} \equiv 0$. The so obtained problem is called random phase approximation. Which physical situation does it model?

1.12 Consider the one-dimensional equation (1.3.13) ($x \in \mathbb{R}_x$, $v \in \mathbb{R}_v$) with $E_{\mathrm{eff}} = E_{\mathrm{eff}}(x)$ given. Show that $F = F(x, v)$ is a steady state solution if and only if F is a function of the Hamiltonian, i.e. if and only if there is a function $\phi: \mathbb{R} \to \mathbb{R}$ such that $F(x, v) = \phi(H(x, v))$, $H(x, v) = mv^2/2 - qV_{\mathrm{eff}}(x)$, holds with $d/dx\, V_{\mathrm{eff}} = -E_{\mathrm{eff}}$.
Remark: This is used to construct steady state solutions of the coupled Vlasov-Poisson problem (see [1.42]).

1.13 Consider the Vlasov equation (1.3.13), (1.3.14) with a smooth interaction field E_{int} (which may be obtained by smoothing the Coulomb field (1.3.19) about $x = y$). Take $F(x, v, t = 0) = \delta(x - x_0, v - v_0)$. Show that the solution $F(x, v, t)$ has the form of a δ-function centered at a point $(x(t), v(t)) \in \mathbb{R}_x^3 \times \mathbb{R}_v^3$. Derive the initial value problem for $(x(t), v(t))$.

1.14 Carry out the calculations to derive the property (1.3.87) of the transition rate s from the principle of detailed balance (1.3.84) by using the Fermi-Dirac statistics (1.3.85), (1.3.86).

1.15 Solve the relaxation time approximation (1.3.101) for a given constant effective field $E_{\mathrm{eff}} \equiv E$, a constant relaxation time τ and $n(x, t) \equiv 1$.

Hint: Use the representation (1.3.102) and invert the characteristic map $(x_0, k_0) \to (x(t), k(t))$.

1.16 Formulate the Boltzmann equation with a magnetic field and the Boltzmann-Maxwell system. Solve the corresponding relaxation time approximation for given constant electric and magnetic fields assuming $n = 1$.

1.17 Prove that initially orthogonal wave functions remain orthogonal for all times, i.e. conclude

$$\int_{\mathbb{R}^{3M}} \psi_I^{(1)}(x)\overline{\psi_I^{(2)}(x)}\,dx = 0 \to \int_{\mathbb{R}^{3M}} \psi^{(1)}(x,t)\overline{\psi^{(2)}(x,t)}\,dx = 0, \qquad t > 0,$$

where $\psi^{(1)}$, $\psi^{(2)}$ are solutions of the M-particle Schrödinger equation with initial data $\psi_I^{(1)}$ and $\psi_I^{(2)}$ resp.

1.18 Let the one-dimensional static potential be given by

$$V(x) = \begin{cases} 0, & x < 0, x > a \\ V_0, & 0 \leqslant x \leqslant a, \end{cases} \qquad a > 0.$$

Solve the eigenvalue problem (1.4.14) for the Schrödinger equation, i.e. find $\varepsilon \in \mathbb{R}$ and $\psi = \psi(x) \in L^2(\mathbb{R})$ such that (1.4.14) holds. Consider the cases $V_0 > 0$, $V_0 < 0$ and the limits $V_0 \to \infty$, $V_0 \to -\infty$.

1.19 Compute the density matrices for the eigenstates of Problem 1.18. Calculate the Wigner functions for the limiting cases $V_0 \to \infty$, $V_0 \to -\infty$.

1.20 Prove that (1.4.15) is equivalent to (1.4.16), (1.4.17), (1.4.18).
Hint: Multiply (1.4.15) by a C^∞-test function φ with compact support, integrate by parts and use

$$\int_{\mathbb{R}} \delta(x)\psi(x)\varphi(x)\,dx = \psi(0)\varphi(0)$$

if ψ is continuous at $x = 0$. Then perform the reverse integration by parts separately for $x < 0$ and $x > 0$.

1.21 Compute the $L^2(\mathbb{R})$-eigenstates of (1.4.15), their density matrices, Wigner functions, particle and current densities.

1.22 Let the one-dimensional static potential be given by

$$V(x) = \begin{cases} V_0, & x > 0 \\ 0, & x \leqslant 0. \end{cases}$$

Compute the reflection and transmission coefficients for a monoenergetic beam of electrons represented by a right-moving wave.

1.23 Derive the quantum Liouville equation for $V(x) = -(m/q)\delta(x)$, $x \in \mathbb{R}$. Simplify the operator $\theta_\hbar[V]$ as much as possible. Write down the Fourier transformed equation and formulate it without δ-functions.
Hint: Derive interface conditions at the lines $x = (\hbar/2m)\eta$, $x = -(\hbar/2m)\eta$ similarly to Problem 1.20.

1.24 Let ε_0, ε_1 be eigenvalues of the Schrödinger equation (1.4.14) with corresponding eigenfunctions $\psi_0, \psi_1 \in L^2(\mathbb{R}^{3M})$. Prove that $i(\varepsilon_0 - \varepsilon_1)/\hbar$ is an eigenvalue of the quantum Liouville equation with eigenfunction

$$\int_{\mathbb{R}^{3M}} \overline{\psi_0\left(x + \frac{\hbar}{2m}\eta\right)\psi_1\left(x - \frac{\hbar}{2m}\eta\right)}\, e^{iv\cdot\eta}\,d\eta.$$

Remark: It is shown in [1.38] that the spectrum of the quantum transport operator $T_{\text{quan}} = v\cdot\text{grad}_x + (q/m)\theta_\hbar[V]$ is the closure of the set consisting of the values $i(\alpha - \beta)/\hbar$, where α, β are spectral values of the corresponding Hamiltonian operator. This result, together with a more detailed analysis of the structure of the spectrum was used in [1.18] to characterize the steady states of the quantum Liouville equation.

1.25 Show that $\theta_\hbar[V]$ maps real valued functions into real valued functions, if V is real valued.

1.26 Let $V = V(x) \in \mathbb{R}$ hold. Show that the pseudo-differential operator $\theta_\hbar[V]$ is formally skew-adjoint on $L^2(\mathbb{R}_x^{3M} \times \mathbb{R}_v^{3M})$, i.e. prove

$$\int_{\mathbb{R}_x^{3M}} \int_{\mathbb{R}_v^{3M}} f\overline{\theta_\hbar[V]g}\, dv\, dx = -\int_{\mathbb{R}_x^{3M}} \int_{\mathbb{R}_v^{3M}} \bar{g}\theta_\hbar[V]f\, dv\, dx.$$

1.27 Compute the potential, which corresponds to the wave function (1.4.85).

1.28 Consider the initial value problem for the quantum Liouville equation (1.4.39) with initial datum $w_I(x, v) = \delta(x - x_0, v - v_0)$ for fixed points $x_0 \in \mathbb{R}_x^{3M}$, $v_0 \in \mathbb{R}_v^{3M}$. What are the conditions on the potential such that the solution is given by $w(x, v, t) = \delta(x - x(t), v - v(t))$, where $(x(t), v(t))$ is a curve in the (x, v)-space with initial value (x_0, v_0)? Is the given δ-initial datum quantum mechanically admissible (uncertainty principle)?

1.29 Consider the quantum Liouville equation (1.4.39) as model for an ultra-integrated semiconductor device of characteristic length $l = 10^{-8}$ m. In a typical operation mode the potential is of the order of magnitude $\bar{V} = 0.5$ eV and a typical simulation time scale is $T = 10^{-14}$ s. Use these values and $m = 0.6 \times 10^{-31}$ kg to scale the quantum Liouville equation by introducing dimensionless variables. Identify the parameter, which plays the role of $\hbar$ in the scaled equation. Is it small (compared to 1)?

References

[1.1] A. Arnold, P. Degond, P. A. Markowich, H. Steinrück: The Wigner-Poisson Equation in a Crystal. Applied Mathematics Letters *2*, 187–191 (1989).

[1.2] A. Arnold, P. A. Markowich: The Periodic Quantum Liouville-Poisson Problem. Boll. U.M.I. (1989) (to appear).

[1.3] A. Arnold, H. Steinrück: The Electromagnetic Wigner Equation for an Electron with Spin. ZAMP (1989) (to appear).

[1.4] N. C. Ashcroft, N. D. Mermin: Solid State Physics. Holt-Sounders, New York (1976).

[1.5] C. Bardos, P. Degond: Global Existence for the Vlasov-Poisson Equation in Three Space Variables with Small Initial Data. Ann. Inst. Henri Poincare, Analyse Non-linéaire *2*, 101–118 (1985).

[1.6] C. Bender, S. Orszag: Advanced Mathematical Methods for Scientists and Engineers. McGraw-Hill, New York (1977).

[1.7] G. F. Bertsch: Heavy Ion Dynamics at Intermediate Energy. Report, Cyclotron Laboratory and Physics Department, Michigan State University, East Lansing, MI 48824, USA (1978).

[1.8] J. S. Blakemore: Semiconductor Statistics. Pergamon Press, Oxford (1962).

[1.9] N. N. Bogoliubov: Problems of a Dynamical Theory in Statistical Physics. In: Studies in Statistical Mechanics, Vol. I (J. de Boer, G. E. Uhlenbeck, eds.). North-Holland, Amsterdam (1962), p. 5.

[1.10] M. Born, H. S. Green: A General Kinetic Theory of Fluids. Cambridge University Press, Cambridge (1949).

[1.11] F. Brezzi, P. A. Markowich: The Three-Dimensional Wigner-Poisson Problem: Existence, Uniqueness and Approximation. Report, Centre de Mathématiques Appliquées, Ecole Polytechnique, F-91128 Palaiseau, France (1989).

[1.12] P. Carruthers, F. Zachariasen: Quantum Collision Theory with Phase-Space Distributions. Reviews of Modern Physics *55*, 245–285 (1983).

[1.13] C. Cercignani: The Boltzmann Equation and Its Applications (Applied Mathematical Sciences, Vol. 67). Springer-Verlag, Berlin (1988).

[1.14] J. Cooper: Galerkin Approximations for the One-Dimensional Vlasov-Poisson Equation. Math. Method. in Appl. Sci. *5*, 516–529 (1983).

[1.15] R. Dautrey, J. L. Lions: Analyse Mathématique et Calcul Numérique pour les Sciences et les Techniques: Tome 3'. Masson, Paris (1985).

[1.16] P. Degond, P. A. Markowich: A Quantum Transport Model for Semiconductors: The Wigner-Poisson Problem on a Bounded Brillouin Zone. M^2AN (to appear).

[1.17] P. Degond, P. A. Markowich: A Mathematical Analysis of Quantum Transport in Three-Dimensional Crystals. Report, Centre de Math. Appl., Ecole Polytechnique, F-91128 Palaiseau, France (1989).

[1.18] F. Nier: Etudes des Solutions Stationnaires de l'Équation de Wigner. Manuscript, Centre de Mathématiques Appliquées, Ecole Polytechnique, F-91128 Palaiseau, France (1989).

[1.19] F. Nier: Existence d'une Solution Pour un Système Mono-dimensionelle d'Equations de Schrödinger et des Poisson Couplées. Manuscript, Centre de Mathématiques Appliqueés, Ecole Polytechnique, F-91128 Palaiseau, France (1989).

[1.20] S. de Groot: La Transformation de Weyl et la Fonction de Wigner: Use Forme Alternative de la Méchanique Quantique. Lex Presses de l'Université de Montréal, Montreal (1974).

[1.21] R. Di Perna, P. L. Lions: Global Solutions of Vlasov-Poisson Type Equations. Report 8824, CEREMADE, Université Paris-Dauphine, F-75775 Paris, France (1989).

[1.22] R. Di Perna, P. L. Lions: Global Weak Solutions of Vlasov-Maxwell Systems. Report 8817, CEREMADE, Université Paris-Dauphine, F-75775 Paris, France (1989).

[1.23] R. Di Perna, P. L. Lions: On the Cauchy Problem for Boltzmann Equations: Global Existence and Weak Stability. Report CEREMADE, Université Paris-Dauphine, F-75775 Paris, France (1988).

[1.24] D. K. Ferry, N. C. Kluksdahl, C. Ringhofer: Absorbing Boundary Conditions for the Simulation of Quantum Tunneling Phenomena. Transport Equ. and Statist. Phys. (to appear).

[1.25] R. P. Feynman: The Feynman Lectures on Physics 3: Quantum Mechanics. Addison-Wesley, Reading (1965).

[1.26] H. Hofmann: Das Elektromagnetische Feld. Springer-Verlag, Wien-New York (1982).

[1.27] R. L. Hudson: When Is the Wigner Quasiprobability Nonnegative?. Reports on Math. Physics *6*, 249–252 (1974).

[1.28] L. P. Kadanoff, G. Baym: Quantum Statistical Mechanics: Green's Function Methods in Equilibrium and Non-Equilibrium Problems. Benjamin, New York (1962).

[1.29] T. Kato: Perturbation Theory for Linear Operators. Springer-Verlag, New York (1966).

[1.30] J. G. Kirkwood: J. Chem. Phys. *14*, 180 (1946).

[1.31] C. Kittel: Introduction to Solid State Physics. J. Wiley & Sons, New York (1968).

[1.32] N. C. Kluksdahl, A. M. Kriman, D. K. Ferry, C. Ringhofer: Self-Consistent Study of the Resonant Tunneling Diode. Phys. Rev. B (to appear).

[1.33] N. A. Krall, A. W. Trivelpiece: Principles of Plasma Physics. McGraw-Hill, New York (1973).

[1.34] L. D. Landau, E. M. Lifschitz: Lehrbuch der Theoretischen Physik, 3: Quantenmechanik. Akademie-Verlag, Berlin (1960).

[1.35] L. D. Landau, E. M. Lifschitz: Lehrbuch der Theoretischen Physik, 1: Mechanik, 2nd edn. Akademie-Verlag, Berlin (1963).

[1.36] I. B. Levinson: Translational Invariance in Uniform Fields and the Equation for the Density Matrix in the Wigner Representation. Soviet Physics JETP *30*, 362–367 (1970).

[1.37] P. A. Markowich: The Stationary Semiconductor Device Equations. Springer-Verlag, Wien-New York (1986).

[1.38] P. A. Markowich: On the Equivalence of the Schrödinger and the Quantum Liouville Equations. Math. Meth. in the Appl. Sci. *11*, 459–469 (1989).
[1.39] P. A. Markowich, C. A. Ringhofer: An Analysis of the Quantum Liouville Equations. ZAMM *69*, 121–127 (1989).
[1.40] A. Messiah: Quantum Mechanics, North-Holland, Amsterdam (1965).
[1.41] H. Neunzert: The Nuclear Vlasov Equation—Methods and Results that Cann(ot) be Taken Over from the 'Classical' Case. Proc. Workshop on Fluid Dynamical Approaches to the Many-Body Problem: Fundamental and Mathematical Aspects. Societa Italiana di Fisica, (1984).
[1.42] H. Neunzert: An Introduction to the Nonlinear Boltzmann-Vlasov Equation. In: Lecture Notes in Math., Vol. 1048. Springer-Verlag, Berlin (1984).
[1.43] B. Niclot, P. Degond, F. Poupaud: Deterministic Particle Simulations of the Boltzmann Transport Equation of Semiconductors. J. Comp. Phys. *78*, 313–350 (1988).
[1.44] F. Poupaud: On a System of Nonlinear Boltzmann Equations of Semiconductor Physics. SIAM J. Math. Anal. (to appear).
[1.45] M. Reed, B. Simon: Methods of Modern Mathematical Physics I: Functional Analysis. Academic Press, New York (1972).
[1.46] M. Reed, B. Simon: Methods of Modern Mathematical Physics II: Fourier Analysis, Self-Adjointness. Academic Press, New York (1975).
[1.47] M. Reed, B. Simon: Methods of Modern Mathematical Physics IV: Analysis of Operators. Academic Press, New York (1978).
[1.48] L. Reggiani (ed.): Hot Electron Transport in Semiconductors. Springer-Verlag, Berlin (1985).
[1.49] C. Ringhofer: A Spectral Method for Numerical Simulation of Quantum Tunneling Phenomena. SIAM J. Num. Anal. (to appear).
[1.50] W. Rudin: Real and Complex Analysis, 2nd ed. McGraw-Hill, New York (1974).
[1.51] S. Selberherr: Analysis and Simulation of Semiconductor Devices. Springer-Verlag, Wien-New York (1984).
[1.52] A. Shubin: Pseudodifferential Operators and Spectral Theory. Springer-Verlag, New York (1986).
[1.53] H. Steinrück: Asymptotic Analysis of the Quantum Liouville Equation. Report, Inst. f. Ang. u. Num. Mathematik, TU-Wien, Austria (1988).
[1.54] H. Steinrück: The Wigner-Poisson Problem in a Crystal: Existence, Uniqueness, Semiclassical Limit in the One-Dimensional Case. Report, Inst. f. Ang. u. Num. Mathematik, TU-Wien, Austria (1989).
[1.55] H. Steinrück: The One-Dimensional Wigner Poisson Problem and Its Relation to the Schrödinger Poisson Problem. Report, Inst. f. Ang. u. Num. Math., TU-Wien, Austria (1989).
[1.56] H. Steinrück: Private communication (1989).
[1.57] S. M. Sze: Physics of Semiconductor Devices. J. Wiley & Sons, New York (1981).
[1.58] V. I. Tatarskii: The Wigner Representation of Quantum Mechanics, Sov. Phys. Usp. *26*, 311–327 (1983).
[1.59] M. E. Taylor: Pseudodifferential Operators. Princeton University Press, Princeton (1981).
[1.60] F. Treves: Introduction to Pseudodifferential and Fourier Integral Operators, Vol. 1: Pseudodifferential Operators. Plenum Press, New York (1980).
[1.61] F. Treves: Introduction to Pseudodifferential and Fourier Integral Operators, Vol. 2: Fourier Integral Operators. Plenum Press, New York (1980).
[1.62] S. Ukai, T. Okabe: On the Classical Solution in the Large in Time of the Two-Dimensional Vlasov Equation. Osaka J. of Math. *15*, 245–261 (1978).
[1.63] E. Wigner: On the Quantum Correction for Thermodynamic Equilibrium. Physical Review *40*, 749–759 (1932).
[1.64] J. Yvon: La Théorie Statistique des Fluides (Actualités Scientifiques et Industrielles, No. 203). Hermann, Paris (1935).

From Kinetic to Fluid Dynamical Models 2

2.1 Introduction

Different approaches to the solution of the kinetic transport models discussed in Chapter 1 are possible. Although several promising attempts towards a numerical solution have been undertaken in the recent past (we only mention particle methods [2.10] and spectral methods [2.13]), the application of numerical methods remains to be a formidable task in general. Apart from that, solutions of the kinetic equations contain in many cases (e.g. close to equilibrium) a good deal of redundant information.

In this Chapter *fluid dynamical models* for semiconductors will be introduced. They represent a reasonable compromise between the contradictory requirements of physical accuracy and computational efficiency. Their common feature is the fact that the number of independent variables is reduced from seven (3 space + 3 velocity coordinates + time) to four (3 space coordinates + time). The dependent variables can usually be interpreted as averages (*moments*) of the phase space number density with respect to the velocity.

Two different approaches for the derivation of fluid dynamical models from kinetic equations exist. They will be presented in the Sections 2.2 and 2.3, respectively. The first is a perturbation argument. It exploits the smallness of a dimensionless parameter, namely the scaled *mean free path*, which appears in an appropriately scaled version of the Boltzmann equation. For the Boltzmann equation of gas dynamics an expansion of the solution in powers of the mean free path has been introduced by Hilbert [2.9] and, accordingly, bears his name. In the context of semiconductors, the Hilbert expansion has been recently carried out and thoroughly analyzed by Poupaud [2.11]. The method is presented in Section 2.2 for a standard bipolar model with the assumptions of low densities and small electric field. In this case the leading terms in the expansion are governed by the standard *drift diffusion equations* for semiconductors which have been derived by van Roosbroeck [2.18] for the first time.

A second way for obtaining fluid dynamical models are *moment methods*. Compared to the Hilbert expansion, their application requires a good deal of physical intuition or a-priori-knowledge about the solution of the Boltzmann equation. Also, the authors are not aware of any rigorous mathematical justification. The main ingredient of a moment method is an ansatz for the phase space density which prescribes the dependence on the velocity and which contains several parameters depending on position and time. After inserting the ansatz, the Boltzmann equation is multiplied by a number of linearly independent functions of velocity and integrated over the velocity space. The result are differential equations for the time and space dependent parameters. In some cases not all integrations can be carried out explicitely. Then the terms in question are usually replaced by phenomenological models. Two different moment methods are presented in Section 2.3. In the first one, the ansatz for the phase space density is motivated by the results of Section 2.2. It leads to a system which can be reduced to the drift diffusion equations by a perturbation argument. Because of the choice of the ansatz all the integrations can be carried out explicitely in this case. The second ansatz [2.2], usually called *shifted Maxwellian*, is motivated by the collision term of the Boltzmann equation for monatomic gases (see [2.3]). It leads to a modified version of the Euler equations of gas dynamics for a gas of charged particles in an electric field. The difference to the Euler equations is the appearance of relaxation terms. In general, these cannot be evaluated explicitely. They are usually replaced by relaxation time approximations. The resulting system (possibly including an extra heat conduction term) is referred to as the *hydrodynamic model* for semiconductors.

The main assumptions in the derivation of the drift diffusion equations are low carrier densities and small fields. The first assumption can be discarded of if a nonlinear collision term is used in the Boltzmann equation. This is necessary when the position space number density is large, which in turn is to be expected for large doping concentrations. The Hilbert expansion [2.11], which differs considerably from that of Section 2.2, is carried out in Section 2.4.

The hydrodynamic model is usually employed to give an appropriate description of high field phenomena. A different approach for the modeling of high field effects is presented in Section 2.5. A Hilbert expansion for a rescaled Boltzmann equation [2.12] leads to a hyperbolic drift equation (compare to Section 3.11). Unfortunately, the mobility coefficient in the drift term, which depends on the electric field, cannot be evaluated explicitely. However, very accurate data from measurements are available which can be used for fitting the coefficients in an ansatz describing the qualitative behaviour.

The recombination-generation terms (1.6.20), (1.6.21) describe direct band-band recombination caused by photon transitions. In practical situations several other recombination-generation mechanisms are important. In Section 2.6 models for band-trap capture and emission, Auger recombination, and impact ionization are presented.

2.2 Small Mean Free Path—The Hilbert Expansion

We consider the bipolar model derived in Section 1.6:

$$\begin{aligned} &\partial_t F_n + v_n(k)\cdot \mathrm{grad}_x F_n - \frac{q}{\hbar} E\cdot \mathrm{grad}_k F_n = Q_n(F_n) + I_n(F_n, F_p) \\ &\partial_t F_p + v_p(k)\cdot \mathrm{grad}_x F_p + \frac{q}{\hbar} E\cdot \mathrm{grad}_k F_p = Q_p(F_p) + I_p(F_n, F_p) \end{aligned} \tag{2.2.1}$$

with the low density approximations (1.3.92):

$$\begin{aligned} Q_n(F_n) &= \int_B \phi_n(x, k, k')\left(\exp\left(\frac{\varepsilon_n(k')}{k_B T}\right) F_n' - \exp\left(\frac{\varepsilon_n(k)}{k_B T}\right) F_n\right) dk' \\ Q_p(F_p) &= \int_B \phi_p(x, k, k')\left(\exp\left(\frac{-\varepsilon_p(k')}{k_B T}\right) F_p' - \exp\left(\frac{-\varepsilon_p(k)}{k_B T}\right) F_p\right) dk' \end{aligned} \tag{2.2.2}$$

for the collision terms and with the models

$$\begin{aligned} I_n(F_n, F_p) &= -\int_B g(x, k, k')\left(\exp\left(\frac{\varepsilon_n(k) - \varepsilon_p(k')}{k_B T}\right) F_n F_p' - 1\right) dk' \\ I_p(F_n, F_p) &= -\int_B g(x, k, k')\left(\exp\left(\frac{\varepsilon_n(k') - \varepsilon_p(k)}{k_B T}\right) F_n' F_p - 1\right) dk' \end{aligned} \tag{2.2.3}$$

for the recombination-generation rates. Note that "$'$" denotes evaluation at k' as in Chapter 1. If we assume that the conduction electrons are located close to the conduction band minimum and the holes close to the valence band maximum a parabolic band approximation for the energy-wave vector relations can be used. It reads (see Section 1.6)

$$\begin{aligned} \varepsilon_n(k) &= E_c + \frac{\hbar^2}{2m_n}|k|^2, \\ \varepsilon_p(k) &= E_v - \frac{\hbar^2}{2m_p}|k|^2, \end{aligned} \tag{2.2.4}$$

where E_c denotes the conduction band minimum, E_v the valence band maximum and m_n and m_p the effective masses of resp. electrons and holes. This gives the velocities

$$\begin{aligned} v_n(k) &= \frac{1}{\hbar}\,\mathrm{grad}_k\, \varepsilon_n(k) = \frac{\hbar}{m_n} k, \\ v_p(k) &= -\frac{1}{\hbar}\,\mathrm{grad}_k\, \varepsilon_p(k) = \frac{\hbar}{m_p} k. \end{aligned} \tag{2.2.5}$$

If the effective masses of electrons and holes are of the same order of magnitude, the exponential terms in (2.2.2) and (2.2.3) suggest the introduction of the reference velocity $\bar{v} = \sqrt{k_B T/m_n}$. For the following the equations (2.2.1) will be written in terms of the scaled wave vector and velocities

$$k_s = \frac{\hbar}{m_n \bar{v}} k,$$

$$v_{ns}(k_s) = k_s, \qquad v_{ps}(k_s) = \frac{m_n}{m_p} k_s.$$

An appropriate scaling of the collision and recombination-generation terms shows that they are proportional to the reciprocals of characteristic time constants, which can be interpreted as average relaxation times. With the average velocity $\bar{v}$, the relaxation times τ_C and τ_R, corresponding to collisions and recombination-generation respectively, can be written as

$$\tau_C = \iota_C/\bar{v}, \qquad \tau_R = \iota_R/\bar{v},$$

where ι_C and ι_R denote the *mean free paths* between two consecutive scattering and, respectively, recombination-generation events.
It is a well known fact (see e.g. [2.11]) that the relaxation times corresponding to the collision terms ($\sim 10^{-12}$ s) are much smaller than those of the recombination-generation terms ($\sim 10^{-9}$ s). Thus,

$$\iota_C \ll \iota_R$$

holds. We denote the ratio ι_C/ι_R by α^2 and introduce a reference length ι_0 by

$$\alpha = \iota_C/\iota_0 .$$

Then α can be interpreted as a scaled version of the mean free path between two scattering events.
The choices of the reference time τ_R and the reference field strength U_T/ι_0 complete the scaling. Here the reference voltage $U_T = k_B T/q$ is the so called *thermal voltage*. At room temperature the reference field strength is of the order of 10^2 V/cm. In VLSI applications, electric fields can be much larger. This shows that the analysis given below does not appropriately account for commonly occurring high field effects.
The scaled version of (2.2.1) is given by

$$\begin{aligned} &\alpha^2 \partial_t F_n + \alpha\{v_n(k)\cdot \operatorname{grad}_x F_n - E\cdot \operatorname{grad}_k F_n\} = Q_n(F_n) + \alpha^2 I_n(F_n, F_p) \\ &\alpha^2 \partial_t F_p + \alpha\{v_p(k)\cdot \operatorname{grad}_x F_p + E\cdot \operatorname{grad}_k F_p\} = Q_p(F_p) + \alpha^2 I_p(F_n, F_p), \end{aligned} \tag{2.2.6}$$

where the index "s" in the scaled quantities is now omitted for reasons of notational convenience. The scaled collision and recombination-generation terms have the same form as the unscaled versions (2.2.2) and (2.2.3) with the integration taken over $\mathbb{R}^3$ and the exponential terms

$$\exp\left(\frac{\varepsilon_n(k)}{k_B T}\right) \quad \text{and} \quad \exp\left(\frac{-\varepsilon_p(k)}{k_B T}\right)$$

replaced by

$$\exp(E_c + |k|^2/2) \quad \text{and} \quad \exp\left(-E_v + \frac{m_n}{m_p}|k|^2/2\right),$$

respectively, where E_c and E_v are the scaled (by $k_B T$) conduction band minimum and valence band maximum, respectively.

The Hilbert expansion is an expansion of solutions of (2.2.6) in terms of powers of the scaled mean free path α:

$$F_n = F_{n0} + \alpha F_{n1} + \cdots,$$
$$F_p = F_{p0} + \alpha F_{p1} + \cdots.$$

Equations for the coefficients in this ansatz are obtained by substitution into (2.2.6) and equating coefficients of equal powers of α. The equations for the leading order terms

$$Q_n(F_{n0}) = Q_p(F_{p0}) = 0$$

have the solutions

$$F_{n0} = n(x, t)M_n(k), \qquad F_{p0} = p(x, t)M_p(k),$$

where M_n, M_p denote the the scaled Maxwellians (see (1.3.95))

$$M_n(k) = \frac{1}{N_c}\exp(-|k|^2/2),$$
$$M_p(k) = \frac{1}{N_v}\exp\left(-\frac{m_n}{m_p}|k|^2/2\right).$$

The constants

$$N_c = (2\pi)^{3/2}, \qquad N_v = \left(\frac{2\pi m_p}{m_n}\right)^{3/2}$$

are chosen such that the integrals of the Maxwellians over the k-space are equal to one. This implies that the as yet unspecified quantities $n(x, t)$, $p(x, t)$ are scaled position space number densities of electrons and holes, respectively.

Equating coefficients of α in (2.2.6) leads to

$$\begin{aligned} M_n v_n \cdot (\operatorname{grad}_x n + nE) &= Q_n(F_{n1}), \\ M_p v_p \cdot (\operatorname{grad}_x p - pE) &= Q_p(F_{p1}). \end{aligned} \tag{2.2.7}$$

The analysis of these equations is facilitated by the following result. In its statement, several technical assumptions concerning the collision cross sections ϕ_n and ϕ_p are omitted.

Lemma (Poupaud [2.11]): *A) A necessary and sufficient condition for the solvability of an equation of the form*

$$Q_{n/p}(f) = g \tag{2.2.8}$$

is

$$\int_{\mathbb{R}^3} g\, dk = 0. \tag{2.2.9}$$

If (2.2.9) *holds,* (2.2.8) *has a one dimensional linear manifold of solutions of the form* $f = f_{n/p} + q_{n/p} M_{n/p}$ *where* $f_{n/p}$ *denotes a particular solution and* $q_{n/p}$ *is a parameter.*
B) The equations

$$Q_n(h_n) = M_n v_n, \qquad Q_p(h_p) = M_p v_p$$

have solutions $h_n(x, k), h_p(x, h) \in \mathbb{R}^3$ *which satisfy*

$$\begin{aligned} \int_{\mathbb{R}^3} v_n \otimes h_n\, dk &= -\mu_n(x) I_3 < 0, \\ \int_{\mathbb{R}^3} v_p \otimes h_p\, dk &= -\mu_p(x) I_3 < 0, \end{aligned} \tag{2.2.10}$$

where I_3 *is the three dimensional unity matrix and* $a \otimes b = ab^T$, *for* $a, b \in \mathbb{R}^3$, *denotes the tensor product. Furthermore, the j-th component of* $h_{n/p}$ *is an odd function of the j-th component of k and has the form*

$$h_{n/p,j}(k) = \bar{h}(k_j, |P_j k|),$$

where $|P_j k|$ *denotes the Euclidian norm of the projection of k onto the plane perpendicular to the* k_j*-direction.*

In terms of the scaled current densities

$$\begin{aligned} J_n(x, t) &= \mu_n(\text{grad}_x\, n + nE), \\ J_p(x, t) &= -\mu_p(\text{grad}_x\, p - pE) \end{aligned} \tag{2.2.11}$$

the solution of (2.2.7) is given by

$$\begin{aligned} F_{n1} &= J_n \cdot h_n/\mu_n + q_n M_n \\ F_{p1} &= -J_p \cdot h_p/\mu_p + q_p M_p, \end{aligned}$$

where $q_n(x, t)$ and $q_p(x, t)$ are as yet unspecified.
Equating coefficients of $\alpha^j, j \geqslant 2$, in (2.2.6) gives

$$\begin{aligned} &\partial_t F_{n,j-2} + v_n \cdot \text{grad}_x\, F_{n,j-1} - E \cdot \text{grad}_k\, F_{n,j-1} \\ &\quad = Q_n(F_{nj}) + I_n(F_{n,j-2}, F_{p,j-2}), \\ &\partial_t F_{p,j-2} + v_p \cdot \text{grad}_x\, F_{p,j-1} + E \cdot \text{grad}_k\, F_{p,j-1} \\ &\quad = Q_p(F_{pj}) + I_p(F_{n,j-2}, F_{p,j-2}). \end{aligned}$$

Assuming that we know the terms up to the order $j-1$, these are equations of the form (2.2.8) for F_{nj} and F_{pj}. The solvability condition (2.2.9) for $j=2$ implies

$$\partial_t n - \operatorname{div}_x J_n = -R, \qquad \partial_t p + \operatorname{div}_x J_p = -R, \tag{2.2.12}$$

where the position space recombination-generation rate R is given by

$$R = A(x)(np - n_i^2). \tag{2.2.13}$$

The quantities n_i and $A(x)$ are defined by

$$n_i = \sqrt{N_c N_v}\exp(-E_g/2), \qquad A(x) = n_i^{-2}\iint_{\mathbb{R}^3\times\mathbb{R}^3} g(x,k,k')\,dk\,dk',$$

where $E_g = E_c - E_v$ denotes the scaled *bandgap* of the semiconductor. The recombination-generation term R has the form of a mass action law with the reaction rate $A(x)$ and the scaled *intrinsic number* n_i.

Unscaled versions of (2.2.11) and (2.2.12) are given by the system of partial differential equations

$$J_n = q\mu_n(U_T \operatorname{grad} n + nE), \qquad q\partial_t n - \operatorname{div} J_n = -qR,$$
$$J_p = -q\mu_p(U_T \operatorname{grad} p + pE), \qquad q\partial_t p + \operatorname{div} J_p = -qR, \tag{2.2.14}$$

called the *drift diffusion equations* of semiconductors. This name originates from the type of dependence of the current densities on the carrier densities and the electric field. The current densities are the sums of drift terms (with the mobilities μ_n and μ_p) and diffusion terms (with the diffusivities $D_n = \mu_n U_T$ and $D_p = \mu_p U_T$). The equations

$$D_n/\mu_n = D_p/\mu_p = U_T \tag{2.2.15}$$

are known as the *Einstein relations.*

As before, the unscaled quantities in (2.2.14) are denoted by the same symbols as their scaled counterparts. In particular, the scaling introduced above implies the reference value $(k_B T m_n)^{3/2}/\hbar^3$ for number densities. Accordingly, the unscaled recombination-generation rate R has the form (2.2.13) with the unscaled intrinsic number given by

$$n_i = \left(\frac{2\pi k_B T\sqrt{m_n m_p}}{\hbar^2}\right)^{3/2} \exp\left(\frac{-E_g}{2k_B T}\right).$$

Another result of the scaling procedure is that the mobilities are inversely proportional to the square roots of the effective masses. The fact that heavy carriers are slower than light ones is responsible for the Gunn effect discussed in Section 4.8.

For a self consistent treatment of the electric field, (2.2.14) has to be supplemented by the Poisson equation (1.6.23), (1.6.25). The resulting system, originally due to van Roosbroeck [2.18], is called the *basic semiconductor*

device equations. It has been the subject of an intensive mathematical, numerical as well as physical scrutiny. A presentation of the most important results is the subject of the Chapters 3 and 4.
Obviously, the leading terms in the Hilbert expansion cannot satisfy general initial and boundary conditions for the Boltzmann equation. If the prescribed data do not have the form of Maxwellians, initial and boundary layers have to be introduced for the construction of a complete formal approximation of the solution. This has been carried out by Poupaud [2.11], who also gave a justification for the formal approximation procedure presented above.

2.3 Moment Methods—The Hydrodynamic Model

A second way for the derivation of fluid dynamical models from the Boltzmann equation are moment methods. As mentioned in the introduction they consist of an ansatz for the distribution function and a system of necessary conditions for solutions of the Boltzmann equation. How "close to sufficiency" these conditions are, can in general only be judged by physical reasoning.
In this Section we consider the classical Boltzmann equation for one type of charge carrier (say electrons):

$$\partial_t F + v \cdot \operatorname{grad}_x F - \frac{q}{m} E \cdot \operatorname{grad}_v F = Q(F) \tag{2.3.1}$$

with a low density collision term:

$$Q(F) = \int_{\mathbb{R}^3} \phi(x, v, v')(MF' - M'F)\, dv',$$

where the Maxwellian is given by

$$M(v) = \left(\frac{m}{2\pi k_B T}\right)^{3/2} \exp\left(\frac{-m|v|^2}{2k_B T}\right).$$

The j-th order moment of the distribution function F is defined as the tensor $M^{(j)}$ of rank j, whose components, which depend on position and time, are given by

$$M^{(j)}_{i_1,\dots,i_j}(x, t) = \int_{\mathbb{R}^3} v_{i_1} \cdots v_{i_j} F(x, v, t)\, dv \qquad \text{for} \qquad j \geqslant 1,$$

$$M^{(0)}(x, t) = \int_{\mathbb{R}^3} F(x, v, t)\, dv.$$

The relevance of the moments is due to the fact that they are related to physical quantities in a simple way. Examples are:

$$\begin{array}{lll} M^{(0)} & n & \text{position space number density,} \\ -qM^{(1)} & J & \text{current density,} \\ \frac{m}{2}\operatorname{tr}(M^{(2)}) & \mathscr{E} & \text{energy density,} \end{array} \tag{2.3.2}$$

where tr denotes the trace (of a matrix).

Equations for the moments can be derived by multiplying the Boltzmann equation by powers of v and by integrating over the velocity space. This leads to the infinite hierarchy

$$\begin{aligned} &\partial_t M^{(0)} + \operatorname{div}_x M^{(1)} = 0, \\ &\partial_t M^{(1)} + \operatorname{div}_x M^{(2)} + \frac{q}{m} M^{(0)} E = \int_{\mathbb{R}^3} vQ(F)\, dv, \\ &\partial_t M^{(2)} + \operatorname{div}_x M^{(3)} + 2\frac{q}{m} M^{(1)} \otimes E = \int_{\mathbb{R}^3} v \otimes v\, Q(F)\, dv, \\ &\ldots \end{aligned} \tag{2.3.3}$$

According to the physical interpretation of the moments these equations represent conservation laws. The first one—already discussed in Chapter 1—represents the conservation of charges. The practical use of the hierarchy (2.3.3) is limited on one hand by the fact that all the moments are coupled, such that truncation of the hierarchy does not give a closed system for a finite number of the moments. On the other hand, the terms originating from the collision integral do not depend on the moments in a simple way in general. These difficulties are overcome by making an ansatz for the distribution function which a priorily fixes its dependence on the velocity. This usually introduces position and time dependent parameters which then are determined by a truncated version of (2.3.3).

Derivation of the Drift Diffusion Model

The first moment method presented here is motivated by the results of the Hilbert expansion. With the particular solution $h(x, v)$ of the equation

$$Q(h) = vM,$$

whose properties are given in the Lemma of the preceding Section, we make the ansatz

$$F(x, v, t) = n(x, t)M(v) + \frac{1}{\mu(x)k_B T} J(x, t) \cdot h(x, v), \tag{2.3.4}$$

where $\mu(x)$ satisfies

$$\int_{\mathbb{R}^3} v \otimes h(x, v)\, dv = -\mu(x) U_T I_3.$$

Straightforward integration gives the relations for the moments of (2.3.4)

$$M^{(0)} = n, \qquad -qM^{(1)} = J, \qquad M^{(2)} = n\frac{k_B T}{m} I_3$$

and a comparison with (2.3.2) shows that the choice of the symbols n and J in (2.3.4) is justified. The energy density is given by

$$\mathscr{E} = \frac{3}{2} k_B T n.$$

The first two equations in (2.3.3) imply

$$q\partial_t n - \operatorname{div}_x J = 0,$$

$$-\frac{\mu m}{q}\partial_t J + q\mu(U_T \operatorname{grad}_x n + nE) = J. \tag{2.3.5}$$

The factor $\mu m/q$ multiplying the time derivative of the current density is the current density relaxation time. It is usually assumed that this relaxation time is small compared to characteristic time constants in the drift diffusion approximation (2.2.14) (see [2.16]). Thus, the term $-(\mu m/q)\partial_t J$ in (2.3.5) is neglected and a unipolar drift diffusion model is obtained from (2.3.5).

The Hydrodynamic Model

A different ansatz for the distribution function is motivated by the collision term for a dilute gas of rigid spheres (see [2.3]). For this case the null manifold of the collision term is five dimensional and its elements can be written as

$$F(x, v, t) = n\left(\frac{m}{2\pi k_B T_e}\right)^{3/2} \exp\left(\frac{-m|v - \bar{v}|^2}{2k_B T_e}\right), \tag{2.3.6}$$

where n, T_e and the three components of $\bar{v}$ are the free parameters (see [2.3, pp. 78ff]). A distribution function of the form (2.3.6) is called *displaced* (or *shifted*) *Maxwellian.* Here, (2.3.6) can be used as an ansatz for a moment method with the parameters depending on position and time. n, T_e, and $\bar{v}$ can be interpreted as number density, effective temperature, and mean velocity, respectively. Since an effective temperature different from the lattice temperature is allowed, it is plausible that certain high field effects are taken into account by (2.3.6). For the moments of (2.3.6) we have

$$M^{(0)} = n, \qquad M^{(1)} = n\bar{v}, \qquad M^{(2)} = n\left(\bar{v} \otimes \bar{v} + \frac{k_B T_e}{m} I_3\right),$$

which implies that the energy density can be written as the sum of a kinetic and a thermal contribution:

$$\mathscr{E} = n\left(\frac{m|\bar{v}|^2}{2} + \frac{3}{2}k_B T_e\right).$$

For the determination of the unknowns the first two equations and the trace of the third equation in (2.3.3) are used. Straightforward but lengthy computations lead to the system, usually referred to as the hydrodynamic semiconductor model:

$$\begin{aligned} &\partial_t n + \operatorname{div}(n\bar{v}) = 0, \\ &\partial_t \bar{v} + (\bar{v}\cdot \operatorname{grad})\bar{v} + \frac{k_B}{mn}\operatorname{grad}(nT_e) + \frac{q}{m}E = (\partial_t \bar{v})_c, \\ &\partial_t T_e + \frac{2}{3}T_e \operatorname{div}\bar{v} + \bar{v}\cdot \operatorname{grad} T_e = (\partial_t T_e)_c, \end{aligned} \tag{2.3.7}$$

where we denoted

$$(\bar{v}\cdot \operatorname{grad})\bar{v} = \bar{v}_1 \frac{\partial \bar{v}}{\partial x_1} + \bar{v}_2 \frac{\partial \bar{v}}{\partial x_2} + \bar{v}_3 \frac{\partial \bar{v}}{\partial x_3}, \qquad \bar{v} = (\bar{v}_1, \bar{v}_2, \bar{v}_3).$$

If the terms on the right-hand sides, stemming from the collision terms, are omitted, then (2.3.7) are the Euler equations of gas dynamics for a gas of charged particles in an electric field. A weakness of the ansatz, when applied to the semiconductor problem, is displayed by exactly these terms. They are given by

$$(\partial_t \bar{v})_c = \frac{1}{n}\int_{\mathbb{R}^3} vQ(F)\,dv,$$

$$(\partial_t T_e)_c = \frac{m}{3k_B n}\int_{\mathbb{R}^3} |v|^2 Q(F)\,dv - \frac{2m}{3k_B n}\bar{v}\cdot \int_{\mathbb{R}^3} vQ(F)\,dv.$$

In general, it is impossible to obtain the dependence of the integrals on the parameters explicitely. For the purpose of simulation the collision terms are often replaced by relaxation time approximations. We refer the reader to [2.1] for a model which seems to meet with approval in the literature.
The problems at the end of this section shed light on the mathematical properties of the hydrodynamic semiconductor model.
In [2.2], where the model (2.3.7) in the context of semiconductors has been introduced, an additional heat conduction term

$$-\frac{2}{3k_B n}\operatorname{div}(\varkappa \operatorname{grad} T_e)$$

was added to the left hand side of the temperature continuity equation. Here, $\varkappa$ denotes the heat conductivity of the electron gas.
The type of the differential equations in (2.3.7) changes at the transition from subsonic flow to supersonic flow. In the supersonic regime the occurance of electron shock waves is possible. The interested reader can find a brief discussion of the nonlinear wave structure in [2.5].
A different model with an account for energy flow has been proposed in [2.7]. In [2.8] a simplified version has been derived by a perturbation argument, which can be interpreted as a modification of the drift diffusion model. Its special appeal lies in the fact that high field effects are modelled in a way compatible with experiment.

2.4 Heavy Doping Effects—Fermi-Dirac Distributions

In this Section we consider cases where the distribution function is not necessarily small compared to one. Therefore a nonlinear collision term has to be used in the Boltzmann equation. A scaled version of the classical unipolar model (2.3.1) reads

$$\alpha^2 \partial_t F + \alpha(v \cdot \operatorname{grad}_x F - E \cdot \operatorname{grad}_v F) = Q(F), \tag{2.4.1}$$

where the collision integral is given by

$$Q(F) = \int_{\mathbb{R}^3} \phi(x, v, v')(MF'(1 - F) - M'F(1 - F'))\, dv'$$

(see Section 1.3) with the Maxwellian

$$M(v) = (2\pi)^{-3/2} \exp(-|v|^2/2).$$

As in Section 2.2, α denotes the scaled mean free path and we introduce a power series expansion of F in terms of α:

$$F = F_0 + \alpha F_1 + \cdots .$$

The equation

$$Q(F_0) = 0$$

implies [2.11] that the leading term is a *Fermi-Dirac distribution*:

$$F_0 = F_D(|v|^2/2 - \Phi),$$

where

$$F_D(u) = \frac{1}{1 + e^u}$$

holds and $\Phi = \Phi(x, t)$ is the Fermi energy (see Chapter 1). Equating coefficients of α in (2.4.1) gives

$$F_0(1 - F_0)v \cdot (\operatorname{grad}_x \Phi + E) = L(\Phi)F_1\,, \tag{2.4.2}$$

where $L(\Phi)$ is the Frechet derivative of Q evaluated at F_0:

$$L(\Phi)f = \int_{\mathbb{R}^3} \phi(x, v, v')(M(f'(1 - F_0) - F_0' f) - M'(f(1 - F_0') - F_0 f'))\, dv'\,.$$

A result [2.11], which is in the spirit of the Lemma in Section 2.2, states that an equation of the form

$$L(\Phi)f = g \tag{2.4.3}$$

has a solution if and only if

$$\int_{\mathbb{R}^3} g\, dv = 0 \tag{2.4.4}$$

holds and that the null space of $L(\Phi)$ is spanned by $F_0(1 - F_0)$. In [2.11] it is shown that the equation

$$L(\Phi)h = vF_0(1 - F_0)$$

has a solution with the property that the matrix

$$\Pi(\Phi) = -\int_{\mathbb{R}^3} v \otimes h \, dv$$

is positive definite. The solution of (2.4.2) can be written as

$$F_1 = (\mathrm{grad}_x \Phi + E)\cdot h + q(x, t)F_0(1 - F_0).$$

Equating coefficients of α^2 in (2.4.1) gives an equation of the form (2.4.3) for F_2. The solvability condition (2.4.4) implies

$$\partial_t n - \mathrm{div}\, J = 0, \tag{2.4.5}$$

where the scaled electron density and current density are given by

$$n(\Phi) = \int_{\mathbb{R}^3} F_0 \, dv, \qquad J = \Pi(\Phi)(\mathrm{grad}\, \Phi + E).$$

Equation (2.4.5) is a nonlinear parabolic equation for the Fermi energy Φ. Since the electric field can be expressed in terms of the electrostatic potential as $E = -\mathrm{grad}\, V$ the current density is proportional to the gradient of the *quasi-Fermi potential* $\varphi_n = \Phi - V$. The continuity equation in terms of the quasi-Fermi potential reads

$$\partial_t n(\varphi_n + V) - \mathrm{div}(\Pi(\varphi_n + V)\, \mathrm{grad}\, \varphi_n) = 0.$$

The perturbation argument leading to the fluid dynamical model (2.4.5) has been justified in [2.11]. A practical application would be facilitated by some, at least qualitative, knowledge of the dependence of the matrix Π on its argument. At present, the authors are not aware of results in that direction.

2.5 High Field Effects—Mobility Models

As mentioned in Section 2.2 the validity of the Hilbert expansions presented so far is restricted to the case of small electric fields. A totally new situation occurs if the scaled electric field is large, say of the order of magnitude of α^{-1}. The appropriately rescaled Boltzmann equation then reads

$$\alpha\partial_t F + \alpha v\cdot \mathrm{grad}_x F - E\cdot \mathrm{grad}_v F = Q(F), \tag{2.5.1}$$

where we also introduced a time scale faster than the one used in Section 2.2. For the collision term we choose a (linear) low density approximation. The leading term of the Hilbert expansion satisfies

$$-E\cdot \mathrm{grad}_v F_0 = Q(F_0). \tag{2.5.2}$$

This equation does not only occur in the context of the Hilbert expansion but is, by itself, of physical interest as a model for stationary, homogeneous situations. It is well known that it does not have an integrable solution in general, which would be necessary for the definition of the position space number density n. The nonexistence of such a solution is called *runaway phenomenon*. For a mathematical analysis of related questions we refer the reader to [2.4]. The occurance of runaway depends on the *collision frequency*

$$\lambda(v) = \int_{\mathbb{R}^3} \varphi(v, v')M(v')\,dv'.$$

The following result can be found in [2.12]:

Lemma: *A necessary condition for the existence of a positive, integrable solution of* (2.5.2) *is*

$$\int_0^\infty \lambda(sE)\,ds = \infty,$$

i.e. the collision frequency does not decay too fast in the direction of the electric field.

For the following we make the assumption that (2.5.2) has a positive solution $M_E(v)$ which satisfies

$$\int_{\mathbb{R}^3} M_E(v)\,dv = 1.$$

Then the leading term in the Hilbert expansion has the form

$$F_0 = n(x, t)M_E(v).$$

Equating coefficients of α in (2.5.1) gives

$$\partial_t(nM_E) + \operatorname{div}(vM_E n) - E\cdot \operatorname{grad}_v F_1 = Q(F_1).$$

Integration in the v-direction implies

$$\partial_t n + \operatorname{div}(\overline{v}(E)n) = 0,$$

with the average velocity defined by

$$\overline{v}(E) = \int_{\mathbb{R}^3} vM_E\,dv.$$

(2.5.3) is a hyperbolic equation for n. Thus, n might have discontinuities. A similar situation occurs in gas dynamics where the Euler equations allow for shocks. These shocks are eliminated by introducing viscosity terms and, thus, considering the Navier Stokes equations, which can be derived from the Boltzmann equation of gas dynamics by the Chapman-Enskog method. A similar approach [2.12] leads to a diffusion term of order α in the present situation:

$$\partial_t n + \operatorname{div}(\overline{v}(E)n - \alpha D(E)\operatorname{grad} n) = 0,$$

where $D(E)$ is the (positive definite) diffusivity tensor.

It is easy to see that

$$\bar{v}(0) = 0, \qquad \operatorname{grad}_E \bar{v}(0) = -\mu_0 I_3 < 0 \tag{2.5.4}$$

holds for the average velocity, where the low field mobility μ_0 can be computed as in Section 2.2. Unfortunately, it is impossible to obtain the dependence of $\bar{v}$ on the field explicitely. For simulation purposes it is common to use an ansatz fitted to experimental results. As a first step, it is certainly reasonable to write $\bar{v}$ in the form

$$\bar{v}(E) = -\mu(|E|)E$$

with $\mu(0) = \mu_0$. Here (2.5.4) is taken into account and it is assumed that the direction of the average velocity is given by the direction of the field.
Experiments show the effect of *velocity saturation* at large electric fields:

$$\lim_{|E| \to \infty} |\bar{v}(E)| = v_{\text{sat}}.$$

A model for the mobility which shows this behaviour has been derived in [2.7] and [2.8]:

$$\mu(|E|) = \frac{2\mu_0}{1 + \sqrt{1 + (2\mu_0|E|/v_{\text{sat}})^2}}. \tag{2.5.5}$$

Several other models which are similar to the above are used (see [2.16] for an overview and references). In numerical simulations it is common to use a drift diffusion model with a mobility like in (2.5.5) and to compute the diffusivity from the Einstein relations (2.2.15). However, it is likely that in reality the Einstein relations are violated for high electric fields. In particular, the diffusivity cannot be expected to decay for large electric fields.
The *transferred electron effect* in two-valley semiconductors (e.g. GaAs) with large effective mass of the electrons in the upper valley leads to a nonmonotone velocity-field relation (see Fig. 2.5.1). It has already been mentioned

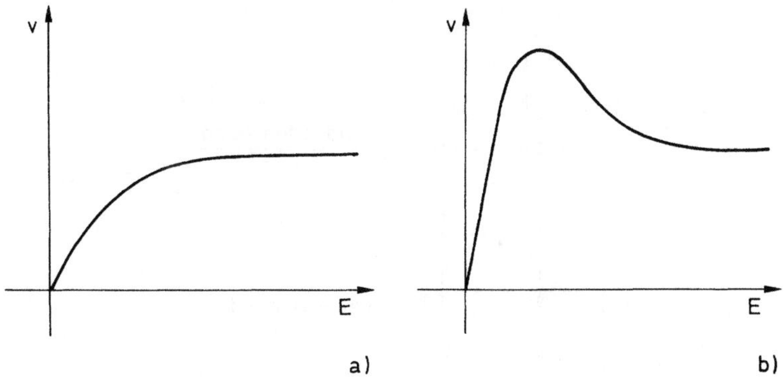

Fig. 2.5.1 Velocity vs. field for (a) Si, (b) GaAs

that large effective mass means low mobility. This explains the effect on the velocity-field relation because for higher electric fields the electron density in the upper band increases (see [2.17] for details). In these cases (2.5.5), which is an acceptable model for Si and Ge, has to be changed accordingly.

2.6 Recombination-Generation Models

The recombination-generation rate

$$R = A(np - n_i^2)$$

derived in Section 2.2 is a model for direct band-band recombination caused by photon transitions. It is well known that other recombination-generation mechanisms are much more important in semiconductor devices. In this Section we discuss three such mechanisms which are usually taken into account in semiconductor device modelling. This will, however, not be done on the level of the Boltzmann equation. Instead, models will be presented which can be used directly in the fluid dynamical equations.

The first mechanism to be considered is *Auger recombination.* Two different processes are shown schematically in Fig. 2.6.1:

a) *Electron capture:* An electron moves from the conduction band to the valence band and recombines with an hole there. Its energy is transferred to another electron in the conduction band.

b) *Hole capture:* An electron moves from the conduction band to the valence band and recombines with an hole there. Its energy is transferred to another hole in the valence band.

The processes acting in the opposite directions:

c) *electron emission,*
d) *hole emission*

are also possible. The rates of these processes in position space are modelled by mass action laws. With the low density assumption they are given by

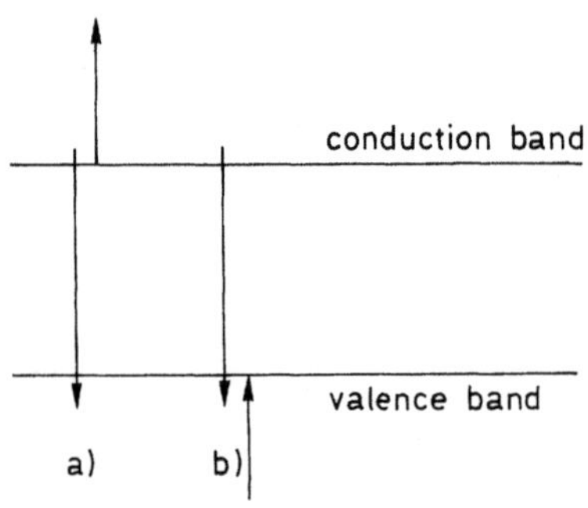

Fig. 2.6.1 Auger recombination

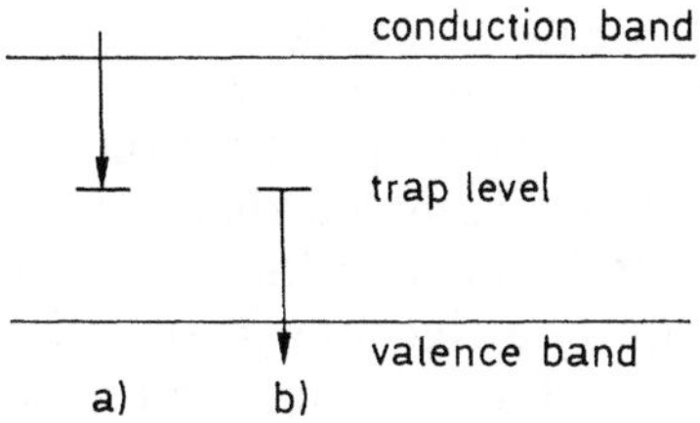

Fig. 2.6.2 Band-trap capture

a) $C_n n^2 p$,
b) $C_p n p^2$,
c) $\tilde{C}_n n$,
d) $\tilde{C}_p p$,

The *principle of detailed balance* (see Section 1.3) states that each process balances its counterpart in thermal equilibrium. The thermal equilibrium condition $np = n_i^2$ derived in Section 2.2 implies

$$\tilde{C}_n = n_i^2 C_n, \qquad \tilde{C}_p = n_i^2 C_p.$$

The total Auger recombination-generation rate can then be written as

$$R_{AU} = (C_n n + C_p p)(np - n_i^2). \tag{2.6.1}$$

Next we consider *band-trap capture* and *emission*. These are important in the presence of impurities which generate additional energy (trap) levels in the forbidden band. The processes shown in Fig. 2.6.2 are

a) *Electron capture:* An electron moves from the conduction band to an unoccupied trap.
b) *Hole capture:* An electron moves from an occupied trap to the valence band. An hole disappears.

Again, the processes in the opposite direction are also possible. The rates are given by

a) $C_a n(N_{tr} - n_{tr})$,
b) $C_b p n_{tr}$,
c) $C_c n_{tr}$,
d) $C_d(N_{tr} - n_{tr})$,

where N_{tr} denotes the density of traps and n_{tr} the density of occupied traps. The densities n, p, and n_{tr} satisfy the differential equations

$$\begin{aligned} \partial_t n - \frac{1}{q} \operatorname{div} J_n &= C_c n_{tr} - C_a n(N_{tr} - n_{tr}), \\ \partial_t p - \frac{1}{q} \operatorname{div} J_p &= C_d(N_{tr} - n_{tr}) - C_b p n_{tr}, \\ \partial_t n_{tr} &= C_a n(N_{tr} - n_{tr}) - C_c n_{tr} + C_b p n_{tr} - C_d(N_{tr} - n_{tr}). \end{aligned} \tag{2.6.2}$$

Since the impurities are assumed to have fixed positions, the trapped electrons do not contribute to the current flow.

In the classical theory of band-trap transitions, due to Shockley, Read [2.15], and Hall [2.6], it was assumed that the relaxation of n_{tr} towards equilibrium happens much faster than the relaxation of n and p. Although the authors are not aware of a rigorous justification of this assumption it will be adopted here because the resulting recombination-generation model has been generally accepted. If a moderate time scale is considered, this assumption justifies setting $\partial_t n_{tr} = 0$ in (2.6.2). From the resulting algebraic equation n_{tr} can be computed:

$$n_{tr} = N_{tr} \frac{C_a n + C_d}{C_a n + C_c + C_b p + C_d} .$$

This implies for the Shockley-Read-Hall (SRH) recombination-generation rate:

$$R_{SRH} = \frac{np - n_1 p_1}{\tau_p(n + n_1) + \tau_n(p + p_1)} ,$$

where the densities n_1 and p_1 are given by

$$n_1 = C_c/C_a , \qquad p_1 = C_d/C_b$$

and the carrier life times τ_n and τ_p by

$$\tau_n = (C_a N_{tr})^{-1} , \qquad \tau_p = (C_b N_{tr})^{-1} .$$

The requirement that the recombination-generation rate vanishes in thermal equilibrium implies $n_1 p_1 = n_i^2$. A more detailed analysis shows that n_1 and p_1 depend on the location of the trap level [2.17]. In particular,

$$n_1 = p_1 = n_i$$

holds if the trap level is in the middle of the forbidden band.

Strictly speaking, the above considerations for Auger and SRH recombination are only valid close to thermal equilibrium and for small electric fields. This observation is of particular importance for the generation process corresponding to Auger recombination at high electric fields. An effect called *impact ionization* which cannot be modelled by (2.6.1) is observed. A phenomenological description is provided by the commonly used model (see [2.16])

$$R_{II} = -\alpha_n |J_n|/q - \alpha_p |J_p|/q , \tag{2.6.4}$$

where the ionization rates α_n and α_p are strongly field dependent. A simple choice is the so called *lucky drift model* [2.14]:

$$\alpha_n = \alpha_n^\infty \exp(-E_n^{crit}/|E|) , \qquad \alpha_p = \alpha_p^\infty \exp(-E_p^{crit}/|E|) , \tag{2.6.5}$$

where α_n^∞ and α_p^∞ are maximal ionization rates, and E_n^{crit} and E_p^{crit} are critical field strengths.

Problems

2.1 Derive the following equation for the rate of change of the energy:

$$\frac{d}{dt}\int_{\mathbb{R}_x^3} \mathscr{E}\,dx - \int_{\mathbb{R}_x^3} J\cdot E\,dx = \frac{m}{2}\int_{\mathbb{R}_x^3}\int_{\mathbb{R}_v^3} |v|^2 Q(F)\,dv\,dx$$

from (2.3.2), (2.3.3). Assume that $M^{(3)}$ vanishes as $|x| \to \infty$.

2.2 Definition: A flow is called irrotational if its velocity vector satisfies curl $\bar{v} = 0$. Simplify the velocity equation of the hydrodynamic model (2.3.7) assuming that the flow of the electron gas is irrotational.
Hint: Prove $(\bar{v}\cdot\text{grad})\bar{v} = \frac{1}{2}\,\text{grad}\,(|\bar{v}|^2) - \bar{v} \times \text{curl}\,\bar{v}$.

2.3 Definition: A flow is called incompressible if its velocity vector satisfies div $\bar{v} = 0$. Simplify the hydrodynamic model (2.3.7) assuming that the flow of the electron gas is incompressible.
Take the relaxation model for the temperature:

$$(\partial_t T_e)_c = -\frac{(T_e - T)}{\tau_T},$$

where $T > 0$ denotes the (constant) lattice temperature and $\tau_T > 0$ the (constant) temperature relaxation time. Solve the electron continuity equation and the temperature equation (in terms of the velocity field $\bar{v}$) with the initial data

$$n(x, t = 0) = n_I(x), \qquad x \in \mathbb{R}^3,$$
$$T_e(x, t = 0) = T_I(x), \qquad x \in \mathbb{R}^3.$$

2.4 Consider the steady state hydrodynamic model (2.3.7), i.e. set $\partial_t n = 0$, $\partial_t \bar{v} = 0$, $\partial_t T_e = 0$. Assume infinitely fast temperature relaxation $(\partial_t T_e)_c = 0$. Prove that $T_e = Kn^{2/3}$, where $K > 0$ is an arbitrary constant, is a solution of the temperature equation.
Remark: $p = k_B n T_e$ is called pressure of the electron gas and a relation of the form $p = p(n)$ is called equation of state. Thus, under the above assumptions, the electron gas has the equation of state $p = Kk_B n^{5/3}$.

2.5 Consider the one-dimensional steady state hydrodynamic model (2.3.7) with an equation of state $p = p(n)$ (see Problem 2.4) and with the velocity relaxation term

$$(\partial_t \bar{v})_c = -\frac{\bar{v}}{\tau_{\bar{v}}},$$

where the velocity relaxation time $\tau_{\bar{v}}$ is positive. Prove that the nonlinear current dependent drift-diffusion equation

$$J = q\mu\left[\left(\frac{m}{q^3}\frac{J^2}{n} + \frac{1}{q}p(n)\right)_x + nE\right]$$

holds with the electron mobility $\mu = \tau_{\bar{v}} q/m$.
Hint: Note that $J \equiv$ const. holds in the one-dimensional steady state case!
Remark: We conclude that the "classical" drift-diffusion model (2.2.14) is obtained from the hydrodynamic model (at least in the steady state one-dimensional case with velocity

relaxation) by neglecting terms of order J^2 (small current limit) and using the equation of state $p(n) = (U_T q)n$ (linear pressure-density relationship).

2.6 Definition: (i) A flow is called subsonic, if its velocity field satisfies

$$|\bar{v}| < \sqrt{p'(n)/m}.$$

$\sqrt{p'(n)}$ is called soundspeed of the flow.
(ii) Let

$$F\colon \mathbb{R}^4 \to \mathbb{R}, \qquad F = F(r, q, u, x)$$

be a function. The equation

$$F(u_{xx}, u_x, u, x) = 0$$

is called (everywhere) elliptic if $\partial F/\partial r > 0$ on $\mathbb{R}^4$.
Prove that the drift-diffusion equation of Problem 2.5 together with the conservation equation $J_x = 0$ leads to an elliptic equation for the density n if and only if the flow is subsonic.

References

[2.1] G. Baccarani, M. R. Wordeman: An Investigation of Steady-State Velocity Overshoot Effects in Si and GaAs Devices. Solid State Electr. *28*, 407–416 (1985).

[2.2] K. Bløtekjær: Transport Equations for Electrons in Two-Valley Semiconductors. IEEE Trans. Electron. Devices *ED-17*, 38–47 (1970).

[2.3] C. Cercignani: The Boltzmann Equation and Its Applications. Springer-Verlag, New York (1988).

[2.4] G. Frosali: Functional-Analytic Techniques in the Study of Time-Dependent Electron Swarms in Weakly Ionized Gases. Preprint, Istituto di Matematica, Università di Ancona (1988).

[2.5] C. L. Gardner, J. W. Jerome, D. J. Rose: Numerical Methods for the Hydrodynamic Device Model: Subsonic Flow. Presented at the meeting on Mathematische Modellierung und Simulation elektrischer Schaltungen, Oberwolfach (1988).

[2.6] R. N. Hall: Electron-Hole Recombination in Germanium. Physical Review *87*, 387 (1952).

[2.7] W. Hänsch, M. Miura-Mattausch: The Hot-Electron Problem in Small Semiconductor Devices. J. Appl. Phys. *80*, 650–656 (1986).

[2.8] W. Hänsch, C. Schmeiser: Hot Electron Transport in Semiconductors. ZAMP *40*, 440–455 (1989).

[2.9] D. Hilbert: Math. Ann. *72*, 562 (1912).

[2.10] B. Niclot, P. Degond, F. Poupaud: Deterministic Particle Simulations of the Boltzmann Transport Equation of Semiconductors. J. Comp. Phys. *78*, 313–350 (1988).

[2.11] F. Poupaud: Etude Mathematique et Simulations Numeriques de Quelques Equations de Boltzmann. Thesis, Univ. Paris 6 (1988).

[2.12] F. Poupaud: Runaway Phenomena and Fluid Approximation Under Strong Electric Field in Semiconductor Theory. Manuscript, Laboratoire de Mathématiques, Université de Nice (1988).

[2.13] C. A. Ringhofer: A Spectral Method for Numerical Simulation of Quantum Tunneling Phenomena. SIAM J. Numer. Anal. (1989) (to appear).

[2.14] W. Shockley: Problems Related to *p-n* Junctions in Silicon. Solid State Electr. *2*, 35–67 (1961).

[2.15] W. Shockley, W. T. Read: Statistics of the Recombinations of Holes and Electrons. Physical Review *87*, 835–842 (1952).

[2.16] S. Selberherr: Analysis and Simulation of Semiconductor Devices. Springer-Verlag, Wien-New York (1984).
[2.17] S. M. Sze: Physics of Semiconductor Devices, 2nd edn. John Wiley & Sons, New York (1981).
[2.18] W. V. van Roosbroeck: Theory of Flow of Electrons and Holes in Germanium and Other Semiconductors. Bell Syst. Techn. J. *29*, 560–607 (1950).

3 The Drift Diffusion Equations

3.1 Introduction

The drift diffusion equations are the most widely used model to describe semiconductor devices today. The bulk of the literature on mathematical models for device simulation is concerned with this nonlinear system of partial differential equations and numerical software for its solution is commonplace at practically every research facility in the field. From an engineering point of view, the interest in the drift diffusion model is to replace as much laboratory testing as possible by numerical simulation in order to minimize costs. To this end, it is important that computations can be performed in a reasonable amount of time. This implies that the involved mathematical models cannot be too complicated, such as, for instance, the higher dimensional transport equations described in Chapter 1. For the current state of technology the drift diffusion equations seem to represent a reasonable compromise between computational efficiency and an accurate description of the underlying device physics. Therefore transport equations are used mainly to compute data for the model parameters in the drift diffusion equations in the engineering environment. It should be pointed out, however, that, with the increased miniaturization of semiconductor devices, one comes closer and closer to the limits of validity of the drift diffusion equations, even in an industrial environment. The reason for this is, on one hand, that in ever smaller devices the assumption that the free carriers can be modelled as a continuum becomes invalid. On the other hand the drift diffusion equations are derived through a limit process where the mean free path of a particle tends to zero. Through miniaturization and the use of materials other than silicon this mean free path becomes larger and larger in comparison to the size of the device. In addition, quantum mechanical effects start to play a more and more important role in novel device structures. For these reasons, and because of the rapid increase in available computing power, transport equations will be used more and more in device simulation in the future. But even then the drift diffusion equations will remain an important tool since the microscopic effects not described by them occur only locally. Thus, the most likely approach will be to use more

sophisticated models only locally, for instance in the channel of a MOS-transistor (see Chapter 4), and to use the drift diffusion equations in the parts of the device where they suffice to describe the physics.
In this Chapter we will discuss the analytical properties of the drift diffusion equations. We will mainly be interested in the structure of their solutions. Of course this structure will strongly depend on the underlying geometry; i.e. on the device under consideration. We will, however, not discuss specific devices in this Chapter but leave their discussion to Chapter 4. So, we will consider only *P-N* junctions and will not concern ourselves with how different types of devices can be made up by configurations of these *P-N* junctions.
We consider the system of partial differential equations

$$\begin{aligned}
&\text{a)}\ \operatorname{div}(\varepsilon \operatorname{grad} V) = q(n - p - C)\\
&\text{b)}\ \operatorname{div} J_n = q(\partial_t n + R)\\
&\text{c)}\ \operatorname{div} J_p = q(-\partial_t p - R)\\
&\text{d)}\ J_n = q(D_n \operatorname{grad} n - \mu_n n \operatorname{grad} V)\\
&\text{e)}\ J_p = q(-D_p \operatorname{grad} p - \mu_p p \operatorname{grad} V),
\end{aligned} \tag{3.1.1}$$

where V denotes the electric potential ($-\operatorname{grad} V$ is the electric field.), n and p are the concentrations of free carriers of negative and positive charge, called electrons and holes, and J_n and J_p are the densities of the electron and the hole current respectively. D_n, D_p, μ_n and μ_p are the diffusion coefficients and mobilities of electrons and holes respectively. ε is the permittivity constant whose approximate value in silicon is 10^{-12} As V^{-1} cm^{-1}. q is the elementary charge whose value is approximately 10^{-19} As. We assume the device geometry to be given by a domain $\Omega \subseteq \mathbb{R}^d$ with $d = 1$, 2 or 3. Physically $d = 3$ holds, of course. For many devices, however, it suffices to consider two dimensional models ($d = 2$) since their extension in one dimension is much larger than in the other two. Even one dimensional models are sometimes used today. The boundary $\partial\Omega$ of the domain Ω is assumed to consist of a Dirichlet part $\partial\Omega_D$ and a Neumann part $\partial\Omega_N$:

$$\partial\Omega = \partial\Omega_D \cup \partial\Omega_N, \qquad \partial\Omega_D \cap \partial\Omega_N = \{\ \}. \tag{3.1.2}$$

The Dirichlet part $\partial\Omega_D$ of the boundary corresponds to *Ohmic contacts.* There the potential V and the concentrations n and p are prescribed. The boundary values are derived from the following considerations. At Ohmic contacts the space charge, given by the right-hand side of (3.1.1)a) vanishes. So

$$n - p - C = 0 \qquad \text{for} \qquad x \in \partial\Omega_D \tag{3.1.3}$$

holds. Furthermore the system is in thermal equilibrium there, which is expressed by the relation

$$np = n_i^2 \qquad \text{for} \qquad x \in \partial\Omega_D\,. \tag{3.1.4}$$

n_i is the intrinsic density ($\cong 10^{10}\ \mathrm{cm}^{-3}$ in silicon at room temperature). Moreover, the quasi Fermi levels ϕ_n and ϕ_p, given by

$$\text{a)}\quad \phi_n = V - U_T \ln\left(\frac{n}{n_i}\right), \qquad \text{b)}\quad \phi_p = V + U_T \ln\left(\frac{p}{n_i}\right), \tag{3.1.5}$$

assume the values of the applied voltage at Ohmic contacts. Here U_T denotes the thermal voltage which, at room temperature, is roughly 0.025 V. From the conditions (3.1.3)–(3.1.5) the boundary values for V, n and p can be uniquely determined. Inserting (3.1.4) into (3.1.3) gives one quadratic equation for n and p each, which have unique positive solutions given by

$$\left.\begin{array}{l}
\text{a)}\quad n(x,t) = n_D(x) = \tfrac{1}{2}\left(C(x) + \sqrt{C(x)^2 + 4n_i^2}\right) \\
\text{b)}\quad p(x,t) = p_D(x) = \tfrac{1}{2}\left(-C(x) + \sqrt{C(x)^2 + 4n_i^2}\right) \\
\qquad \text{for} \quad x \in \partial\Omega_D . \\
\text{(3.1.5) gives the boundary values for the potential } V\text{:} \\
\text{c)}\quad V(x,t) = V_D(x,t) = U(x,t) + V_{bi}(x) \\
\qquad V_{bi}(x) = U_T \ln\left(\dfrac{n_D(x)}{n_i}\right) \qquad \text{for} \qquad x \in \partial\Omega_D .
\end{array}\right\} \tag{3.1.6}$$

$U(x, t)$ denotes the applied potential. So, differences in $U(x, t)$ between different segments of $\partial\Omega_D$ correspond to the applied bias between these contacts. Note, that (3.1.4) immediately implies that ϕ_n equals ϕ_p at Ohmic contacts. The Neumann parts $\partial\Omega_N$ of the boundary model insulating or artificial surfaces. Thus a zero current flow and a zero electric field in the normal direction are prescribed.

$$\begin{array}{l}
\text{a)}\quad \dfrac{\partial V}{\partial \nu}(x,t)\ (:= \operatorname{grad} V \cdot \nu) = 0 \\
\text{b)}\quad J_n(x,t)\cdot \nu = 0, \\
\text{c)}\quad J_p(x,t)\cdot \nu = 0 \qquad \text{for} \qquad x \in \partial\Omega_N .
\end{array} \tag{3.1.7}$$

In this Chapter ν will always denote the unit outward normal vector on the boundary $\partial\Omega$. In addition the concentrations of the free carriers n and p at time $t = 0$ are prescribed.

$$n(x,0) = n^I(x), \qquad p(x,0) = p^I(x) \qquad \text{for} \qquad x \in \Omega \tag{3.1.8}$$

holds and the complete initial boundary value problem is given by the equations (3.1.1), the boundary conditions (3.1.6), (3.1.7) and the initial conditions (3.1.8).

This setting is not sufficient to describe devices like MOS transistors which, in addition, contain an oxide layer attached to the semiconductor. In the oxide a different set of equations holds and interface conditions are given at the semiconductor oxide interface. We will leave the discussion of this case to Chapter 4.

Various models for the recombination rate R in (3.1.1)b)c) can be found in the literature (see [3.34]). In this Chapter we will, for the sake of simplicity, only consider the Shockley Read Hall term which is of the form

$$R = \frac{np - n_i^2}{\tau_p(n + n_i) + \tau_n(p + n_i)}. \tag{3.1.9}$$

Here, again, n_i denotes the intrinsinc density. τ_n and τ_p are the lifetimes of electrons and holes respectively. Other models, such as the Auger- or the impact ionization model can be found in Chapter 2. We will always assume the mobilities and the diffusion coefficients to satisfy the Einstein relations

$$D_n = \mu_n U_T, \qquad D_p = \mu_p U_T, \tag{3.1.10}$$

where U_T ($\cong 0.025$ V) is the thermal voltage. There is a variety of models for the mobilities μ_n and μ_p. They can be grouped into two different categories which have to be treated in different ways analytically. In one case they are simply functions of position. In the other case they are modelled as dependent on the electric field $-\text{grad}\ V$ in order to represent so called velocity saturation effects. The field dependent mobility models will be discussed in detail in the corresponding Section of this Chapter. We refer the reader to [3.34] for a discussion of different mobility models.

In this Chapter we will restrict ourselves to the analysis of P-N junctions. The term P-N junction refers to the sign change of the doping concentration $C(x)$ in (3.1.1)a). This doping concentration is produced by diffusion of different materials into the silicon crystal and by implantation with an ion beam. This produces a preconcentration of ions in the crystal which is modelled by the function $C(x)$. So $C(x) = C_+(x) - C_-(x)$ holds where C_- and C_+ are the concentrations of negative and positive ions respectively. Where the preconcentration of negative ions predominates in Ω, $C(x) < 0$ holds and these subregions of Ω are called P-regions. Similarly, in the N-region, where the preconcentration of positive ions dominates, $C(x) > 0$ holds. The boundaries between the N- and the P-regions, i.e. the manifolds where $C(x)$ changes its sign, are called P-N junctions. For the simplest P-N junction device, the P-N diode, the geometrical configuration in the two-dimensional case is depicted in Fig. 3.1.1.

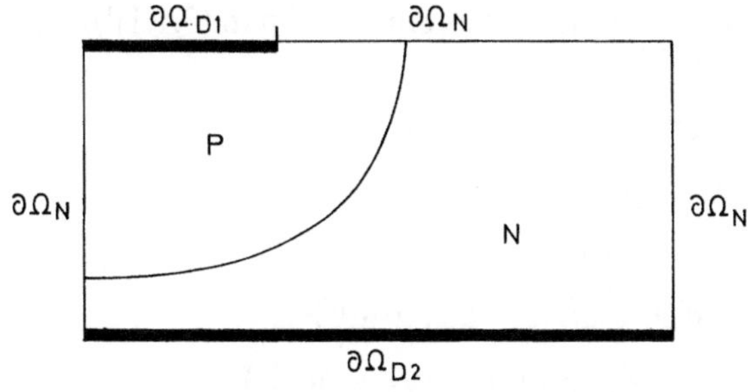

Fig. 3.1.1 *P-N* diode

The function of a *P-N* junction diode is, roughly speaking, that of a valve. If a potential difference of an appropriate sign, that means a difference in the values of $U(x, t)$ in (3.1.6)c) between the Dirichlet boundary parts $\partial\Omega_{D1}$ and $\partial\Omega_{D2}$ in Fig. 3.1.1 is applied to the *P-N* junction, a so called depletion region will form around the *P-N* junction. There, very few free carriers will exist and $n \cong 0$ and $p \cong 0$ will hold in (3.1.1), inside the depletion region. The depletion region will effectively act as an insulator and no, or very little, current will flow. If the potential difference is applied the other way around the depletion region will vanish and current will flow. This is the simplest type of semiconductor device and will be used in this Chapter to explain the analytical features of the drift diffusion equations.
The main difficulties in the treatment of the drift diffusion equations are, on one hand, their nonlinear nature and, on the other hand, the extreme differences in the magnitude of the involved quantities. These differences lead to an almost singular type of behaviour of solutions of the drift diffusion equations. Solutions of the drift diffusion equations will exhibit an extreme layer behaviour in the spatial as well as in the time direction and will, therefore, be amenable to singular perturbation analysis. In this Chapter we will analyze the drift diffusion equations by means of such an asymptotic analysis. This will provide an understanding of the types of mechanisms which are important in different subregions of the device domain Ω and on different time scales. Different approaches have to be used depending on the biasing conditions and the geometries under consideration. We will keep the discussion in rather broad terms in order for the results to be applicable to as wide a range of geometrical configurations, and therefore as wide a range of devices, as possible. Excerpts of the analytical machinery developed in this Chapter will be used to treat specific devices in Chapter 4.

3.2 The Stationary Drift Diffusion Equations

The majority of the analytical and computational work on the drift diffusion equations so far has been concerned with the stationary problem. This means that the drift diffusion equations (3.1.1) are considered at a steady state and that the time derivatives $\partial_t n$ and $\partial_t p$ in (3.1.1)b)c) are neglected. So we consider the problem

$$\begin{aligned}
&\text{a)} \quad \operatorname{div}(\varepsilon \operatorname{grad} V) = q(n - p - C) \\
&\text{b)} \quad \operatorname{div} J_n = qR \\
&\text{c)} \quad \operatorname{div} J_p = -qR \\
&\text{d)} \quad J_n = q(D_n \operatorname{grad} n - \mu_n n \operatorname{grad} V) \\
&\text{e)} \quad J_p = q(-D_p \operatorname{grad} p - \mu_p p \operatorname{grad} V),
\end{aligned} \tag{3.2.1}$$

together with the boundary conditions

a) $n(x) = n_D(x) = \frac{1}{2}(C(x) + \sqrt{C(x)^2 + 4n_i^2})$

b) $p(x) = p_D(x) = \frac{1}{2}(-C(x) + \sqrt{C(x)^2 + 4n_i^2})$

c) $V(x) = V_D(x) = U(x) + V_{bi}(x)$

$$V_{bi}(x) = \ln\left(\frac{n_D(x)}{n_i}\right) \quad \text{for} \quad x \in \partial\Omega_D \tag{3.2.2}$$

d) $\frac{\partial V}{\partial \nu}(x)\ (:= \operatorname{grad} V \cdot \nu) = 0$

e) $J_n(x) \cdot \nu = 0,$ f) $J_p(x) \cdot \nu = 0$ for $x \in \partial\Omega_N$.

From a practical point of view, the interest in the stationary drift diffusion equations lies in the dependence of the current densities J_n and J_p on the applied bias, the doping concentration $C(x)$ and the geometry. For instance, one tries, through variation of these parameters, to minimize the so called leakage current. That is the small current still flowing through a reverse biased *P-N* junction. While the tools to treat such a complicated problem, as the drift diffusion equations in higher dimensions, have to be computational, a great deal of insight can be gained by analysis. The development of numerical methods for the drift diffusion equations benefits greatly from the analytical understanding of the solutions, and might even be impossible without it. In the following Sections we are interested in four basic questions about the stationary drift diffusion equations:

1. Does a solution exist, and, if yes, how smooth is it?
2. What is the structure of solutions of the stationary drift diffusion equations?
3. What are the stability properties of these solutions?

Question 1 is of a general mathematical interest. It turns out that solutions exist and lie in the function spaces one would expect them to. Question 2 is of extreme importance for the development of numerical methods. As mentioned earlier, solutions of the drift diffusion equations exhibit layer structure. That means that they have large gradients locally, usually near the *P-N* junctions. The performance of numerical methods can be improved a great deal if they are adapted to this layer behaviour. As had to be expected, the structure of solutions to the drift diffusion equations looks quite differently for forward and reverse biased *P-N* junctions. Thus, these two cases are treated separately in the subsequent Sections. The stability properties of the stationary drift diffusion equations are extremely dependent on the geometry and not analyzed satisfactorily at this point. We will present the available results in Section 3.6. There it turns out that the conditioning of the drift diffusion equations is acceptable if every *N*- and *P*-region contains a contact. If this is not the case, i.e. if so called floating regions are present, the stability properties can worsen drastically.

3.3 Existence and Uniqueness for the Stationary Drift Diffusion Equations

Before analyzing the structure and the properties of solutions to the equations obtained by the steady state drift diffusion approximation in the subsequent Chapters, we will first briefly discuss the existence of these solutions. From a practical point of view one would prefer existence results based on the implicit function theorem. Such results would provide information about the norm of the inverse of the linearized problem. They would show the existence of an isolated solution and, more importantly, give an indication of the conditioning of the problem and of the convergence properties of Newton's method. The conditioning of the steady state drift diffusion equations has been investigated in [3.3] and various papers in the literature ([3.5], [3.17], [3.28], [3.29]) deal with the convergence of iterative methods. However, they do not treat the whole coupled system of equations or they assume the existence of a suitable a priori bound on the inverse of the linearization. At the end of this Section we will briefly discuss these results. The only existence results available for the coupled problem and arbitrarily large bias are not constructive since the arguments are based on the Schauder Fixed Point Theorem in one way or the other. In this Chapter we will give an example for an existence theorem for a simplified case (Theorem (3.3.16)) in order to outline the basic approach. We refer the reader to the literature ([3.4], [3.10], [3.19], [3.24]) for results in more general cases.

Since the scaling is of no particular importance for the purposes of this Section we will treat the drift diffusion equations in an unscaled form for the moment. So we consider the system

$$\begin{aligned}
&\text{a)}\quad \varepsilon\,\Delta V = q(n - p - C(x))\\
&\text{b)}\quad \operatorname{div} J_n = qR, \qquad \text{c)}\quad J_n = q(D_n \operatorname{grad} n - \mu_n n \operatorname{grad} V)\\
&\text{d)}\quad \operatorname{div} J_p = -qR, \qquad \text{e)}\quad J_p = q(-D_p \operatorname{grad} p - \mu_p p \operatorname{grad} V).
\end{aligned} \tag{3.3.1}$$

Equations (3.3.1) have the disadvantage of containing the convection terms $-n \operatorname{grad} V$ and $-p \operatorname{grad} V$ which prohibit the use of the maximum principle in a simple way. If the Einstein relations

$$D_n = U_T \mu_n, \qquad D_p = U_T \mu_p \tag{3.3.2}$$

can be assumed, with U_T the thermal voltage (see [3.35]), it is beneficial to change from the concentrations n and p to the so called Slotboom variables u and v given by

$$\text{a)}\quad n = n_i e^{V/U_T} u, \qquad \text{b)}\quad p = n_i e^{-V/U_T} v. \tag{3.3.3}$$

The current relations then become

$$\begin{aligned}
&\text{a)}\quad J_n = q U_T n_i \mu_n e^{V/U_T} \operatorname{grad} u,\\
&\text{b)}\quad J_p = -q U_T n_i \mu_p e^{-V/U_T} \operatorname{grad} v.
\end{aligned} \tag{3.3.4}$$

After inserting the current densities J_n and J_p into the continuity equations (3.3.1)b)d) one obtains the elliptic system

$$\begin{aligned} &\text{a)} \quad \varepsilon \, \Delta V = q n_i (e^{V/U_T} u - e^{-V/U_T} v) - qC(x) \\ &\text{b)} \quad U_T n_i \operatorname{div}(\mu_n e^{V/U_T} \operatorname{grad} u) = R \\ &\text{c)} \quad U_T n_i \operatorname{div}(\mu_p e^{-V/U_T} \operatorname{grad} v) = R . \end{aligned} \tag{3.3.5}$$

In this form the continuity equations (3.3.5)b)c) are self adjoint. In the Slotboom variables u and v the boundary conditions (3.2.2)d)e)f) at artificial or insulating surfaces become pure Neumann conditions

$$\left.\frac{\partial V}{\partial \nu}\right|_{\partial\Omega_N} = \left.\frac{\partial u}{\partial \nu}\right|_{\partial\Omega_N} = \left.\frac{\partial v}{\partial \nu}\right|_{\partial\Omega_N} = 0. \tag{3.3.6}$$

For Ohmic contacts we obtain from (3.2.2)a)b)c)

$$V|_{\partial\Omega_D} = V_D|_{\partial\Omega_D}, \qquad u|_{\partial\Omega_D} = u_D|_{\partial\Omega_D}, \qquad v|_{\partial\Omega_D} = v_D|_{\partial\Omega_D}, \tag{3.3.7}$$

with $u_D = n_i^{-1} e^{-V_D/U_T} n_D$ and $v_D = n_i^{-1} e^{V_D/U_T} p_D$. We remark that, since n and p represent physical concentrations, the Slotboom variables u and v have to remain positive.

Existence theorems for the problem (3.3.5)–(3.3.7) usually employ the Schauder Fixed Point Theorem. The construction of the fixed point map depends on the form of the recombination rate, the mobilities, the geometry and so on. To outline the basic idea we will give an existence proof following the approach in [3.19] but we will use some simplifying assumptions. We will consider the Shockley Read Hall recombination term only. So after changing variables to (V, u, v) the recombination rate R in (3.3.5) is of the form

$$R = n_i \frac{uv - 1}{\tau_p(e^{V/U_T} u + 1) + \tau_n(e^{-V/U_T} v + 1)} . \tag{3.3.8}$$

We assume that the mobilities μ_n and μ_p are uniformly bounded functions of position only and that

$$0 < \underline{\mu}_n \leqslant \mu_n(x) \leqslant \bar{\mu}_n, \qquad 0 < \underline{\mu}_p \leqslant \mu_p(x) \leqslant \bar{\mu}_p \qquad \forall x \in \Omega \tag{3.3.9}$$

hold. Furthermore we will take the boundary $\partial\Omega$ and the boundary data V_D, u_D, and v_D in (3.3.7) to be as smooth as necessary.

We reiterate that existence results for more general situations can be found in the literature (see [3.19, Chapter 3]). For instance the form of the recombination rate can be taken so as to include the Auger recombination term without any additional problems. The boundedness of the mobilities away from zero is a more severe restriction. It excludes the field dependent mobilities used to model velocity saturation effects. A condition of the form (3.3.9) is necessary, however, to guarantee the uniform ellipticity of the continuity equations. Therefore most existence proofs do assume an a priori bound on the mobilities even when modeling them as dependent on the field $-\operatorname{grad} V$. On the other hand the inclusion of an oxide layer, where Laplace's equation has to be solved for V, does not pose a major problem. The fixed point map is constructed such that its evaluation only involves the solution of semilinear or linear scalar boundary value problems. Let G be given by $G(u_0, v_0) = (u_1, v_1)$, where (u_1, v_1) is computed from (u_0, v_0) as follows.

Step 1: Solve Poisson's equation

$$\begin{gathered} -\varepsilon\,\Delta V + qn_i(e^{V/U_T}u_0 - e^{-V/U_T}v_0) - qC(x) = 0 \\ \left.\frac{\partial V}{\partial \nu}\right|_{\partial\Omega_N} = 0, \qquad V|_{\partial\Omega_D} = V_D|_{\partial\Omega_D} \end{gathered} \tag{3.3.10}$$

for $V = V_1$.

Step 2: Solve

$$\begin{gathered} \text{a)} \quad -U_T \operatorname{div}(\mu_n e^{V_1/U_T} \operatorname{grad} u) + \frac{uv_0 - 1}{\tau_p(e^{V_1/U_T}u_0 + 1) + \tau_n(e^{-V_1/U_T}v_0 + 1)} = 0 \\ \text{b)} \quad \left.\frac{\partial u}{\partial \nu}\right|_{\partial\Omega_N} = 0, \qquad u|_{\partial\Omega_D} = u_D|_{\partial\Omega_D} \end{gathered} \tag{3.3.11}$$

for $u = u_1$.

Step 3: Solve

$$\begin{gathered} \text{a)} \quad -U_T \operatorname{div}(\mu_n e^{-V_1/U_T} \operatorname{grad} v) + \frac{u_0 v - 1}{\tau_p(e^{V_1/U_T}u_0 + 1) + \tau_n(e^{-V_1/U_T}v_0 + 1)} = 0 \\ \text{b)} \quad \left.\frac{\partial v}{\partial \nu}\right|_{\partial\Omega_N} = 0, \qquad v|_{\partial\Omega_D} = v_D|_{\partial\Omega_D} \end{gathered} \tag{3.3.12}$$

for $v = v_1$.

By solving the boundary value problems (3.3.10)–(3.3.12) we mean a solution in the usual weak sense (see [3.12]). Obviously a fixed point of the nonlinear operator G is a weak solution of the coupled problem (3.3.5)–(3.3.7). The existence of such a fixed point is established by showing that the map G is completely continuous and by applying the Schauder Fixed Point Theorem. Of course, for this approach one has to choose an appropriate space for defining G. We will leave this question for later (for Theorem (3.3.16)) and first convince ourselves that the map G is well defined; that means that the involved boundary value problems are uniquely solvable. All three problems (3.3.10)–(3.3.12) can be written in the general form

$$\begin{gathered} -\operatorname{div}(a(x) \operatorname{grad} w) + f(x, w) = 0, \qquad x \in \Omega \\ \left.\frac{\partial w}{\partial \nu}\right|_{\partial\Omega_N} = 0, \qquad w|_{\partial\Omega_D} = w_D|_{\partial\Omega_D}, \end{gathered} \tag{3.3.13}$$

where w takes the place of V, u and v respectively. The coefficient $a(x)$ in (3.3.13) is either the constant ε or equal to $\mu_n e^{V_1/U_T}$ or $\mu_p e^{-V_1/U_T}$. In any case it is uniformly bounded away from zero if μ_n and μ_p are and if V_1 is bounded, which makes the semilinear equation (3.3.13) uniformly elliptic. $f(x, w)$ is a monotone increasing function of w in all three cases (3.3.10)–(3.3.12) if u_0 and v_0 are positive. In (3.3.11) and (3.3.12) f is linear in w. The existence of a unique solution of semilinear partial differential equations of the type as in (3.3.13) is, under certain assumptions, a standard result in the theory of elliptic partial differential equations. We will state in Lemma 3.3.14 such a result in a form suitable for our purposes and refer the reader to [3.19] for the proof. The result of Lemma 3.3.14 requires that the coefficient $a(x)$ in (3.3.13) is in the space $L^\infty(\Omega)$, i.e. that $a(x)$ is bounded uniformly in Ω. The solution $w(x)$ will lie in the intersection of the spaces $L^\infty(\Omega)$ and $H^1(\Omega)$. $H^1(\Omega)$ is the space of functions which are square integrable and whose gradient is square integrable as well. So

$$\int_\Omega (w(x)^2 + |\nabla w(x)|^2)\, dx < \infty$$

holds.

Lemma 3.3.14: *Let the following assumptions hold:*
A1) *The function $f(x, w)$ is monotonically increasing in w for all $x \in \Omega$.*
A2) *$a(x) \in L^\infty(\Omega)$ and $a(x) \geqslant \underline{a} > 0$ holds for some constant $\underline{a}$.*
A3) *There exist functions $\underset{\sim}{g}(w)$ and $\tilde{g}(w)$ such that $\underset{\sim}{g}(w) \leqslant f(x, w) \leqslant \tilde{g}(w)$ hold $\forall\, x \in \Omega$, $\forall\, w$.*
A4) *There exist solutions $\underset{\sim}{w}$ and $\tilde{w}$ of $\underset{\sim}{g}(\tilde{w}) = 0$ and $\tilde{g}(\underset{\sim}{w}) = 0$.*
Then there exists a unique solution w of the problem (3.3.13) in $H^1(\Omega) \cap L^\infty(\Omega)$. This solution satisfies

$$\underline{w} \leqslant w(x) \leqslant \overline{w}$$

$$\underline{w} = \min\left\{\inf_{\partial\Omega_D} w_D,\ \underset{\sim}{w}\right\}, \qquad \overline{w} = \max\left\{\sup_{\partial\Omega_D} w_D,\ \tilde{w}\right\}. \tag{3.3.15}$$

The proof can be found in [3.19].
Using Lemma 3.3.14 we can now, by showing that the map G is well defined and completely continuous, employ the Schauder theorem to establish the existence of a weak solution to (3.3.5).

Theorem 3.3.16: *Let $K \geqslant 1$ be a constant satisfying*

$$\frac{1}{K} \leqslant u_D(x),\ v_D(x) \leqslant K \qquad \forall\, x \in \partial\Omega_D.$$

Then the problem

a) $\varepsilon\, \Delta V = q n_i (e^{V/U_T} u - e^{-V/U_T} v) - qC(x)$

b) $U_T n_i \operatorname{div}(\mu_n e^{V/U_T} \operatorname{grad} u) = R$

c) $U_T n_i \operatorname{div}(\mu_p e^{-V/U_T} \operatorname{grad} v) = R$

d) $$\left.\frac{\partial V}{\partial \nu}\right|_{\partial\Omega_N} = \left.\frac{\partial u}{\partial \nu}\right|_{\partial\Omega_N} = \left.\frac{\partial v}{\partial \nu}\right|_{\partial\Omega_N} = 0$$

e) $V|_{\partial\Omega_D} = V_D|_{\partial\Omega_D}, \qquad u|_{\partial\Omega_D} = u_D|_{\partial\Omega_D}, \qquad v|_{\partial\Omega_D} = v_D|_{\partial\Omega_D}$

has a solution $(V^*, u^*, v^*) \in (H^1(\Omega) \cap L^\infty(\Omega))^3$ *which satisfies the* L^∞-*estimate*

$$\frac{1}{K} \leqslant u(x), v(x) \leqslant K \qquad \text{in } \Omega,$$

$$\min\left(\inf_{\partial\Omega_D} V_D, U_T \ln\left[\frac{1}{2Kn_i}(\underline{C} + (\underline{C}^2 + 4n_i^2)^{1/2})\right]\right) \leqslant V(x)$$

$$V(x) \leqslant \max\left(\sup_{\partial\Omega_D} V_D, U_T \ln\left[\frac{K}{2n_i}(\bar{C} + (\bar{C}^2 + 4n_i^2)^{1/2})\right]\right) \qquad \text{in } \Omega \tag{3.3.19}$$

where $\underline{C} \leqslant C(x) \leqslant \bar{C}$ *holds.*

Proof: First we choose an appropriate space for the fixed point map G. Let $\mathcal{N}$ be defined by

$$\mathcal{N} = \left\{(u, v) \in L^2(\Omega): \frac{1}{K} \leqslant u, v \leqslant K \text{ a.e. in } \Omega\right\}, \tag{3.3.20}$$

where $L^2(\Omega)$ is the space of square integrable functions; i.e. the space of functions (u, v) for which

$$\int_\Omega |(u(x), v(x))|^2 \, dx < \infty$$

holds. We show that G maps $\mathcal{N}$ into itself and is completely continuous. Given $(u_0, v_0) \in \mathcal{N}$, by virtue of Lemma 3.3.14, there exists a solution V_1 of (3.3.10). $\underset{\sim}{g}$ and $\tilde{g}$ can be chosen as

$$\underset{\sim}{g}(V) = n_i q\left(\frac{1}{K} e^{V/U_T} - K e^{-V/U_T}\right) - q\bar{C} \tag{3.3.21}$$

$$\tilde{g}(V) = n_i q\left(K e^{V/U_T} - \frac{1}{K} e^{-V/U_T}\right) - q\underline{C}.$$

Solving $\underset{\sim}{g}(\tilde{V}) = 0$ and $\tilde{g}(\underset{\sim}{V}) = 0$ gives

$$\underset{\sim}{V} = U_T \ln\left[\frac{K}{2n_i}(\bar{C} + (\bar{C}^2 + 4n_i^2)^{1/2})\right] \tag{3.3.22}$$

$$\tilde{V} = U_T \ln\left[\frac{1}{2Kn_i}(\underline{C} + (\underline{C}^2 + 4n_i^2)^{1/2})\right].$$

Applying Lemma 3.3.14 to equation (3.3.11) we use

$$\tilde{g}(u) = \frac{Ku - 1}{\tau_p\left(\dfrac{e^{\underline{V}/U_T}}{K} + 1\right) + \tau_n\left(\dfrac{e^{-\bar{V}/U_T}}{K} + 1\right)},$$

where $\underline{V} \leqslant V_1(x) \leqslant \bar{V}$ holds, and obtain $\underset{\sim}{u} = 1/K$. Analogously we obtain $\tilde{u} = K$ which implies

$$\frac{1}{K} \leqslant u_1(x) \leqslant K. \tag{3.3.24}$$

In the same way we obtain $1/K \leqslant v_1(x) \leqslant K$. Thus, G maps $\mathcal{N}$ into itself. The continuity of $\mathcal{N}$ is a simple consequence of the well posedness of uniformly elliptic boundary value problems. On the other hand the continuous dependence of u_1 and v_1 on the data of the corresponding boundary value problems implies

$$\|u_1\|_{1,2,\Omega} + \|v_1\|_{1,2,\Omega} \leqslant F(\|u_0\|_{2,\Omega}, \|v_0\|_{2,\Omega}, \|u_D\|_{1,2,\Omega}, \|v_D\|_{1,2,\Omega}) \tag{3.3.25}$$

for some positive and continuous function F. Here, the symbols $\|\cdot\|_{2,\Omega}$ and $\|\cdot\|_{1,2,\Omega}$ denote the norms in $L^2(\Omega)$ and $H^1(\Omega)$. So

$$\|f\|_{2,\Omega} = \left(\int_\Omega |f(x)|^2\,dx\right)^{1/2},$$

$$\|f\|_{1,2,\Omega} = \left(\int_\Omega (|f(x)|^2 + |\nabla f(x)|^2)\,dx\right)^{1/2}$$

holds. Thus, $\|u_1\|_{1,2,\Omega} + \|v_1\|_{1,2,\Omega} \leqslant \text{const}$ holds for all (u_0, v_0) in $\mathcal{N}$. The Rellich Kondrachov Theorem (see [3.1]) now assures that $G(\mathcal{N})$ is precompact in $(L^2(\Omega))^2$. This, together with the continuity of G, gives complete continuity and the Schauder Fixed Point Theorem (see [3.12]) assures the existence of a fixed point of G which is a solution of (3.3.5)–(3.3.7). □

Theorem 3.3.16 serves as a typical example of an existence result for the steady state drift diffusion problem. Various other results of this type treat more complicated geometries or parameter models, which affects the structure of the fixed point map and introduces additional technical complications. Such results can be found in [3.16], [3.10], [3.18] or [3.19].

Global uniqueness of the solution of (3.3.5)–(3.3.7) cannot be expected in the general case since there are devices, such as thyristors, whose performance is based explicitely on the existence of multiple solutions (see Chapter 4). One can, however, obtain a uniqueness result in the case that the applied bias, and therefore the current densities J_n and J_p, are sufficiently small. In the case of thermal equilibrium (no voltage applied, $J_n = J_p = 0$) the system (3.3.5) reduces to the scalar problem

$$\varepsilon\,\Delta V_E = n_i q(e^{V_E/U_T}u_E - e^{-V_E/U_T}v_E) - qC(x)$$
$$\left.\frac{\partial V_E}{\partial \nu}\right|_{\partial\Omega_N} = 0, \qquad V_E|_{\partial\Omega_D} = V_D|_{\partial\Omega_D}, \tag{3.3.26}$$

where $u_E = u_D$ and $v_E = v_D$ are constant. One can show easily the isolatedness of the solution V_E of (3.3.26) and estimate the norm of the inverse of the linearization of (3.3.5) at the solution (V_E, u_E, v_E). The implicit function theorem then implies a unique solution of (3.3.5)–(3.3.7) for sufficiently small voltages and (consequently) current densities.

3.4 Forward Biased *P*-*N* Junctions

We now turn to analyzing the structure and the quantitative properties of solutions of the drift diffusion equations (3.2.1). The main tool for this analysis is singular perturbation theory. It is well known that the solutions of the drift diffusion equations behave differently in different subregions of the device. For instance, steep gradients occur locally across *P*-*N* junctions and in narrow regions underneath semiconductor-oxide interfaces, i.e. in the channel of a MOS transistor (see Chapter 4). The basic idea of singular perturbation analysis is to replace the Basic Semiconductor Equations locally, in different regions of the device, by simpler problems whose solutions contain all the essential qualitative features of the original solution. These approximations are then used to gain insights into the behaviour of the solution which could not be achieved by looking at the structurally more complicated original system. Since we restrict ourselves in this Chapter to simple *P*-*N* junctions, and leave more complicated devices with more than one junction to Chapter 4, we only have to distinguish between two basic situations. In the case of a reverse biased *P*-*N* junction one observes the formation of a depletion region with no, or very few, free carriers. so $n \cong p \cong 0$ in (3.2.1)a) holds. This region acts effectively as an insulator and only a very small current, the so called leakage current, flows. If the potential difference is applied the other way around, i.e. if a forward bias is applied, the depletion region vanishes. The free carriers tend to neutralize the doping concentration and current flows. So $p - n + C \cong 0$ will hold except in narrow layers around the *P*-*N* junctions where large electric fields and concentration gradients will occur. These two situations require two different types of perturbation analysis. In this Section we will concentrate on the forward bias case where, except in the above mentioned layer regions, the space charge $q(p - n + C)$ is very small.
We start by bringing the drift diffusion equations (3.2.1) into an appropriate scaled and dimensionless form. Suppose the geometry under consideration has a characteristic length scale (for instance the diameter) L. We use the scaling

$$x = Lx_s \tag{3.4.1}$$

for the position variable x. x_s, the scaled position variable, is then at most of order $O(1)$ and dimensionless. For the dependent variables in (3.2.1) we use the following scaling which has turned out to be the most useful for the singular perturbation analysis of forward biased *P-N* junctions (see [3.21], [3.18])

$$\text{a)}\quad V = U_T V_s, \qquad \text{b)}\quad n = \tilde{C} n_s, \qquad \text{c)}\quad p = \tilde{C} p_s,$$
$$\text{d)}\quad J_n = \frac{qU_T \tilde{C}\tilde{\mu}}{L} J_{n_s}, \qquad \text{e)}\quad J_p = \frac{qU_T \tilde{C}\tilde{\mu}}{L} J_{p_s}, \tag{3.4.2}$$

where the subscript s denotes the scaled and dimensionless variable. U_T in (3.4.2) is the thermal voltage which, at room temperature, has a value of 0.0259 V and $\tilde{C}$ is the maximal absolute value of the doping concentration $C(x)$. For VLSI applications typical values of $\tilde{C}$ range from 10^{15} cm^{-3} to 10^{21} cm^{-3}. $\tilde{\mu}$ in (3.4.2)d)e) is a characteristic value for the mobilities μ_n and μ_p and is, for silicon, usually of the order of 1000 cm^2 V^{-1} s^{-1}. We will only consider situations in this Section where we can assume the Einstein relations

$$D_n = U_T \mu_n, \qquad D_p = U_T \mu_p \tag{3.4.3}$$

to hold. Thus we set

$$\mu_n = \tilde{\mu}\mu_{n_s}, \qquad \mu_p = \tilde{\mu}\mu_{p_s}, \qquad D_n = U_T \tilde{\mu}\mu_{n_s}, \qquad D_p = U_T \tilde{\mu}\mu_{p_s}. \tag{3.4.4}$$

Using this scaling, the drift diffusion equations (3.2.1) become

$$\text{a)}\quad \lambda^2 \, \Delta V_s = n_s - p_s - C_s(x)$$
$$\text{b)}\quad \operatorname{div} J_{n_s} = R_s, \qquad \text{c)}\quad J_{n_s} = \mu_{n_s}(\operatorname{grad} n_s - n_s \operatorname{grad} V_s) \tag{3.4.5}$$
$$\text{d)}\quad \operatorname{div} J_{p_s} = -R_s, \qquad \text{e)}\quad J_{p_s} = \mu_{p_s}(-\operatorname{grad} p_s - p_s \operatorname{grad} V_s)$$

with $C(x) = \tilde{C} C_s(x)$. R_s in (3.4.5)b)d) is the appropriately scaled recombination rate. If the Shockley Read Hall term is taken as a model for the recombination rate, R_s is of the form

$$R_s = \frac{n_s p_s - \delta^4}{\tau_{p_s}(n_s + \delta^2) + \tau_{n_s}(p_s + \delta^2)}. \tag{3.4.6}$$

The boundary conditions (3.2.2) read in scaled form

$$\text{a)}\quad V_s(x) = V_{D_s}(x) = U_s(x) + V_{bi_s}(x)$$
$$V_{bi_s}(x) = \ln\left[\frac{1}{2\delta^2}\left(C_s(x) + \sqrt{C_s(x)^2 + 4\delta^4}\right)\right]$$
$$\text{b)}\quad n_s(x) = n_{D_s}(x) = \tfrac{1}{2}\left(C_s(x) + \sqrt{C_s(x)^2 + 4\delta^4}\right)$$
$$\text{c)}\quad p_s(x) = p_{D_s}(x) = \tfrac{1}{2}\left(-C_s(x) + \sqrt{C_s(x)^2 + 4\delta^4}\right) \quad \text{for} \quad x \in \partial\Omega_{D_s}$$
$$\text{d)}\quad \frac{\partial V_s}{\partial \nu}(x) = 0, \qquad J_{n_s}(x)\cdot \nu = 0,$$
$$J_{p_s}(x)\cdot \nu = 0 \qquad \text{for} \qquad x \in \partial\Omega_{N_s}. \tag{3.4.7}$$

From here on we will omit the subscript s for notational convenience. The parameters λ in (3.4.5)a) and δ in (3.4.7) are

$$\lambda = \left(\frac{\varepsilon U_T}{qCL^2}\right)^{1/2}, \qquad \delta = \left(\frac{n_i}{C}\right)^{1/2}. \tag{3.4.8}$$

λ is the scaled minimal normed Debye length of the device (see [3.35]). λ will act as a singular perturbation parameter in the forward bias case as well as in the reverse bias case in the next Section. However, for reasons that will be explained in the corresponding Section, a different form of scaling has to be used for P-N junctions under extreme reverse biasing conditions.

The Equilibrium Case

The derivation of an approximation to the solution of a given problem via singular perturbation analysis follows a certain recipe. The steps in this recipe become technically more involved the more structurally complex the original problem is. As a matter of fact, for a nonlinear system of partial differential equations, such as the drift diffusion equations, we will in general not be able to carry out some of these steps. So it is for instance still an open problem to show that the asymptotic expansion derived in the next Section actually approximates the solution of (3.4.5) except in some special cases. This does not diminish the value of the singular perturbation approach since one can always resort to numerical computations to 'convince' oneself of the validity of the expansions. In order to demonstrate the basic techniques involved in singular perturbation analysis, and to give a flavor of the type of results obtained, we will consider the case of a P-N junction in thermal equilibrium first, using the scaling (3.4.2) for the forward bias case. This has the advantage that parts of the system (3.4.5) can be integrated exactly and (3.4.5) reduces to a nonlinear Poisson equation for the potential V. We will also restrict ourselves to the two dimensional case. So we assume that the device occupies a region $\Omega \subseteq \mathbb{R}^2$ with a Dirichlet boundary $\partial\Omega_D$ and a Neumann boundary $\partial\Omega_N$ ($\partial\Omega = \partial\Omega_D \cup \partial\Omega_N$, $\partial\Omega_D \cap \partial\Omega_N = \{\ \}$). The drift diffusion equations can be reduced to a nonlinear Poisson equation with a certain set of boundary data corresponding to zero applied bias. This set is given by

$$\text{a)} \quad V(x) = V_{bi} = \ln\left[\frac{1}{2\delta^2}\left(C(x) + \sqrt{C(x)^2 + 4\delta^4}\right)\right]$$

$$\text{b)} \quad n(x) = n_D(x) = \tfrac{1}{2}\left(C(x) + \sqrt{C(x)^2 + 4\delta^4}\right)$$

$$\text{c)} \quad p(x) = p_D(x) = \tfrac{1}{2}\left(-C(x) + \sqrt{C(x)^2 + 4\delta^4}\right) \qquad \text{for} \qquad x \in \partial\Omega_D$$

$$\text{d)} \quad \frac{\partial V}{\partial \nu}(x) = 0, \qquad J_n(x)\cdot \nu = 0,$$

$$J_p(x)\cdot \nu = 0 \qquad \text{for} \qquad x \in \partial\Omega_N. \tag{3.4.9}$$

So $U = 0$ in (3.4.7) holds. Here, again Ω denotes the domain of the device and $\partial\Omega_D$ and $\partial\Omega_N$ denote the Dirichlet and Neumann parts of the boundary. ν is the unit outward normal vector on the domain boundary $\partial\Omega$.
A solution n, p, J_n and J_p of the continuity equations and current relations (3.4.5)b)–e) is given by

$$n = \delta^2 e^V, \qquad p = \delta^2 e^{-V}, \qquad J_n = 0, \qquad J_p = 0. \tag{3.4.10}$$

(3.4.10) is a solution of the continuity equations as long as there is no generation-recombination in thermal equilibrium, i.e. as long as $R = 0$ holds whenever $np = \delta^4$. This condition is satisfied for the Shockley Read Hall term (3.4.6). The Poisson equation then reduces to

$$\text{a)} \quad \lambda^2 \Delta V = \delta^2(e^V - e^{-V}) - C(x)$$

$$\text{b)} \quad V|_{\partial\Omega_D} = V_{bi}(x)|_{\partial\Omega_D} = \ln\left[\frac{1}{2\delta^2}(C(x) + \sqrt{C(x)^2 + 4\delta^4})\right]\bigg|_{\partial\Omega_D}$$

$$\text{c)} \quad \frac{\partial V}{\partial \nu}\bigg|_{\partial\Omega_N} = 0. \tag{3.4.11}$$

To keep matters simple, we will take Ω to be a rectangle with $\partial\Omega_D$ consisting of the two vertical sides and $\partial\Omega_N$ of the two horizontal sides. The *P-N* junction shall be given by a curve $\Gamma = \{(x, y) = (X(s), Y(s))\}$ which intersects $\partial\Omega_N$ in the two points $(X(s_1), Y(s_1))$ and $(X(s_2), Y(s_2))$. The situation is depicted schematically in Fig. 3.4.1.
We assume the *P-N* junction to be abrupt. So $C(x)$ has a jump discontinuity at Γ and is as smooth as we like (say constant) away from Γ. We can assume without any loss of generality that the tangent vector $(dX/ds, dY/ds) =: (\dot{X}, \dot{Y})$ on Γ is normalized. Also, to avoid technical difficulties, we assume that Γ intersects $\partial\Omega_N$ in a right angle and without curvature. So

$$\begin{aligned} &\dot{X}(s)^2 + \dot{Y}(s)^2 = 1 \qquad \forall s \\ &\dot{X}(s_i) = \ddot{X}(s_i) = \ddot{Y}(s_i) = 0, \qquad i = 1, 2 \end{aligned} \tag{3.4.12}$$

holds.

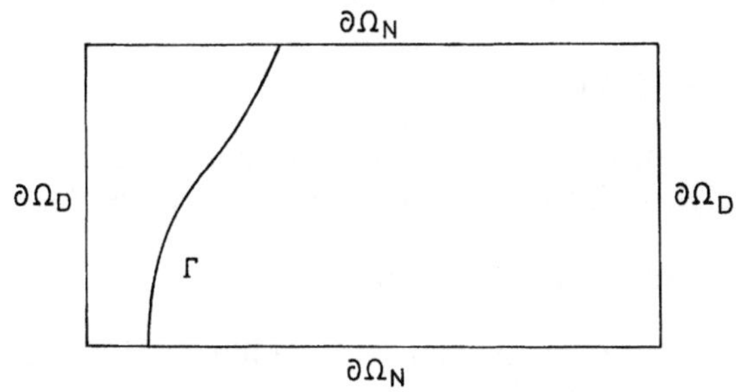

Fig. 3.4.1 Simplified geometry

λ in (3.4.11)a) will be very small in practice. So, for instance, for $L = 10^{-4}$ cm and $\tilde{C} = 10^{19}$ cm^{-3} in (3.4.2) $\lambda = O(10^{-3})$ holds. If the term ΔV in (3.4.11)a) is of moderate size this implies that the right hand side of (3.4.11)a), the space charge, is small (of order λ^2). If the scaling (3.4.2) was correct ΔV will not be large except, maybe, in small subregions (layers) where V has a steep gradient. It is therefore reasonable to assume that, away from these layers, V will be approximated by $\bar{V}_0$, the solution of the zero space charge approximation

$$0 = 2\delta^2 \sinh(\bar{V}_0) - C(x). \tag{3.4.13}$$

(3.4.13) is a simple problem since the differential equation (3.4.11)a) has been reduced to an algebraic relation. In the language of singular perturbation theory (3.4.13) is called the reduced problem and $\bar{V}_0$ is called the outer, or reduced, solution (see [3.25]). We use the index 0 for the outer solution since $\bar{V}_0$ will be the zeroth order term of an asymptotic expansion in powers of λ at the end of this Section. Solving (3.4.13) we obtain

$$\bar{V}_0(x) = \ln\left[\frac{C(x) + \sqrt{C(x)^2 + 4\delta^4}}{2\delta^2}\right]. \tag{3.4.14}$$

$\bar{V}_0(x)$ also satisfies the boundary conditions (3.4.11)b)c) assuming that $\partial C/\partial r|_{\partial\Omega_N} = 0$ holds. On the other hand, since the doping profile $C(x)$ is discontinuous at Γ, so is $\bar{V}_0$. Because of the regularizing effect of the Laplace operator we would expect the solution V to be continuous, and so $\bar{V}_0$ is probably not a good approximation to V in the vicinity of Γ. The reason for this is that our original premise, namely that div(grad V) is of moderate size, is not valid there. In order to investigate the solution in a neighbourhood of Γ we employ a local coordinate transformation. For a point (x, y) close to Γ let s denote the parameter value of the nearest curve point $(X(s), Y(s))$ and ξ the perpendicular distance of (x, y) to Γ divided by λ as shown in Fig. 3.4.2 ($\xi > 0$ on one side of Γ and $\xi < 0$ on the other side). Using the normalized tangent and normal vectors $(\dot{X}, \dot{Y})$ and $(\dot{Y}, -\dot{X})$, the transformation $(x, y) \leftrightarrow (\xi, s)$ is given by

$$\begin{aligned} x &= X(s) + \lambda\xi\, \dot{Y}(s) \\ y &= Y(s) - \lambda\xi\, \dot{X}(s). \end{aligned} \tag{3.4.15}$$

This coordinate transformation is one to one as long as $|\lambda\xi|$ is sufficiently small, i.e. in a neighborhood of Γ. In the (ξ, s) variables (3.4.11)a) reads

$$\partial_\xi^2 V = 2\delta^2 \sinh(V) - C + O(\lambda). \tag{3.4.16}$$

In order to obtain an approximation to V, which is valid in a neighbourhood of Γ as well, we will derive the layer term $\hat{V}_0$ which is a function of the

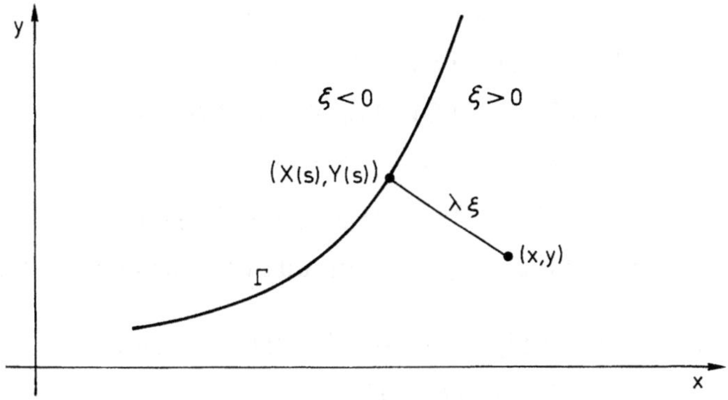

Fig. 3.4.2 Local coordinates

transformed variables (ξ, s). So we approximate V by $\hat{V}_0(\xi, s)$ inside the layer.
So, letting λ go to zero, we obtain

$$\partial_\xi^2 \hat{V}_0(\xi, s) = 2\delta^2 \sinh(\hat{V}_0(\xi, s)) - C(X(s)\mp, Y(s)\mp). \tag{3.4.17}$$

$C(X(s)\mp, Y(s)\mp)$ in (3.4.17) means the corresponding onesided limit of the function C at the *P-N* junction Γ. Equation (3.4.17) is called the layer equation in the language of singular perturbation theory and $\hat{V}_0$ is called the inner solution (see [3.25]). The layer equation is now also simpler than the original equation since it is only a family of ordinary differential equations in ξ parameterized by s. For $\xi \to \mp\infty$ the layer term $\hat{V}_0$ should match with the outer solution $\bar{V}_0$. So we require that

$$\hat{V}_0(\infty, s) = \bar{V}(X(s)+, Y(s)+), \qquad \hat{V}_0(-\infty, s) = \bar{V}(X(s)-, Y(s)-) \tag{3.4.18}$$

holds. Let us first check whether $\hat{V}_0$ can be really used as a layer term in (3.4.16), i.e. whether there is a solutive $\hat{V}_0$ of (3.4.19) which converges to $\bar{V}_0$ for ξ to $\pm\infty$. Equation (3.4.17) can be integrated in terms of the inverse of a monotone function. Multiplying both sides of (3.4.17) by $\partial_\xi \hat{V}_0$ and integrating once gives

$$\begin{aligned}\tfrac{1}{2}\partial_\xi \hat{V}_0(\xi, s)^2 &= 2\delta^2 \cosh(\hat{V}_0(\xi, s)) \\ &\quad - C(X(s)-, Y(s)-)\hat{V}_0(\xi, s) - A_-(s), \quad \text{for} \quad \xi < 0 \\ \tfrac{1}{2}\partial_\xi \hat{V}_0(\xi, s)^2 &= 2\delta^2 \cosh(\hat{V}_0(\xi, s)) \\ &\quad - C(X(s)+, Y(s)+)\hat{V}_0(\xi, s) - A_+(s), \quad \text{for} \quad \xi > 0.\end{aligned} \tag{3.4.19}$$

If $\hat{V}_0$ converges to $\bar{V}_0$ for ξ to $\pm\infty$, $\hat{V}_0(\pm\infty, s) = \bar{V}_0(X(s)\pm, Y(s)\pm)$,

$\partial_\xi \hat{V}_0(\pm\infty, s) = 0$ holds. This gives for $A_-(s)$ and $A_+(s)$ in (3.4.19)

$$\begin{aligned} A_+(s) &= 2\delta^2 \cosh(\bar{V}_0(X(s)+, Y(s)+) \\ &\quad - C(X(s)+, Y(s)+)\bar{V}_0(X(s)+, Y(s)+) \\ A_-(s) &= 2\delta^2 \cosh(\bar{V}_0(X(s)-, Y(s)-) \\ &\quad - C(X(s)-, Y(s)-)\bar{V}_0(X(s)-, Y(s)-). \end{aligned} \tag{3.4.20}$$

Moreover, the sign of the jump discontinuity of $\bar{V}_0$ determines the sign σ of $\partial_\xi \hat{V}_0$.

$$\begin{aligned} \sigma(s) := \operatorname{sign}(\partial_\xi \bar{V}_0(\xi, s)) = \operatorname{sign}[\bar{V}_0(X(s)+, Y(s)+) \\ - \bar{V}_0(X(s)-, Y(s)-)] \end{aligned} \tag{3.4.21}$$

holds. So by taking square roots we obtain

$$\begin{aligned} \sqrt{2}\sigma(s) = \partial_\xi \hat{V}_0(\xi, s)\{2\delta^2 \cosh(\hat{V}_0(\xi, s)) \\ - C(X(s)\pm, Y(s)\pm)\hat{V}_0(\xi, s) - A_\pm(s)\}^{-1/2}. \end{aligned} \tag{3.4.22}$$

Integrating (3.4.22) with respect to ξ now gives

$$\begin{aligned} &\text{a)} \quad F_+(\hat{V}_0(\xi, s), s) = \sqrt{2}\sigma\xi \qquad \text{for} \quad \xi > 0 \\ &\text{b)} \quad F_-(\hat{V}_0(\xi, s), s) = \sqrt{2}\sigma\xi \qquad \text{for} \quad \xi < 0 \\ &\text{c)} \quad F_\mp(z, s) = \int_{B(s)}^{z} \frac{dy}{\sqrt{2\delta^2 \cosh(y) - C_\mp y - A_\mp(s)}} \\ &\text{d)} \quad B(s) = \hat{V}_0(0, s) = \frac{A_+(s) - A_-(s)}{C(X(s)-, Y(s)-) - C(X(s)+, Y(s)+)}, \end{aligned} \tag{3.4.23}$$

where $\bar{V}_{0\mp}$ and $C_\mp$ denote the onesided limits at Γ. $B(s)$ in (3.4.23)d) has been computed by setting $\xi = 0$ in (3.4.19). Since the integrand in (3.4.23)c) is positive $F_\mp(z, s)$ is a monotone function of z and therefore the equations (3.4.23)a)b) are solvable for all ξ. So the layer problem (3.4.17) has a solution $\hat{V}_0$ which converges exponentially towards $\bar{V}_0$ for ξ to $\mp\infty$. To obtain a uniform $O(\lambda)$ approximation of the solution V we define the composite expansion V_0^c (see [3.25]) by

$$V_0^c = \begin{cases} \bar{V}_0(x, y) + [\hat{V}_0(\xi, s) - \bar{V}_0(X(s)+, Y(s)+)]\phi(\sqrt{\lambda}\xi) \\ \text{for} \quad \xi > 0 \\ \bar{V}_0(x, y) + [\hat{V}_0(\xi, s) - \bar{V}_0(X(s)-, Y(s)-)]\phi(\sqrt{\lambda}\xi) \\ \text{for} \quad \xi > 0. \end{cases} \tag{3.4.24}$$

In order to not having to deal with exponentially small terms at the boundary $\partial\Omega$ the function ϕ serves to eliminate the layer term away from Γ. It satisfies

$$\phi \in C^\infty, \qquad \phi(r) = 1 \quad \text{for} \quad |r| \leqslant 1, \qquad \phi(r) = 0 \quad \text{for} \quad |r| \geqslant 2. \tag{3.4.25}$$

The composite zero order approximation V_0^c then satisfies the differential equations together with the boundary conditions up to terms of order $O(\lambda)$ (see Problem 3.4).
So far we have only derived necessary conditions for the terms in our asymptotic approximation. The question which arises is of course whether, and if yes in what sense, the composite approximation V_0^c does approximate the solution V. In this simple case this question can be answered precisely by using the maximum principle (see [3.12]). If we define the remainder Q_0 by

$$Q_0 = V - V_0^c, \tag{3.4.26}$$

we derive the following remainder equation for Q_0

$$\lambda^2 \Delta Q_0 = F(x, y, Q_0) - \lambda G(x, y) \tag{3.4.27}$$

$$Q_0|_{\partial\Omega_D} = 0, \qquad \left.\frac{\partial Q_0}{\partial \nu}\right|_{\partial\Omega_N} = 0 \tag{3.4.28}$$

$$F(x, y, Q_0) = 2\delta^2 \sinh(V_0^c(x, y) + Q_0) - 2\delta^2 \sinh(V_0^c(x, y)). \tag{3.4.29}$$

The function $G(x, y)$ in (3.4.27) is of a complicated form involving the lower order partial derivatives of $\hat{V}_0$ w.r.t. ξ and s which arise from the coordinate transformation $(x, y) \leftrightarrow (\xi, s)$. However, $G(x, y)$ is a bounded function (see Problem 3.1). (3.4.27) implies that the derived approximation, at least, solves the differential equation up to terms of order $O(\lambda)$. Since $F(x, y, Q_0)$ is monotonically increasing in Q_0, it follows from the maximum principle, that the point $(\bar{x}, \bar{y})$, where Q_0 attains its maximum $\bar{Q}$, either lies on the Dirichlet boundary $\partial\Omega_D$ or that

$$F(\bar{x}, \bar{y}, \bar{Q}) - \lambda G(\bar{x}, \bar{y}) \leqslant 0 \tag{3.4.30}$$

holds. Since F is monotone in Q_0, $\bar{Q} \leqslant \tilde{Q}$ holds, where $\tilde{Q}$ is the solution of

$$F(\bar{x}, \bar{y}, \tilde{Q}) - \lambda G(\bar{x}, \bar{y}) = 0. \tag{3.4.31}$$

Thus $\max_\Omega Q_0(x, y) = O(\lambda)$ holds. In the same way one shows $\min_\Omega Q(x, y) = O(\lambda)$ and thus

$$\max_\Omega |Q_0(x, y)| = O(\lambda) \tag{3.4.32}$$

holds and V_0^c is an $O(\lambda)$ approximation to V. Since the remainder term Q_0 is of order $O(\lambda)$ it is reasonable to assume that, for $\lambda = 10^{-3}$ for instance, which corresponds to a device diameter of 1μ in silicon and a maximal doping $\tilde{C}$ of roughly $10^{19}\ \mathrm{cm}^{-3}$, V_0^c is a good approximation to the solution V. If even better approximations are desired these can be obtained by further expansion in powers of λ. We set, formally,

$$V = V(x, y, \lambda) \approx \sum_{n=0}^{\infty} V_n^c(x, y)\lambda^n. \tag{3.4.33}$$

(3.4.33) is an asymptotic expansion and the sum on the right hand side will in general not converge. If the sum is truncated after the Nth term the remainder Q_N will be of order $O(\lambda^N)$. V_0^c, $\bar{V}_0$, $\hat{V}_0$ and Q_0 were already derived above. The higher order terms are determined by inserting (3.4.33) into the equation (3.4.11)a) and expanding in powers of λ. This gives

$$
\begin{aligned}
&\text{a)}\quad \bar{V}_1 = 0\\
&\text{b)}\quad \Delta\bar{V}_{n-2} = 2\delta^2\cosh(\bar{V}_0)\bar{V}_n + F_n(\bar{V}_0,\ldots,\bar{V}_{n-1}), \qquad n = 2, 3\ldots\\
&\text{c)}\quad \partial_\xi^2\hat{V}_n = 2\delta^2\cosh(\hat{V}_0)\hat{V}_n + G_n(\bar{V}_0,\ldots,\bar{V}_n,\hat{V}_0,\ldots,\hat{V}_{n-1}),\\
&\qquad n = 1, 2, \ldots\\
&\text{d)}\quad V_n^c = \begin{cases} \bar{V}_n(x,y) + [\hat{V}_n(\xi,s) - \bar{V}_n(X(s)+, Y(s)+)]\phi(\sqrt{\lambda}\xi) \\ \text{for}\quad \xi > 0 \\ \bar{V}_n(x,y) + [\hat{V}_n(\xi,s) - \bar{V}_n(X(s)-, Y(s)-)]\phi(\sqrt{\lambda}\xi) \\ \text{for}\quad \xi > 0. \end{cases}
\end{aligned}
\tag{3.4.34}
$$

and so on. So, approximations of any order could be obtained theoretically. Note, however, that $\bar{V}_n$ in (3.4.34)b) is a function of $\Delta\bar{V}_{n-2}$. Thus, with each additional term in the expansion regularity is lost depending on the geometry of Ω and Γ. If $\bar{V}_0$ is smooth enough so that $\bar{V}_0, \ldots, \bar{V}_n$ are reasonably well defined, it can be shown by using the same maximum principle argument as above that the remainder term Q_N is of order λ^N in the L^∞-norm. If we reconsider the procedure to obtain an asymptotic expansion of the solution of the equilibrium problem (3.4.11), the steps involved are

A) Determine the reduced equation by setting λ equal to zero.
B) Check which parts of the problem (i.e. boundary conditions, continuity requirements etc.) can be satisfied with the outer solution obtained in Step A.
C) Add a layer correction term to the outer solution where the outer solution is apparently insufficient (in this case in the vicinity of the P-N junction Γ).
D) Show that the so derived reduced problem and the layer problem actually have solutions with the desired properties (regularity, decay etc.).
E) Show, that the composite expansion actually is an asymptotic approximation to the solution.

In the case of the equilibrium problem, when everything can be reduced to a single differential equation, we could actually carry out all the five steps above. However, when currents are present, and one has to deal with the whole coupled system (3.4.5), this will in general not be the case. One can always set the perturbation parameter to zero and obtain a reduced equation. So Steps A and B above are usually no problem. Also the derivation of the layer equations in Step C is possible if one finds the right coordinate transformation. In the case of the steady state Basic Semiconductor Equations this is no problem either. To find the solution of the reduced problem is more tricky. In the equilibrium case the reduced problem consists of the transcendental equation (3.4.13) which obviously had a unique solution. In the non equilibrium case the reduced problem is still a system of nonlinear partial differential equations. The layer equations usually degenerate into ordinary differential equations, and so finding solutions to them is easier. To show in Step E that the derived expansion actually is an asymptotic approximation of the solution is generally a very difficult task for systems of nonlinear partial differential equations. Since the whole approach consists of finding an approximate solution which satisfies the boundary value problem up to small terms, Step E involves the estimation of the inverse of the involved nonlinear differential operator. Using the implicit function theorem this task can be reduced to estimating the inverse of the linearization at the solution. However, even to show that this linearization has an inverse whose norm is bounded uniformly in the perturbation parameter λ, is usually impossible for complicated systems of partial differential equations, except in special cases.

The Non-Equilibrium Case

When the boundary data V_D in (3.4.9) are not given by the built-in-potential V_{bi} the densities n and p in (3.4.10) do not satisfy the boundary conditions and therefore do not constitute a solution of the boundary value problem. So, since we cannot simply integrate out the continuity equations and the current relations, the singular perturbation analysis has to be carried out on the whole coupled system (3.4.5). As pointed out before, the derivation of the reduced equations and the layer equations is straightforward if one follows the Steps A–C above. To prove that the asymptotic expansions actually constitute an approximation of the solution is not possible in general except in special cases such as in one dimension. The approximation property has been verified, however, numerically (see [3.21]).
As it already was the case for the existence and uniqueness results in Section 3.3, it is more convenient to write the drift diffusion equations in the Slotboom variables V, u and v of (3.3.3). Using the scaling introduced at the beginning of this Section, they are given by

$$u = \delta^{-2} n e^{-V}, \qquad v = \delta^{-2} p e^{V}. \tag{3.4.35}$$

The transformed problem now reads

$$\begin{aligned}
&\text{a)}\quad \lambda^2 \Delta V = \delta^2 (e^V u - e^{-V} v) - C(x)\\
&\text{b)}\quad \operatorname{div} J_n = R, \qquad J_n = \mu_n e^V \operatorname{grad} u\\
&\text{c)}\quad \operatorname{div} J_p = -R, \qquad J_p = -\mu_p e^{-V} \operatorname{grad} v\\
&\text{d)}\quad V|_{\partial\Omega_D} = V_D|_{\partial\Omega_D}, \qquad u|_{\partial\Omega_D} = u_D|_{\partial\Omega_D}, \qquad v|_{\partial\Omega_D} = v_D|_{\partial\Omega_D},
\end{aligned} \tag{3.4.36}$$

$$\text{e)}\quad \frac{\partial V}{\partial \nu}\bigg|_{\partial\Omega_N} = \frac{\partial u}{\partial \nu}\bigg|_{\partial\Omega_N} = \frac{\partial v}{\partial \nu}\bigg|_{\partial\Omega_N} = 0.$$

$$\text{f)}\quad u_D(x) = e^{-U(x)}, \qquad \text{d)}\quad v_D(x) = e^{U(x)}$$

$$\text{g)}\quad V_D(x) = U(x) + \ln\left[\frac{1}{2\delta^2}\left(C(x) + \sqrt{C(x)^2 + 4\delta^4}\right)\right]$$

$$= U(x) + V_{bi}(x), \qquad x \in \partial\Omega_D.$$

Again $U(x)$ denotes the applied potential at the contacts. Note, that $u(x) \equiv v(x) \equiv 1$ holds in the equilibrium case discussed above. In the non equilibrium case the form of the mobilities μ_n and μ_p impacts the form of the asymptotic expansions significantly. The two basic cases to be distinguished are the presence and the absence of velocity saturation effects. If, in the presence of velocity saturation, μ_n and μ_p behave like $1/|\text{grad}\, V|$ for large values of the electric field $-\text{grad}\, V$ this implies that inside the layers the mobilities are of order $O(\lambda)$. We will leave this case for later on and consider the absence of velocity saturation first. So we simply assume for the moment that μ_n and μ_p depend on the position x only (and not on λ). Since we will not derive the higher order terms in the asymptotic expansion we will drop the subscript 0 for the zero order term from now on.

So, for $\lambda \to 0$, the reduced problem for the outer solution $\bar{w} = (\bar{V}, \bar{u}, \bar{v}, \bar{J}_n, \bar{J}_p)$ is given by

$$\begin{aligned}
&\text{a)}\quad 0 = \delta^2(e^{\bar{V}}\bar{u} - e^{-\bar{V}}\bar{v}) - C(x)\\
&\text{b)}\quad \text{div}\, \bar{J}_n = \bar{R}, \qquad \bar{J}_n = \mu_n e^{\bar{V}}\, \text{grad}\, \bar{u}\\
&\text{c)}\quad \text{div}\, \bar{J}_p = -\bar{R}, \qquad \bar{J}_p = -\mu_p e^{-\bar{V}}\, \text{grad}\, \bar{v}\\
&\text{d)}\quad \bar{V}|_{\partial\Omega_D} = V_D|_{\partial\Omega_D}, \qquad \bar{u}|_{\partial\Omega_D} = u_D|_{\partial\Omega_D}, \qquad \bar{v}|_{\partial\Omega_D} = v_D|_{\partial\Omega_D}\\
&\text{e)}\quad \frac{\partial \bar{V}}{\partial \nu}\bigg|_{\partial\Omega_N} = \frac{\partial \bar{u}}{\partial \nu}\bigg|_{\partial\Omega_N} = \frac{\partial \bar{v}}{\partial \nu}\bigg|_{\partial\Omega_N} = 0,
\end{aligned} \tag{3.4.37}$$

and it is understood that $\bar{w} = (\bar{V}, \bar{u}, \bar{v}, \bar{J}_n, \bar{J}_p)$ denotes the zero order term of the outer solution. $\bar{R}$ in (3.4.37)b)c) denotes the recombination rate evaluated at the reduced solution $\bar{w}$. (The form of the recombination term is not important for the moment.) Firstly we observe that we have three boundary conditions at each part of the boundary while we have only two second order differential equations (3.4.37)b) and (3.4.37)c). However, since the boundary conditions at Ohmic contacts are derived from the condition of vanishing space charge, they are consistent with the zero space charge approximation (3.4.37)a). In other words, if we solve (3.4.37)a) for $\bar{V}$ and insert into (3.4.37)b) and (3.4.37)c), we obtain the boundary value problem

$$\begin{aligned}
&\text{a)}\quad \text{div}\, \bar{J}_n = \bar{R}, \qquad \bar{J}_n = \mu_n e^{\bar{V}(x,\bar{u},\bar{v})}\, \text{grad}\, \bar{u}\\
&\text{b)}\quad \text{div}\, \bar{J}_p = -\bar{R}, \qquad \bar{J}_p = -\mu_p e^{-\bar{V}(x,\bar{u},\bar{v})}\, \text{grad}\, \bar{v}\\
&\text{c)}\quad \bar{u}|_{\partial\Omega_D} = u_D|_{\partial\Omega_D}, \qquad \bar{v}|_{\partial\Omega_D} = v_D|_{\partial\Omega_D},
\end{aligned} \tag{3.4.38}$$

d) $\left.\dfrac{\partial \bar{u}}{\partial \nu}\right|_{\partial\Omega_N} = \left.\dfrac{\partial \bar{v}}{\partial \nu}\right|_{\partial\Omega_N} = 0.$

e) $\bar{V}(x, \bar{u}, \bar{v}) = \ln\left[\dfrac{C(x) + \sqrt{C(x)^2 + 4\delta^4\bar{u}\bar{v}}}{2\delta^2\bar{u}}\right]$

for $\bar{u}$ and $\bar{v}$. $\bar{V} = \bar{V}(x, \bar{u}, \bar{v})$ then satisfies the boundary condition (3.4.37)d)e) automatically if $\partial C/\partial\nu|_{\partial\Omega_D} = 0$ holds. (3.4.38) constitutes the reduced problem. Again, as in the equilibrium case, the necessity for a layer term arises from the fact that $\bar{V}$, defined by (3.4.38)e), is discontinuous at the P-N junction Γ because of the discontinuity of $C(x)$ there. Performing the same local coordinate transformation (3.4.15) as in the equilibrium case, we derive the layer term $\hat{w} = (\hat{V}, \hat{u}, \hat{v}, \hat{J}_n, \hat{J}_p)$ which is a function of the layer variables ξ and s in (3.4.15). Again we change from the (x, y) variables to the (ξ, s) variables. This yields, for $\lambda \to 0$,

$$\begin{aligned} &\text{a)} \quad \partial_\xi^2 \hat{V} = \delta^2[e^{\hat{V}(\xi,s)}\hat{u}(\xi, s) - e^{\hat{V}(\xi,s)}\hat{v}(\xi, s)] - C_\mp(s) \\ &\text{b)} \quad \partial_\xi \hat{J}_n \cdot \vec{n} = 0, \qquad 0 = \mu_{n_\mp}(s) e^{\hat{V}(\xi,s)} \partial_\xi \hat{u}(\xi, s) \\ &\text{c)} \quad \partial_\xi \hat{J}_p \cdot \vec{n} = 0, \qquad 0 = -\mu_{p_\mp}(s) e^{-\hat{V}(\xi,s)} \partial_\xi \hat{v}(\xi, s). \end{aligned} \tag{3.4.39}$$

Here $\vec{n} = (\dot{Y}(s), -\dot{X}(s))^T$ denotes the normal vector on the P-N junction Γ. In (3.4.39) $f_\mp(s)$ denotes the appropriate one sided limit of the function f at Γ. So $f_+(s) = f(X(s)+, Y(s)+)$ for $\xi > 0$ and $f_-(s) = f(X(s)-, Y(s)-)$ for $\xi < 0$ holds. Since the layer terms have to match the outer solution $\bar{w}$ at $\xi = \pm\infty$, we immediately obtain

$$\begin{aligned} &\text{a)} \quad \hat{u}(\xi, s) \equiv \bar{u}(X(s), Y(s)), \qquad \hat{v}(\xi, s) \equiv v(X(s), Y(s)) \\ &\text{b)} \quad \hat{J}_n(\xi, s) \cdot \vec{n}(s) \equiv \bar{J}_n(X(s), Y(s)) \cdot \vec{n}(s), \\ &\text{c)} \quad \hat{J}_p(\xi, s) \cdot \vec{n}(s) \equiv \bar{J}_p(X(s), Y(s)) \cdot \vec{n}(s). \end{aligned} \tag{3.4.40}$$

A layer term occurs in the potential V and in the tangential components of the current densities J_n and J_p only, while $\bar{u}, \bar{v}, \bar{J}_n \cdot \vec{n}$ and $\bar{J}_p \cdot \vec{n}$ are continuous across Γ. The layer equation is given by

$$\begin{aligned} &\text{a)} \quad \partial_\xi^2 \hat{V} = \delta^2[e^{\hat{V}(\xi,s)}\bar{u}(s) - e^{-\hat{V}(\xi,s)}\bar{v}(s)] - C_+(s) \qquad \text{for} \quad \xi > 0 \\ &\text{b)} \quad \partial_\xi^2 \hat{V} = \delta^2[e^{\hat{V}(\xi,s)}\bar{u}(s) - e^{-\hat{V}(\xi,s)}\bar{v}(s)] - C_-(s) \qquad \text{for} \quad \xi < 0, \end{aligned} \tag{3.4.41}$$

where $\bar{u}(s)$ stands for $\bar{u}(X(s), Y(s))$ and so on. As in the equilibrium problem, we require $\hat{V}$ to match the outer solution $\bar{V}$ at $\xi = \mp\infty$. So we require

$$\hat{V}(\infty, s) = \bar{V}(X(s)+, Y(s)+), \qquad \hat{V}(-\infty, s) = \bar{V}(X(s)-, Y(s)-). \tag{3.4.42}$$

After solving the reduced problem and the layer problem (3.4.41)–(3.4.42) the tangential components of the layer current densities are given by

$$\begin{aligned} &\text{a)} \quad \hat{J}_n = \vec{t}(s)\mu_{n\pm}(s)e^{\hat{V}}\partial_s\bar{u}(s) \\ &\text{b)} \quad \hat{J}_p = -\vec{t}(s)\mu_{p\pm}(s)e^{-\hat{V}}\partial_s\bar{v}(s), \qquad t(s) = (\dot{X}(s), \dot{Y}(s))^T. \end{aligned} \tag{3.4.43}$$

Since the layer problem (3.4.41)–(3.4.42) around the *P-N* junction is of the same form as in the equilibrium case the same technique can be applied to show that (3.4.41)–(3.4.42) has a solution $\hat{V}$ which converges monotonically towards the one sided limit of the outer solution $\bar{V}$ for ξ to $\pm\infty$. If no velocity saturation effects are considered, and the mobilities μ_n and μ_p are simply functions of position, bounded uniformly away from zero, the same approach as in Section 3.3 can be used to show the existence of a solution of the reduced problem (3.4.38). If the recombination rate is given by the scaled Shockley Read Hall term

$$R = \frac{uv - 1}{\tau_p(e^V u + 1) + \tau_n(e^{-V} v + 1)}, \tag{3.4.44}$$

then the corresponding fixed point map G is given by $G(u_0, v_0) = (u_1, v_1)$ where u_1 and v_1 satisfy the linear boundary value problems

$$\begin{aligned} &\text{a)} \quad \operatorname{div}[\mu_n e^{V_0} \operatorname{grad} u_1] = \frac{u_1 v_0 - 1}{\tau_p(e^{V_0}u_0 + 1) + \tau_n(e^{-V_0}v_0 + 1)} \\ &\text{b)} \quad \operatorname{div}[\mu_p e^{-V_1} \operatorname{grad} v_1] = \frac{u_1 v_1 - 1}{\tau_p(e^{V_1}u_1 + 1) + \tau_n(e^{-V_1}v_0 + 1)} \\ &\text{c)} \quad u_1|_{\partial\Omega_D} = u_D|_{\partial\Omega_D}, \qquad v_1|_{\partial\Omega_D} = v_D|_{\partial\Omega_D}, \\ &\text{d)} \quad \left.\frac{\partial u_1}{\partial \nu}\right|_{\partial\Omega_N} = \left.\frac{\partial v_1}{\partial \nu}\right|_{\partial\Omega_N} = 0. \\ &\text{e)} \quad V_0 = \ln\left[\frac{C(x) + \sqrt{C(x)^2 + 4\delta^4 u_0 v_0}}{2\delta^2 u_0}\right] \\ &\text{f)} \quad V_1 = \ln\left[\frac{C(x) + \sqrt{C(x)^2 + 4\delta^4 u_1 v_0}}{2\delta^2 u_1}\right]. \end{aligned} \tag{3.4.45}$$

We refer the reader to [3.19] for a more detailed existence proof for the reduced problem. As in the equilibrium case (3.4.24), the zero order composite expansion is given by

$$w_0^c = \begin{cases} \bar{w}(x, y) + [\hat{w}(\xi, s) - \bar{w}(X(s)+, Y(s)+)]\phi(\sqrt{\lambda}\xi) \\ \qquad\qquad\qquad \text{for} \quad \xi > 0, \\ \bar{w}(x, y) + [\hat{w}(\xi, s) - \bar{w}(X(s)-, Y(s)-)]\phi(\sqrt{\lambda}\xi) \\ \qquad\qquad\qquad \text{for} \quad \xi < 0, \end{cases} \tag{3.4.46}$$

where the vector w denotes (V, u, v, J_n, J_p). The derivation of the higher order terms in the asymptotic expansion is again straightforward (see Problem 3.5).

Asymptotic Validity in the One-Dimensional Case

To show that the obtained approximation is asymptotically valid, i.e. to show that the solution of (3.4.36) converges in some sense to w_0^c for $\lambda \to 0$, is, in full generality, a yet unresolved problem. Ideally, one would like an explicit estimate of the difference between the exact solution and the asymptotic approximation obtained above. Unfortunately, such estimates are available only in special cases (no generation terms, one dimensional problems etc.). In more than one dimension it has been established that the solution of the full problem (3.4.36) converges to w_0^c as λ tends to zero, in the absence of recombination and generation terms (see [3.15]). However, this result says nothing about the rate of convergence, and therefore about the approximation quality for finite λ. Results which do also provide convergence rates are only available in the one dimensional case. They can be found in c.f. [3.2] and [3.23].

If we consider the one dimensional model of a *P-N* junction the system (3.4.36) reduces to

$$\begin{aligned} &\text{a)}\quad \lambda^2 \frac{d^2V}{dx^2} = \delta^2(e^V u - e^{-V} v) - C(x) \\ &\text{b)}\quad \frac{d}{dx} J_n = R, \qquad J_n = \mu_n e^V \frac{du}{dx} \\ &\text{c)}\quad \frac{d}{dx} J_p = -R, \qquad J_p = -\mu_p e^{-V} \frac{dv}{dx}, \end{aligned} \tag{3.4.47}$$

where x is the position variable which varies, after scaling, between $x = -1$ and $x = 1$. The boundary conditions in the one dimensional case are of the form

$$\begin{aligned} &u(x) = e^{-U(x)}, \qquad v(x) = e^{U(x)} \\ &V(x) = V_D(x) = U(x) + \ln\left[\frac{1}{2\delta^2}(C(x) + \sqrt{C(x)^2 + 4\delta^4})\right] \\ &\qquad = U(x) + V_{bi}(x), \qquad \text{for} \quad x = -1 \quad \text{and} \quad x = 1. \end{aligned} \tag{3.4.48}$$

In [3.23] an abrupt *P-N* junction in the center of the device is assumed. So $C(x)$ in (3.4.47)a) has a jump discontinuity at $x = 0$ and $C(x) > 0$ for $x > 0$ and $C(x) < 0$ for $x < 0$ holds. Also the absence of recombination-generation effects is assumed; so $R \equiv 0$ holds. The authors show that if the boundary potential V_D varies within a certain range then

$$\begin{aligned} &\max_{[-1,1]} |V - V_0^c| + \max_{[-1,1]} |u - u_0^c| + \max_{[-1,1]} |v - v_0^c| \\ &\quad \leqslant \text{const}\, \lambda \delta^2 |V_D(1) - V_D(-1) + 1|^3 e^{0.5(U(1) - U(-1))} \end{aligned} \tag{3.4.49}$$

holds. Since the proof is technically quite involved it will be omitted here. It should be pointed out, however, that the proof makes explicit use of the fact, that in the one dimensional case J_n and J_p are constant in the absence of recombination-generation. Thus it is unlikely that it can easily be generalized to higher dimensions. Although this is unsatisfactory for practical purposes, the result contains an interesting feature. If we choose typical values for the device parameters, say 10^{-4} cm for the length L of the device and 10^{19} cm^{-3} for the maximum doping concentration, and calculate the resulting values for λ and δ in silicon, the bounds on the applied potential difference, which have to be assumed in order for the estimate (3.4.49) to be valid imply that the applied bias satisfies

$$-0.8\ \mathrm{V} \leqslant U_T[U(1) - U(-1)] \leqslant 0.2\ \mathrm{V}. \tag{3.4.50}$$

In this setting a negative potential difference corresponds to forward bias and a positive one to reverse bias. For such a diode -0.8 V represents quite a large forward bias. On the other hand one would expect the drift diffusion equations to give a reasonable description of the device for reverse bias values of up to a couple of volts. So this is consistent with the original premise of this Section to give an asymptotic analysis for forward biased *P-N* junctions. It also is an indication that in general the validity of the above derived approximations will break down for some moderate values of the reverse bias. Thus, the reverse bias case will be treated by a different kind of asymptotic analysis in Section 3.5.

Velocity Saturation Effects—Field Dependent Mobilities

It is a well known fact of device physics, that the proportionality of the drift velocities μ_n grad V and $-\mu_p$ grad V to the electric field $-$grad V only holds at moderate field strengths. Due to carrier heating, the velocities saturate for high electric fields, i.e.

$$\lim_{|\mathrm{grad}\, V| \to \infty} |\mu_n \,\mathrm{grad}\, V| = v_n \quad \text{and} \quad \lim_{|\mathrm{grad}\, V| \to \infty} |\mu_p \,\mathrm{grad}\, V| = v_p \tag{3.4.51}$$

holds, where v_n and v_p are the saturation velocities. In order to reflect this property, the mobilities μ_n and μ_p are modeled as dependent on the electric field such that the saturation relation (3.4.51) holds. One way to do this is to choose the mobilities as

$$\mu_n = \mu_n^s = \frac{v_n \bar{\mu}_n}{v_n + \bar{\mu}_n |\mathrm{grad}\, V|}, \qquad \mu_p = \mu_p^s = \frac{v_p \bar{\mu}_p}{v_p + \bar{\mu}_p |\mathrm{grad}\, V|}, \tag{3.4.52}$$

where $\bar{\mu}_n$ and $\bar{\mu}_p$ are field independent, mobility models (see [3.34]). If the original drift velocities $\mu_{n,p}$ grad V are small compared to $v_{n,p}$ (3.4.52) represents only a small perturbation of the original model and

$$\mu_n^s \sim \bar{\mu}_n, \qquad \mu_p^s \sim \bar{\mu}_p \tag{3.4.53}$$

holds. On the other hand the new drift velocities μ_n^s grad V and $-\mu_p^s$ grad V satisfy the saturation relation (3.4.51). Of course there is a multitude of possible formulas for μ_n^s and μ_p^s which would do the same job. We refer the reader to [3.34] for other formulas used in practice. It should be pointed out, however, that the form of μ_n^s and μ_p^s is based on no other physical consideration than the saturation relation (3.4.51), and different models are obtained only by fitting experimental data. If we use the scaling of this Section together with (3.4.52) we obtain

$$\begin{aligned}
&\text{a)}\quad \lambda^2 \Delta V = \delta^2(e^V u - e^{-V} v) - C(x)\\
&\text{b)}\quad \operatorname{div} J_n = R, \qquad J_n = \mu_n^s e^V \operatorname{grad} u\\
&\text{c)}\quad \operatorname{div} J_p = -R, \qquad J_p = -\mu_p^s e^{-V} \operatorname{grad} v\\
&\text{d)}\quad \mu_n^s = \frac{\alpha_n \mu_n}{\alpha_n + \mu_n |\operatorname{grad} V|}, \qquad \mu_p^s = \frac{\alpha_p \mu_p}{\alpha_p + \mu_p |\operatorname{grad} V|},
\end{aligned} \tag{3.4.54}$$

with $\alpha_{n,p}$, the scaled saturation velocities, given by

$$\alpha_n = \frac{v_n l}{\mu_n U_T}, \qquad \alpha_p = \frac{v_p l}{\mu_p U_T}. \tag{3.4.55}$$

Inside the layer, where grad V will be large ($\cong O(1/\lambda)$), this will obviously change the structure of the solution and of the singular perturbation approximation. In addition, as we will see, the different layer behaviour also impacts the reduced problem and therefore the total current flow through the *P-N* junction. The reduced equations remain the same, however, with μ_n and μ_p in (3.4.37) replaced by μ_n^s and μ_p^s. Following the expansion procedure gives, for the layer equations

$$\begin{aligned}
&\text{a)}\quad \partial_\xi^2 \hat{V} = \delta^2[e^{\hat{V}(\xi,s)}\hat{u}(\xi,s) - e^{-\hat{V}(\xi,s)}\hat{v}(\xi,s)] - C_\pm(s)\\
&\text{b)}\quad \partial_\xi \hat{J}_n(\xi,s)\cdot \vec{n}(s) = 0, \qquad \partial_\xi \hat{J}_p(\xi,s)\cdot \vec{n}(s) = 0\\
&\text{c)}\quad \hat{J}_n \cdot \vec{n} = \frac{\alpha_n}{|\partial_\xi \hat{V}|} e^{\hat{V}(\xi,s)} \partial_\xi \hat{u}(\xi,s)\\
&\text{d)}\quad \hat{J}_p \cdot \vec{n} = \frac{\alpha_p}{|\partial_\xi \hat{V}|} e^{-\hat{V}(\xi,s)} \partial_\xi \hat{v}(\xi,s)
\end{aligned} \tag{3.4.56}$$

with $\vec{n}(s) = (\dot{Y}(s), -\dot{X}(s))$ as the normalized normal vector on the *P-N* junction Γ. Again, we obtain

$$\begin{aligned}
&\text{a)}\quad \hat{J}_n \cdot \vec{n} = \bar{J}_n(X(s), Y(s)) \cdot \vec{n}\\
&\text{b)}\quad \hat{J}_p \cdot \vec{n} = \bar{J}_p(X(s), Y(s)) \cdot \vec{n}.
\end{aligned} \tag{3.4.57}$$

The difference to the unsaturated case comes from the current relations (3.4.56)c)d). From (3.4.56)c) we obtain

$$\begin{aligned}
&\text{a)}\quad \partial_\xi \hat{u} = e^{-\hat{V}}[\bar{J}_n(s)\cdot\vec{n}(s)]\frac{1}{\alpha_n}\sigma\partial_\xi\hat{V}\\
&\text{b)}\quad \partial_\xi \hat{v} = e^{\hat{V}}[\bar{J}_p(s)\cdot\vec{n}(s)]\frac{1}{\alpha_p}\sigma\partial_\xi\hat{V}\\
&\text{c)}\quad \sigma = \operatorname{sign}[\partial_\xi\hat{V}(\xi,s)].
\end{aligned}\tag{3.4.58}$$

Here $\bar{J}_n(s)$ denotes $\bar{J}_n(X(s), Y(s))$. If $\hat{V}$ is monotone in ξ for fixed s then σ is independent of ξ and (3.4.58) can be integrated. This gives

$$\begin{aligned}
\hat{u} &= -e^{-\hat{V}}[\bar{J}_n(s)\cdot\vec{n}(s)]\frac{1}{\alpha_n}\sigma(s) + A_n(s)\\
\hat{v} &= -e^{\hat{V}}[\bar{J}_p(s)\cdot\vec{n}(s)]\frac{1}{\alpha_p}\sigma(s) + A_p(s)
\end{aligned}\tag{3.4.59}$$

with integration constants A_n and A_p. Because of the matching conditions $\hat{u}(\pm\infty) = \bar{u}(X(s)\pm, Y(s)\pm)$, $\hat{v}(\pm\infty) = \bar{v}(X(s)\pm, Y(s)\pm)$ the integration constants are given by

$$\begin{aligned}
&\text{a)}\quad A_n(s) = \bar{u}_+(s) + \frac{\sigma}{\alpha_n}(\bar{J}_n(s)\cdot\vec{n}(s))e^{-\bar{V}_+(s)}\\
&\qquad\qquad = \bar{u}_-(s) + \frac{\sigma}{\alpha_n}(\bar{J}_n(s)\cdot\vec{n}(s))e^{-\bar{V}_-(s)}\\
&\text{b)}\quad A_p(s) = \bar{v}_+(s) + \frac{\sigma}{\alpha_p}(\bar{J}_p(s)\cdot\vec{n}(s))e^{\bar{V}_+(s)}\\
&\qquad\qquad = \bar{v}_-(s) + \frac{\sigma}{\alpha_p}(\bar{J}_p(s)\cdot\vec{n}(s))e^{\bar{V}_-(s)}.
\end{aligned}\tag{3.4.60}$$

Here $\bar{u}_\pm(s)$ denote the onesided limits $\bar{u}(X(s)\pm, Y(s)\pm)$ and so on. So the layer problem is given by (3.4.56)a) together with the boundary conditions

$$\hat{V}(\infty, s) = \bar{V}(X(s)+, Y(s)+), \qquad \hat{V}(-\infty, s) = \bar{V}(X(s)-, Y(s)-).\tag{3.4.61}$$

One can again show, as in the equilibrium case, that the solution $\hat{V}$ exists and is indeed monotone for fixed s. So σ in (3.4.58) is really independent of ξ. The difference in the reduced problem comes from the fact that now a layer term arises in the u and v variables as well. This implies that $\bar{V}$, $\bar{u}$ and $\bar{v}$ will be discontinuous across the P-N junction Γ. Thus the reduced problem can not be formulated as in (3.4.38) but in addition interface conditions have to be applied for u, v across Γ. From (3.4.60) we obtain that the functions

$$\begin{aligned}
&\text{a)}\quad \bar{u} + e^{-\bar{V}}(\bar{J}_n\cdot\vec{n}(s))\frac{\sigma(s)}{\alpha_n}\\
&\text{b)}\quad \bar{v} + e^{\bar{V}}(\bar{J}_p\cdot\vec{n}(s))\frac{\sigma(s)}{\alpha_p}
\end{aligned}\tag{3.4.62}$$

remain continuous. Thus the reduced problem is of the form

$$\left.\begin{aligned}
&\text{a)}\quad \operatorname{div}\bar{J}_n = \bar{R}, \quad \bar{J}_n = \mu_n^s e^{\bar{V}(x,\bar{u},\bar{v})} \operatorname{grad}\bar{u}\\
&\text{b)}\quad \operatorname{div}\bar{J}_p = -\bar{R}, \quad \bar{J}_p = -\mu_p^s e^{-\bar{V}(x,\bar{u},\bar{v})} \operatorname{grad}\bar{v}\\
&\text{c)}\quad \bar{V}(x,\bar{u},\bar{v}) = \ln\left[\frac{C(x) + \sqrt{C(x)^2 + 4\delta^4\bar{u}\bar{v}}}{2\delta^2\bar{u}}\right]\\
&\text{for } (x,y)\in\Omega \text{ together with the boundary conditions}\\
&\text{d)}\quad \bar{u}|_{\partial\Omega_D} = u_D|_{\partial\Omega_D}, \qquad \bar{v}|_{\partial\Omega_D} = v_D|_{\partial\Omega_D},\\
&\text{e)}\quad \left.\frac{\partial\bar{u}}{\partial\nu}\right|_{\partial\Omega_N} = \left.\frac{\partial\bar{v}}{\partial\nu}\right|_{\partial\Omega_N} = 0\\
&\text{on } \partial\Omega, \text{ and the interface conditions which say that the terms}\\
&\text{f)}\quad \bar{u} + e^{-\bar{V}}(\bar{J}_n\cdot\vec{n}(s))\frac{\sigma(s)}{\alpha_n}, \qquad \sigma(s) = \operatorname{sign}(\bar{V}_+ - \bar{V}_-)\\
&\text{g)}\quad \bar{v} + e^{\bar{V}}(\bar{J}_p\cdot\vec{n}(s))\frac{\sigma(s)}{\alpha_p}\\
&\text{h)}\quad \bar{J}_n\cdot\vec{n}(s)\\
&\text{i)}\quad \bar{J}_p\cdot\vec{n}(s)
\end{aligned}\right\} \quad (3.4.63)$$

remain continuous across Γ.

Since the reduced problem has a different form, the effect of the field dependent mobilities is not only noticable locally, inside the layers, but also globally in the outer solution. It will therefore influence the current flow through the *P-N* junction although μ_n^s and μ_p^s will be close to $\bar{\mu}_n$ and $\bar{\mu}_p$ for the outer solution. Note, that, since the effect of the field dependent mobilities on the outer solution occurs through the interface conditions (3.4.63)f)g), this effect will be the stronger the larger the jump in $\bar{V}$ at the *P-N* junction Γ, and therefore $\partial_\xi \hat{V}$, is. So the locally large electric fields at the *P-N* junctions influence the solution globally.

3.5 Reverse Biased *P-N* Junctions

We now turn to the study of *P-N* junctions under reverse biasing conditions. In the reverse bias case the singular perturbation scaling, introduced in the previous Section, describes the solution of the drift diffusion equations only up to a moderate value of the reverse bias. The reason for this is that, if a potential difference of the appropriate sign is applied to the *P-N* junction, one observes the formation of a depletion region in the semiconductor where no free carriers are present. This region acts as an insulator and, ideally, no current flows. In Section 3.4 equation (3.4.5) implied that the space charge $\rho = -n + p + C(x)$ is of order λ^2 except in narrow layer regions. Therefore, if that scaling is used, the depletion region has to lie entirely within the *P-N* junction layer. On the other hand the limits of validity of this asymptotic analysis which we have briefly discussed in the previous Paragraph (see (3.4.50)), indicate that the approximation breaks down anyway for large reverse bias values. Thus, for large reverse bias, the layer region becomes so bloated that the scaling (3.4.2) has to be modified. We will at first analyze the formation of the depletion region for moderate reverse bias values, using the scaling (3.4.2). This can be done by an additional asymptotic analysis of the layer term $\hat{V}$ in (3.4.41) for $\delta \to 0$.

Moderately Reverse Biased P-N Junctions

If we compute the layer term $\hat{V}$ in (3.4.41) in the same way as in the equilibrium case, we obtain $\hat{V}$ in terms of the inverse of a monotone function.

$$\begin{aligned}
&\text{a)}\quad F_+(\hat{V}(\xi, s), s) = \sqrt{2\sigma\xi} \qquad \text{for} \qquad \xi \geqslant 0 \\
&\text{b)}\quad F_-(\hat{V}(\xi, s), s) = \sqrt{2\sigma\xi} \qquad \text{for} \qquad \xi \leqslant 0 \\
&\text{c)}\quad F_\mp(V, s) = \int_{\hat{V}(0)}^{V} \frac{dz}{\sqrt{\delta^2(e^z\bar{u}(s) + e^{-z}\bar{v}(s)) - C_\mp(s)z - A_\mp(s)}} \\
&\text{d)}\quad A_\mp(s) = \delta^2(e^{\bar{V}_\mp(s)}\bar{u}(s) + e^{-\bar{V}_\mp(s)}\bar{v}(s)) - C_\mp(s)\bar{V}_\mp(s) \\
&\text{e)}\quad \hat{V}(0) = \frac{A_-(s) - A_+(s)}{C_+(s) - C_-(s)} .
\end{aligned} \tag{3.5.1}$$

For fixed $\hat{V} \neq \bar{V}_\mp$ and $\delta \to 0$, $F_\mp(\hat{V}, s)$ will tend to $F_{\mp_0}(\hat{V}, s)$, given by

$$F_{\mp_0}(\hat{V}, s) = \int_{\hat{V}(0)}^{\hat{V}} \frac{dz}{\sqrt{C_\mp(\bar{V}_\mp - z)}} . \tag{3.5.2}$$

Solving $F_\mp(\hat{V}, s) = \sqrt{2\sigma\xi}$ yields

$$\hat{V} = \bar{V}_\mp - \frac{1}{C_\mp}\left[C_\mp \frac{\sigma\xi}{\sqrt{2}} - \sqrt{C_\mp(\bar{V}_\mp - V(0))}\right]^2 , \tag{3.5.3}$$

which gives $\partial_\xi^2 \hat{V} \cong -C$ and therefore $\hat{n} \cong \hat{p} \cong 0$ inside the layer.
The limit $\delta \to 0$ in $F_\mp$ is not uniform for $\hat{V}$close to $\bar{V}_\mp$. This raises the general question whether performing an additional asymptotic analysis for small δ in the layer equations is exact, i.e. whether the two limits $\lambda \to 0$ and $\delta \to 0$ commute. We will leave the discussion of the interdependence of these two limits to Chapter 4. At this point the precise transition mechanism between the forward bias case and the situation under extreme reverse biasing conditions, treated in the next Section, is not understood completely.

P-N Junctions Under Extreme Reverse Bias Conditions

If we increase the reverse bias further, the depletion region, and with it the *P-N* junction layer, will cover a significant part of the device and the approximation, derived in the previous Section, breaks down. For *P-N* junctions under extreme reverse biasing conditions an asymptotic analysis based on a different scaling can be performed. This scaling was first introduced in [3.6] and is given by

$$V = \frac{qL^2\tilde{C}}{\varepsilon} V_s, \qquad n = \tilde{C} n_s, \qquad p = \tilde{C} p_s$$
$$J_n = \frac{q\tilde{\mu}\, U_T \tilde{C}}{L} J_{n_s}, \qquad J_p = \frac{q\tilde{\mu}\, U_T \tilde{C}}{L} J_{p_s}. \tag{3.5.4}$$

The scaled boundary value problem then becomes

$$\text{a)}\quad \Delta V_s = n_s - p_s - C_s$$
$$\text{b)}\quad \operatorname{div} J_{n_s} = R_s, \qquad \lambda^2 J_{n_s} = \mu_n(\lambda^2 \operatorname{grad} n_s - n_s \operatorname{grad} V_s)$$
$$\text{c)}\quad \operatorname{div} J_{p_s} = -R_s, \qquad \lambda^2 J_{p_s} = \mu_p(-\lambda^2 \operatorname{grad} p_s - p_s \operatorname{grad} V_s)$$
$$\text{d)}\quad V_s|_{\partial\Omega_D} = U_s + V_{bi_s}|_{\partial\Omega_D}, \qquad n_s|_{\partial\Omega_D} = n_{D_s}|_{\partial\Omega_D}, \qquad p_s|_{\partial\Omega_D} = p_{D_s}|_{\partial\Omega_D}$$
$$\text{e)}\quad \left.\frac{\partial V_s}{\partial v}\right|_{\partial\Omega_N} = J_n\cdot\vec{n}|_{\partial\Omega_N} = J_p\cdot\vec{n}|_{\partial\Omega_N} = 0 \tag{3.5.5}$$
$$\text{f)}\quad n_s = \tfrac{1}{2}\left(C_s(x) + \sqrt{C_s(x)^2 + 4\delta^4}\right),$$
$$p_s = \tfrac{1}{2}\left(-C_s(x) + \sqrt{C_s(x)^2 + 4\delta^4}\right.$$
$$\text{g)}\quad U_{D_s}(x) = \frac{U(x)\varepsilon\delta^2}{qL^2 n_i},$$
$$V_{bi_s}(x) = \frac{U_T\delta^2\varepsilon}{qL^2 n_i}\ln\left[\frac{C_s(x) + \sqrt{C_s(x)^2 + 4\delta^4}}{2\delta^2}\right]$$
$$\text{h)}\quad \lambda = \sqrt{\frac{U_T\varepsilon}{qL^2\tilde{C}}}, \qquad \delta^2 = \frac{n_i}{\tilde{C}}.$$

In this scaling the depletion region, i.e. the region where $\Delta V \cong -C$ holds, is not necessarily small. So, in particular so called punch through effects can be considered. (3.5.5) is, however, a scaling for very large bias values, since in order to keep $U_s(x)$ in (3.5.5)d) of order $O(1)$, the unscaled bias $U(x)$ has to grow with the maximal doping concentration $\tilde{C}$. The analysis of the drift diffusion equations in the scaling (3.5.4) leads to the solution of free boundary problems with the edges of the depletion region as the free boundaries. We will first treat the one dimensional case, where these edges are just points, and leave the discussion of the results known for the higher dimensional cases for later on.

The One-Dimensional Problem

The asymptotic analysis for extreme reverse bias is technically more complicated than for the forward bias case. The solution of the reduced problem behaves differently in different subregions of the device, namely inside the depletion region and outside. The layers occur now not directly at the *P-N* junction but at the edges of the depletion region. So, in addition to finding the reduced solution, one has to determine the location of these edges. In general this will result in the reduced problem becoming a free boundary problem. We will consider the one dimensional case first. In this case, the determination of the boundary of the depletion region becomes particularly simple since it consists only of two points. Aside from being a free boundary problem the reduced problem is also structurally more complex since, as we will see, setting the perturbation parameter λ to zero in (3.5.5) yields an underdetermined problem. In the language of singular perturbation theory (3.5.5) is called a 'singular' singular perturbation problem (see [3.31]). In the one dimensional case we assume that, after scaling, the semiconductor is positioned in the interval [0, 1]. The one dimensional equivalent to (3.5.5) is then of the form

$$\begin{aligned}
&\text{a)}\quad V'' = n - p - C(x) \\
&\text{b)}\quad J_n' = R, \quad \lambda^2 J_n = \mu_n(\lambda^2 n' - nV') \\
&\text{c)}\quad J_p' = -R, \quad \lambda^2 J_p = \mu_p(-\lambda^2 p' - pV')
\end{aligned} \Bigg\} \quad \text{for} \quad x \in [0, 1]$$

$$\begin{aligned}
&\text{d)}\quad V(x) = V_D(x) = U(x) + V_{bi}(x), \qquad n(x) = n_D(x), \\
&\qquad p(x) = p_D(x) \qquad \text{for} \quad x = 0 \quad \text{and} \quad x = 1 \\
&\text{e)}\quad n_D = \tfrac{1}{2}(C + \sqrt{C^2 + 4\delta^4}), \qquad p_D = \tfrac{1}{2}(-C + \sqrt{C^2 + 4\delta^4}).
\end{aligned} \tag{3.5.6}$$

Here $'$ denotes differentiation with respect to x. The location of the *P-N* junction is at some point $\gamma \in (0, 1)$. So, again employing the simplification of an abrupt *P-N* junction, we assume that $C(x)$ has a jump discontinuity at $x = \gamma$ and is as smooth as we like elsewhere. Since this analysis shall hold for large reverse bias, we have to agree on a sign convention for the doping profile C and the bias. So we assume that

$$\operatorname{sign}(C(x)) = \operatorname{sign}(x - \gamma) \tag{3.5.7}$$

holds. A large reverse bias then corresponds to $V_D(1) - V_D(0) > 0$. Letting λ go to zero in (3.5.6) gives the reduced problem

$$\begin{aligned}
&\text{a)}\quad \bar{V}'' = \bar{n} - \bar{p} - C(x) \\
&\text{b)}\quad \bar{J}_n' = \bar{R}, \qquad 0 = -\mu_n \bar{n} \bar{V}' \\
&\text{c)}\quad \bar{J}_p' = -\bar{R}, \qquad 0 = -\mu_p \bar{p} \bar{V}'
\end{aligned} \Bigg\} \quad \text{for} \quad x \in [0, 1]$$

$$\begin{aligned}
&\text{d)}\quad \bar{V}(x) = V_D(x), \qquad \bar{n}(x) = n_D(x), \\
&\qquad \bar{p}(x) = p_D(x) \qquad \text{for} \quad x = 0 \quad \text{and} \quad x = 1.
\end{aligned} \tag{3.5.8}$$

The equations (3.5.8) allow two possible solutions:

Case 1: $\bar{V}' \not\equiv 0, \quad \bar{n} \equiv 0, \quad \bar{p} \equiv 0.$

Case 2: $\bar{V}' \equiv 0.$

In the first case, inside the depletion region, $\bar{V}'' = -C(x)$ will hold. In the second case $\bar{n}$ and $\bar{p}$ cannot be determined immediately from the equations (3.5.8). Since $\bar{n} \equiv \bar{p} \equiv 0$ contradicts the boundary conditions, and we do not expect the depletion region to form at the contact, we derive the following problem for $\bar{V}$

$$\begin{aligned} &\bar{V}' \equiv 0 && \text{for} \quad 0 \leqslant x \leqslant x_1 \\ &\bar{V}''(x) = -C(x) && \text{for} \quad x_1 \leqslant x \leqslant x_2 \\ &\bar{V}' \equiv 0 && \text{for} \quad x_2 \leqslant x \leqslant 1, \end{aligned} \tag{3.5.9}$$

where x_1 and x_2, the endpoints of the depletion region, have yet to be determined. If the concentrations n and p stay bounded for $\lambda \to 0$, then so do the derivatives of V and no zeroth order layer term can form in the potential variable. Thus, the reduced potential $\bar{V}$ and its derivative have to remain continuous across x_1 and x_2. Since $\bar{V}' \equiv 0$ holds for $0 \leqslant x \leqslant x_1$, we have

$$\bar{V}(x_1) = V_D(0), \qquad \bar{V}'(x_1) = 0. \tag{3.5.10}$$

Inside the depletion region $\bar{V}$ is then given by

$$\bar{V}(x) = V_D(0) - \int_{x_1}^{x} \int_{x_1}^{\zeta} C(\eta)\, d\eta\, d\zeta. \tag{3.5.11}$$

Since $\bar{V}' \equiv 0$ holds in $[x_2, 1]$ we have

$$\bar{V}(x_2) = V_D(1), \qquad \bar{V}'(x_2) = 0. \tag{3.5.12}$$

Matching (3.5.11) with the condition (3.5.12) yields

$$\begin{aligned} &\text{a)} \quad V_D(0) - V_D(1) = \int_{x_1}^{x_2} \int_{x_1}^{\zeta} C(\eta)\, d\eta\, d\zeta, \\ &\text{b)} \quad 0 = \int_{x_1}^{x_2} C(\eta)\, d\eta. \end{aligned} \tag{3.5.13}$$

Let us first convince ourselves that (3.5.13) has a solution (x_1, x_2). A family of solutions $(x_1(t), x_2(t))$ of (3.5.13)b) alone, which depends continuously differentiably on the parameter t, has to satisfy

$$C(x_2(t))\dot{x}_2(t) - C(x_1(t))\dot{x}_1(t) \equiv 0. \tag{3.5.14}$$

If we add the equation $\dot{x}_2 = 1$ and the initial conditions $x_1(0) = x_2(0) = \gamma$ to (3.5.14) we obtain a system of ordinary differential equations whose solution satisfies (3.5.13)b). Moreover $\dot{x}_1 \leqslant 0$ and $\dot{x}_2 \geqslant 0$ holds and so x_1 stays to the left of the *P-N* junction γ and x_2 stays to the right of γ. If we define $f(t)$ by

$$f(t) = \int_{x_1(t)}^{x_2(t)} \int_{x_1(t)}^{\zeta} C(\eta)\, d\eta\, d\zeta, \tag{3.5.15}$$

then $f(0) = 0$ holds and we have to solve $f(t) = V_D(0) - V_D(1)$ in order to obtain a solution of (3.5.13). Differentiating f, we see that

$$f'(t) = -C(x_1)\dot{x}_1(x_2 - x_1) = -C(x_2)(x_2 - x_1). \tag{3.5.16}$$

Since, because of (3.5.14), $\dot{x}_2 - \dot{x}_1 \geqslant 1$, and therefore $x_2 - x_1 \geqslant t$ holds, we obtain

$$f'(t) \leqslant -t \min_{x \geqslant \gamma} \{C(x)\}. \tag{3.5.17}$$

So $f(t)$ is monotone from $(-\infty, 0]$ onto $(-\infty, 0]$ and for reverse bias, that means for $V_D(0) - V_D(1) < 0$, (3.5.13) will have a unique solution. After finding x_1 and x_2, the reduced potential $\bar{V}$ is uniquely determined by equation (3.5.9) together with the boundary conditions (3.5.6)d) and the requirement that $\bar{V}$ and the reduced field $\bar{V}'$ remain continuous. So far, $\bar{n}$, $\bar{p}$, $\bar{J}_n$ and $\bar{J}_p$ are known only within the depletion region $[x_1, x_2]$ where, because of (3.5.8),

$$\begin{aligned} &\text{a)} \quad \bar{n}(x) = \bar{p}(x) = 0 \\ &\text{b)} \quad \bar{J}_n(x) = \bar{J}_n(x_1) + \int_{x_1}^{x} R_0(\zeta)\, d\zeta, \\ &\qquad \bar{J}_p(x) = \bar{J}_p(x_1) - \int_{x_1}^{x} R_0(\zeta)\, d\zeta \qquad \text{for} \quad x_1 \leqslant x \leqslant x_2 \end{aligned} \tag{3.5.18}$$

holds. Here $R_0(x)$ denotes the recombination rate R evaluated at $n = p = 0$. Outside the depletion region, where $\bar{V}' \equiv 0$, the reduced equations are insufficient to determine $\bar{n}$ and $\bar{p}$. The only information about the reduced carrier concentrations is given by the condition of vanishing space charge

$$\bar{n} - \bar{p} - C = 0 \qquad \text{for} \qquad x \in [0, x_1) \cup (x_2, 1]. \tag{3.5.19}$$

In this sense (3.5.6) is a singular singular perturbation problem (see [3.31]). The standard approach to deal with such a situation is to find a transformation of the dependent variables which reduces the problem to a regular singular perturbation problem. In our case such a transformation has been given in [3.30] and is of the form $(n, p) \leftrightarrow (\rho, w)$,

$$\begin{aligned} &\text{a)} \quad \rho = n - p, \qquad w = np \\ &\text{b)} \quad n = \tfrac{1}{2}(\rho + \sqrt{\rho^2 + 4w}), \qquad p = \tfrac{1}{2}(-\rho + \sqrt{\rho^2 + 4w}). \end{aligned} \tag{3.5.20}$$

In the transformed variables the equations (3.5.6) read

$$\begin{aligned} &\text{a)} \quad V'' = \rho - C \\ &\text{b)} \quad J_n' = R, \qquad J_p' = -R \\ &\text{c)} \quad w' = \frac{1}{2}\sqrt{\rho^2 + 4w}\left[\frac{J_n}{\mu_n} - \frac{J_p}{\mu_p}\right] - \frac{1}{2}\rho\left[\frac{J_n}{\mu_n} + \frac{J_p}{\mu_p}\right] \\ &\text{d)} \quad \lambda^2 \rho' = \sqrt{\rho^2 + 4w}\, V' + \lambda^2\left[\frac{J_n}{\mu_n} + \frac{J_p}{\mu_p}\right] \qquad \text{for} \qquad x \in (0, 1) \\ &\text{e)} \quad \rho(x) = C(x), \qquad w(x) = \delta^4, \qquad V(x) = V_D(x) \\ &\qquad \text{for} \quad x = 0 \quad \text{and} \quad x = 1. \end{aligned} \tag{3.5.21}$$

Outside the depletion region, where $\bar{V}' = 0$ holds, the reduced equations in the transformed variables are given by

$$\begin{aligned}
&\text{a)}\quad \bar{V}' = 0, \qquad \text{b)}\quad \bar{\rho} = C\\
&\text{c)}\quad \bar{J}_n' = \bar{R}, \qquad \bar{J}_p' = -\bar{R}\\
&\text{d)}\quad \bar{w}' = \frac{1}{2}\sqrt{\bar{\rho}^2 + 4\bar{w}}\left[\frac{\bar{J}_n}{\mu_n} - \frac{\bar{J}_p}{\mu_p}\right] - \frac{1}{2}\bar{\rho}\left[\frac{\bar{J}_n}{\mu_n} + \frac{\bar{J}_p}{\mu_p}\right]\\
&\qquad \text{for} \quad x \in [0, x_1) \cup (x_2, 1]
\end{aligned} \tag{3.5.22}$$

subject to the boundary conditions

$$\bar{w}(x) = \delta^4, \quad \bar{V}(x) = V_D(x) \quad \text{for} \quad x = 0 \quad \text{and} \quad x = 1. \tag{3.5.23}$$

Since the right-hand side of (3.5.22) stays bounded, w cannot exhibit layer behaviour. This gives the conditions on the reduced solution at $x = x_1$ and $x = x_2$.

$$\begin{aligned}
&\text{a)}\quad \bar{w}(x_1) = \bar{w}(x_2) = 0\\
&\text{b)}\quad \bar{J}_n(x_2) = \bar{J}_n(x_1) + \int_{x_1}^{x_2} R_0(\zeta)\, d\zeta,\\
&\qquad \bar{J}_p(x_2) = \bar{J}_p(x_1) + \int_{x_1}^{x_2} R_0(\zeta)\, d\zeta.
\end{aligned} \tag{3.5.24}$$

For $\delta = 0$ the problem (3.5.22)–(3.5.24) has the trivial solution. One can show with a perturbation argument around this trivial solution that for δ sufficiently small (3.5.22)–(3.5.24) has a unique solution (see [3.30]).
Thus we have completely determined the solution of the reduced problem and can now turn to the calculation of the layer terms. Under strong reverse bias conditions the layers do not occur at the *P-N* junction γ, as in the forward bias case, but at x_1 and x_2, the edges of the depletion region. From (3.5.20) we calculate

$$\begin{aligned}
\bar{n}(x) &= \begin{cases} \frac{1}{2}(C(x) + \sqrt{C(x)^2 + 4\bar{w}}) & \text{for} \quad x \in [0, x_1) \cup (x_2, 1],\\ 0 & \text{for} \quad x \in (x_1, x_2),\end{cases}\\
\bar{p}(x) &= \begin{cases} \frac{1}{2}(-C(x) + \sqrt{C(x)^2 + 4\bar{w}}) & \text{for} \quad x \in [0, x_1) \cup (x_2, 1],\\ 0 & \text{for} \quad x \in (x_1, x_2).\end{cases}
\end{aligned} \tag{3.5.25}$$

Since $\bar{w}(x_1) = \bar{w}(x_2) = 0$ holds, $\bar{n}$ is discontinuous at $x = x_2$, where $C > 0$ holds, and $\bar{p}$ is discontinuous at $x = x_1$, where $C < 0$ holds and the corresponding layer terms are needed at the boundaries of the depletion region. The layer equations involve the higher order terms in the expansion. At the right edge of the depletion region ($x = x_2$) we introduce the fast layer variable $\xi = (x - x_2)/\lambda$. Expanding the Poisson equation we see that, close to $x = x_2$, the potential V is of the form

$$\hat{V}(\xi) = \bar{V}(x_2 + \lambda\xi) + \lambda^2\hat{\phi}(\xi) + O(\lambda^3), \tag{3.5.26}$$

where $\hat{\phi}$ denotes the second order layer term. (3.5.26) implies that, close to $x = x_2$, the field $(1/\lambda)\partial_\xi \hat{V}$ is given by $[\bar{V}'(x_2 + \lambda\xi) + \lambda\partial_\xi\hat{\phi}(\xi)]$. Expanding $\bar{V}'(x_2 + \lambda\xi)$ around $\lambda = 0$ gives

$$\partial_\xi \hat{V}(\xi) = \begin{cases} \lambda^2[-\xi C(x_2) + \partial_\xi\hat{\phi}(\xi)] + O(\lambda^3) & \text{for} \quad \xi < 0, \\ \lambda^2\partial_\xi\hat{\phi}(\xi) + O(\lambda^2) & \text{for} \quad \xi > 0. \end{cases} \tag{3.5.27}$$

Inserting this into the equations and letting λ go to zero while keeping ξ fixed gives the following layer equations and matching conditions:

$$\begin{aligned} &\text{a)} \quad \partial_\xi^2\hat{\phi} = \begin{cases} \hat{n} - \hat{p} & \text{for} \quad \xi < 0 \\ \hat{n} - \hat{p} - C(x_2) & \text{for} \quad \xi > 0 \end{cases} \\ &\text{b)} \quad \partial_\xi \hat{J}_n = 0, \qquad \partial_\xi \hat{J}_p = 0 \\ &\text{c)} \quad \partial_\xi \hat{n} = \begin{cases} \hat{n}[-\xi C(x_2) + \partial_\xi\hat{\phi}] & \text{for} \quad \xi < 0 \\ \hat{n}\partial_\xi\hat{\phi} & \text{for} \quad \xi > 0 \end{cases} \\ &\text{d)} \quad \partial_\xi \hat{p} = \begin{cases} -\hat{p}[-\xi C(x_2) + \partial_\xi\hat{\phi}] & \text{for} \quad \xi < 0 \\ -\hat{p}\partial_\xi\hat{\phi} & \text{for} \quad \xi > 0 \end{cases} \\ &\text{e)} \quad \hat{\phi}(\infty) = \hat{\phi}(-\infty) = 0, \qquad \hat{n}(\infty) = C(x_2), \qquad \hat{n}(-\infty) = 0, \\ &\qquad \hat{p}(\infty) = \hat{p}(-\infty) = 0. \end{aligned} \tag{3.5.28}$$

Again, the equations for the carrier concentrations can be integrated exactly and we obtain

$$\begin{aligned} &\text{a)} \quad \hat{n} = \begin{cases} C(x_2)e^{\hat{\phi} - (\xi^2/2)C(x_2)} & \text{for} \quad \xi < 0 \\ C(x_2)e^{\hat{\phi}} & \text{for} \quad \xi > 0 \end{cases} \\ &\text{b)} \quad \hat{p} \equiv 0. \end{aligned} \tag{3.5.29}$$

Thus the only boundary value problem to solve for the layer terms is the equation (3.5.28)a) together with the boundary conditions (3.5.28)e). Using the same methods as in the previous Section, one can show that this boundary value problem has a unique solution (see Problem 3.6). A similar expansion has to be performed around the left edge of the depletion region, where p has a layer term $\check{p}(\eta)$ which depends on the layer variable $\eta = (x - x_1)/\lambda$. This layer problem is of the form

$$\begin{aligned} &\text{a)} \quad \partial_\eta^2\check{\phi} = \begin{cases} -\check{p} & \text{for} \quad \eta > 0 \\ -\check{p} - C(x_1) & \text{for} \quad \eta < 0 \end{cases} \\ &\text{b)} \quad \check{p} = \begin{cases} -C(x_1)e^{-\check{\phi} + (\eta^2/2)C(x_1)} & \text{for} \quad \eta > 0 \\ -C(x_1)e^{-\check{\phi}} & \text{for} \quad \eta < 0. \end{cases} \end{aligned} \tag{3.5.30}$$

In order to obtain a uniform $O(\lambda)$ approximation of the solution one can again derive the composite expansion $(V^c, n^c, p^c, J_n^c, J_p^c)$ given by

$$\text{a)}\quad V^c = \bar{V} + O(\lambda^2), \qquad J_n^c = \bar{J}_n + O(\lambda), \qquad J_p^c = \bar{J}_p + O(\lambda)$$

$$\text{b)}\quad n^c = \begin{cases} \bar{n}(x) + \hat{n}(\xi) + O(\lambda) & \text{for} \quad 0 \leqslant x \leqslant x_2 \\ \bar{n}(x) + \hat{n}(\xi) - C(x_2) + O(\lambda) & \text{for} \quad x \leqslant x_2 \leqslant 1 \end{cases} \tag{3.5.31}$$

$$\text{c)}\quad p^c = \begin{cases} \bar{p}(x) + \check{p}(\eta) + C(x_1) + O(\lambda) & \text{for} \quad 0 \leqslant x \leqslant x_1 \\ \bar{p}(x) + \check{p}(\eta) + O(\lambda) & \text{for} \quad x_1 \leqslant x \leqslant 1. \end{cases}$$

The derivation of higher order approximations is straight forward and is left to the reader (see Problem 3.7). The asymptotic validity of (3.5.31) can again be shown in special cases. However the known results here are of a different flavor than in the forward bias case (see [3.6], [3.7], [3.8], [3.9]). They are based on compactness arguments and do not give any convergence rates. So they only say that, for $\lambda \to 0$, the solutions converge to the reduced solution $(\bar{V}, \bar{n}, \bar{p}, \bar{J}_n, \bar{J}_p)$ derived above. Since the one-dimensional case is in no way particular, as far as asymptotic validity is concerned, we will defer discussion of these results to the next paragraph, where we will deal with the two-dimensional case.

The Two-Dimensional Case

In more than one dimension the approach to finding an asymptotic approximation of the solution follows the same pattern as in the one dimensional case. However, the solution of the resulting free boundary problem is much more difficult since the boundary of the depletion region is a curve in the two dimensional case and a surface in the three-dimensional case. For the sake of simplicity we will restrict ourselves to the case of two dimensions and leave generalizations to the three-dimensional case to the reader. (They are straight forward.) We will use the same assumptions on the device geometry as in the forward biased case. So the device occupies again a region $\Omega \subset \mathbb{R}^2$ with a boundary $\partial\Omega = \partial\Omega_D \cup \partial\Omega_N$ where the boundary conditions

$$\text{a)}\quad V|_{\partial\Omega_D} = V_D|_{\partial\Omega_D}, \qquad n|_{\partial\Omega_D} = n_D|_{\partial\Omega_D}, \qquad p|_{\partial\Omega_D} = p_D|_{\partial\Omega_D}$$
$$\text{b)}\quad \left.\frac{\partial V}{\partial \nu}\right|_{\partial\Omega_N} = J_n \cdot \nu|_{\partial\Omega_N} = J_p \cdot \nu|_{\partial\Omega_N} = 0 \tag{3.5.32}$$

hold. Setting λ to zero in the equations (3.5.5) gives the reduced problem

$$\text{a)}\quad \Delta\bar{V} = \bar{n} - \bar{p} - C$$
$$\text{b)}\quad 0 = \bar{n}\,\text{grad}\,\bar{V}, \qquad \text{c)}\quad 0 = -\bar{p}\,\text{grad}\,\bar{V} \tag{3.5.33}$$
$$\text{d)}\quad \text{div}\,\bar{J}_n = \bar{R}, \qquad \text{e)}\quad \text{div}\,J_p = -\bar{R}.$$

$\bar{J}_n$ and $\bar{J}_p$ have to be determined from the higher order terms in the expansion. (3.5.33) suggests that the device domain Ω splits into a region where $\text{grad}\,\bar{V} \equiv 0$ ($\bar{V} \equiv$ constant) holds, and the depletion region where $\bar{n} \equiv \bar{p} \equiv 0$ holds (see Fig. 3.5.1).

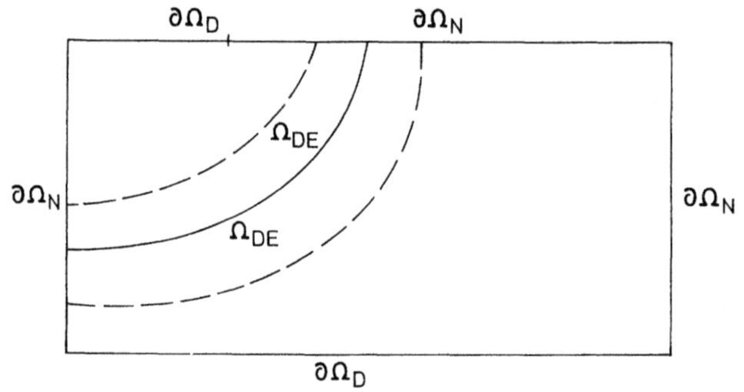

Fig. 3.5.1 Depletion layer

In Ω_{DE}, i.e. in the depletion region, the solution is given by

$$\bar{n} = \bar{p} = 0, \qquad \Delta \bar{V} = -C \qquad \text{for} \qquad x \in \Omega_{DE} \tag{3.5.34}$$

and outside the reduced solution is given by

$$\begin{aligned} &\text{a)} \quad \operatorname{grad} \bar{V} = 0, \qquad \text{b)} \quad \bar{n} = \tfrac{1}{2}(C + \sqrt{C^2 + 4\bar{w}}), \\ &\text{c)} \quad \bar{p} = \tfrac{1}{2}(-C + \sqrt{C^2 + 4\bar{w}}), \qquad x \in \Omega_1 \cup \Omega_2, \end{aligned} \tag{3.5.35}$$

where $\bar{w}$ is the transformed variable according to (3.5.20). $\bar{w}$ is then given as the solution of the two-dimensional equivalent of (3.5.22) (see Problem 3.8). In order for this approach to be feasible, we have to assume that the boundary data V_D are piecewise constant on $\partial\Omega_D$ so as not to contradict (3.5.35)a). So we assume

$$V_D = U_1 \quad \text{for} \quad x \in \partial\Omega_{D_1}, \qquad V_D = U_2 \quad \text{for} \quad x \in \partial\Omega_{D_2}. \tag{3.5.36}$$

For the same reason as in the one-dimensional case the approximation to the potential V does not have a zero order layer term, and so $\bar{V}$ and grad $\bar{V}$ are continuous across X_1 and X_2, the boundaries of the depletion region. This gives the reduced problem for $\bar{V}$:

$$\begin{aligned} &\text{a)} \quad \Delta \bar{V} = -C \qquad \text{for} \qquad x \in \Omega_{DE}, \\ &\text{b)} \quad \bar{V}|_{X_1} = U_1, \qquad \operatorname{grad} \bar{V} \cdot \nu_1|_{X_1} = 0, \\ &\qquad \bar{V}|_{X_2} = U_2, \qquad \operatorname{grad} \bar{V} \cdot \nu_2|_{X_2} = 0 \\ &\text{c)} \quad \bar{V} \equiv U_1 \quad \text{for} \quad x \in \Omega_1, \qquad \bar{V} \equiv U_2 \quad \text{for} \quad x \in \Omega_2, \end{aligned} \tag{3.5.37}$$

where ν_1 and ν_2 are the unit normal vectors on X_1 and X_2 respectively. The asymptotic validity of the derived approximation has been shown in [3.7] and [3.8] for the one-dimensional case and in [3.9] for two and three dimensions. At least for the higher-dimensional cases the available results are of a very weak type. They only say, that the solutions of the full problem (3.5.5) converge to the solution of the reduced problem (3.5.34)–(3.5.37) in the L^∞ weak star sense. This means, that for any test function $\phi \in L^1(\Omega)$

$$\int_\Omega V\phi\,dx \to \int_\Omega \bar{V}\phi\,dx\,, \qquad \int_\Omega n\phi\,dx \to \int_\Omega \bar{n}\phi\,dx\,,$$
$$\int_\Omega p\phi\,dx \to \int_\Omega \bar{p}\phi\,dx \tag{3.5.38}$$

holds. In [3.9] this result has been shown for a one carrier model (that means $C(x) < 0$ everywhere and $n \equiv 0$ everywhere) and for the special case $\delta = 0$. In this case the smooth variable w vanishes identically and $\bar{p} \equiv 1$ holds outside the depletion region.

3.6 Stability and Conditioning for the Stationary Problem

When solving any kind of equation one of the most important questions from a practical point of view is that of stability and conditioning. These terms refer to the sensitivity of the computed solution on the data. In our case the input data are given by the boundary data on the Dirichlet part of the boundary, the doping concentration and the geometry of the device. The question of stability and conditioning becomes particularly important when the drift diffusion equations are solved numerically. Errors, introduced by a numerical solution, can usually be analyzed by regarding the computed solution as the exact solution of a perturbed problem. Stability hinges on an estimate of the inverse of the linearization of the involved differential operator. As pointed out earlier such an estimate, which is independent of the perturbation parameter λ, is not available in the general case. In this Paragraph we will give estimates of approximations to the inverse of the linearized operator and briefly sketch how the performance of iterative solution methods for the nonlinear problem can be influenced by using different sets of dependent variables.
If we formulate the problem to be solved as

$$F(z) = g\,, \tag{3.6.1}$$

where z is the solution and g denotes the data, we consider the perturbed problem

$$F(z') = g'\,. \tag{3.6.2}$$

The terms stability and conditioning now refer to the absolute and relative effect of the perturbation in g on the solution z. So one looks for constants K_s and K_c such that

$$\text{a)}\quad \|z - z'\| \leqslant K_s \|g - g'\|$$
$$\text{b)}\quad \frac{\|z - z'\|}{\|z\|} \leqslant K_c \frac{\|g - g'\|}{\|g\|} \tag{3.6.3}$$

holds, where $\|\cdot\|$ is some suitable norm. K_c is called the condition number and K_s the stability bound. Even if the data of the problem are known precisely the stability and conditioning of a problem are of great significance since errors made in numerical computations can usually be estimated through backward error analysis (see [3.38]). If we solve (3.6.1) numerically the operator F is replaced by some approximate, discrete operator $\tilde{F}$ and

$$\tilde{F}(\tilde{z}) = g \tag{3.6.4}$$

is solved instead. (Usually $\tilde{F}$ also includes the effect of roundoff errors.) The concept of backward error analysis now implies that $\tilde{z}$, the solution of the discretized equation with the exact data, can be interpreted as the solution of the exact equation with perturbed data. That means that there exists a small perturbation δg of the data g, such that

$$F(\tilde{z}) = g + \delta g \tag{3.6.5}$$

holds. So errors in numerically obtained solutions can be measured in terms of the stability bounds. In turn it can be said that problems, where K_c and K_s, are too large, are practically very difficult to solve. They are called ill posed problems. Stability and conditioning are, by virtue of the mean value theorem, closely related to the linearization F' of F since

$$\delta g = F'(z)(\tilde{z} - z) + o(\|\tilde{z} - z\|) \tag{3.6.6}$$

holds. On the other hand the conditioning of the linearized operator F' is also of great importance for the performance of iterative methods for the solution of the drift diffusion equations since they are usually based on a linearization technique. The most prominent of these methods is Newton's method, which is of the form

$$\begin{aligned} &\text{a)}\quad w_{k+1} = w_k + dw \\ &\text{b)}\quad F'(w_k)\,dw = -F(w_k), \end{aligned} \tag{3.6.7}$$

where w_k and w_{k+1} denote the old and the new iterates and (3.6.7)b) has to be solved for the increment dw in each iteration. If we apply Newton's method to the drift diffusion equations we obtain the system

$$\begin{pmatrix} -\lambda^2\Delta^* & 1 & -1 \\ a_{21} & a_{22} & \partial R/\partial p \\ a_{31} & \partial R/\partial n & a_{33} \end{pmatrix} \begin{pmatrix} dV \\ dn \\ dp \end{pmatrix} = \begin{pmatrix} f_1 \\ f_2 \\ f_3 \end{pmatrix}, \tag{3.6.8}$$

$$a_{21} = \operatorname{div}(\mu_n n_k \operatorname{grad} *), \qquad a_{31} = \operatorname{div}(-\mu_p p_k \operatorname{grad} *),$$

$$a_{22} = \operatorname{div}[\mu_n(-\operatorname{grad} * + * \operatorname{grad} V_k)] + \frac{\partial R}{\partial n},$$

$$a_{33} = \operatorname{div}[\mu_p(-\operatorname{grad} * - * \operatorname{grad} V_k)] + \frac{\partial R}{\partial p},$$

where $*$ is a placeholder symbol. We have used the forward bias scaling of Section 3.4 here since it is of greater practical significance (because it is relevant for a much larger bias range). Also, for the sake of simplicity, it is assumed that the mobilities μ_n and μ_p do not depend on V, n and p and that the recombination rate R does not depend on V. The current densities J_n and J_p have been eliminated by inserting them into the continuity equations. Of course (3.6.8) is the linearization of the continuous problem and has to be replaced by the jacobian of the corresponding difference operator when solving the drift diffusion equations numerically. Reasonable stability bounds (which are independent of λ) for the coupled system (3.6.8) are not available yet. One can however, in lieu of such a rigorous approach, carry out a 'decoupled' analysis along the lines of [3.3]. For this purpose it is beneficial to perform a variable transformation in (3.6.8). This transformation is given by

$$\begin{pmatrix} dV \\ dn \\ dp \end{pmatrix} = T \begin{pmatrix} dV \\ y \\ z \end{pmatrix}, \qquad T = \begin{pmatrix} 1 & 0 & 0 \\ n & 1 & 0 \\ -p & 0 & 1 \end{pmatrix}. \tag{3.6.9}$$

Note, that the transformation (3.6.9) is the linearization of the nonlinear transformation to the Slotboom variables $n = e^V u$, $p = e^{-V} v$. The transformed equations then read

$$J \begin{pmatrix} dV \\ y \\ z \end{pmatrix} = \begin{pmatrix} f_1 \\ f_2 \\ f_3 \end{pmatrix}, \qquad J = \begin{pmatrix} a_{11} & 1 & -1 \\ a_{21} & a_{22} & \partial R/\partial p \\ a_{31} & \partial R/\partial n & a_{33} \end{pmatrix}, \tag{3.6.10}$$

$$a_{11} = -\lambda^2 \Delta * + (n_k + p_k) *,$$

$$a_{21} = \operatorname{div}(-J_{n_k} *) + \left(n_k \frac{\partial R}{\partial n} - p_k \frac{\partial R}{\partial p} \right) *,$$

$$a_{31} = \operatorname{div}(-J_{p_k} *) + \left(p_k \frac{\partial R}{\partial p} - n_k \frac{\partial R}{\partial n} \right) *,$$

$$a_{22} = \operatorname{div}[\mu_n(-\operatorname{grad} * + * \operatorname{grad} V_k)] + \frac{\partial R *}{\partial n},$$

$$a_{33} = \operatorname{div}[\mu_p(-\operatorname{grad} * - * \operatorname{grad} V_k)] + \frac{\partial R *}{\partial p},$$

and the update is given by the formula

$$V_{k+1} = V_k + dV, \qquad n_{k+1} = (1 + dV) n_k + y, \tag{3.6.11}$$

$$p_{k+1} = (1 + dV) p_k + z.$$

Note, that, if the drift diffusion equations are written in the Slotboom variables (V, u, v) and Newton's method is applied to the system (3.4.36), up to a transformation the same linear system has to be solved for the increments (dV, du, dv). In this case (dV, du, dv) are related to (dV, y, z) via

$$y = e^{V_k} du, \qquad z = e^{-V_k} dv. \tag{3.6.12}$$

Thus the difference between applying Newton's method to the system (3.4.5) and (3.4.36) is only in the updating strategies, which are of the form

$$
\begin{aligned}
&\text{a)}\quad V_{k+1} = V_k + dV, && n_{k+1} = (1 + dV)n_k + y, \\
& && p_{k+1} = (1 - dV)p_k + z \\
&\text{b)}\quad V_{k+1} = V_k + dV, && e^{V_{k+1}}u_{k+1} = e^{dV}(e^{V_k}u_k + y), \\
& && e^{-V_{k+1}}v_{k+1} = e^{-dV}(e^{-V_k}v_k + z),
\end{aligned}
\tag{3.6.13}
$$

respectively (see Problem 3.2). In practice the matrix J in (3.6.10) is frequently replaced by a, in some sense, simpler matrix in what is called approximate Newton methods (see [3.5]). The most widely used approximate Newton method consists of neglecting the subdiagonal terms in (3.6.10) and solving the system

$$
G\begin{pmatrix} dV \\ y \\ z \end{pmatrix} = \begin{pmatrix} f_1 \\ f_2 \\ f_3 \end{pmatrix}, \qquad G = \begin{pmatrix} a_{11} & 1 & -1 \\ 0 & a_{22} & R_p \\ 0 & 0 & a_{33} \end{pmatrix}
\tag{3.6.14}
$$

in each iteration, where the diagonal blocks a_{ii} are the same as in (3.6.10). If this approach is used together with the update (3.6.13)b) one obtains a version of the so called Gummel method (see [3.14]). Clearly, neglecting the subdiagonal terms in (3.6.10) assumes small currents and recombination rates. We refer the reader to [3.5], [3.17], [3.28] and [3.29] for a convergence analysis and various acceleration techniques for Gummel type methods. The iteration (3.6.14) has the advantage that only three scalar linear differential equations have to be solved at each step instead of a coupled system. For each of these separate equations stability bounds are easily obtainable. So, for instance, for the (1, 1) block of (3.6.14) a boundary value problem of the form

$$
\begin{aligned}
&\text{a)}\quad -\lambda^2 \Delta\, dV + (n + p)\, dV = f \\
&\text{b)}\quad dV|_{\partial\Omega_D} = 0, \qquad \left.\frac{\partial dV}{\partial \nu}\right|_{\partial\Omega_N} = 0
\end{aligned}
\tag{3.6.15}
$$

has to be solved. (Without loss of generality we can assume that the last iterate satisfies the boundary conditions and therefore the boundary conditions for the increment are homogeneous.) Dividing (3.6.15)a) by $(n + p)$ and applying the maximum principle yields

$$
\max_{\Omega} |dV| \leqslant \max_{\Omega} \left| \frac{f}{n + p} \right|,
\tag{3.6.16}
$$

which suggests a row scaling of the first row of the Newton equation (3.6.10) by $(n + p)$. After this row scaling the stability bound for (3.6.15) equals one. So the Poisson equation is generally well conditioned.

The conditioning of the linearized continuity equations is a more subtle problem since here the stability bounds depend on the type of the device. If we consider the (2, 2) block of (3.6.14) in the case $R = 0$, a boundary value problem of the form

$$\begin{aligned}&\operatorname{div}[\mu_n(-\operatorname{grad} y + y \operatorname{grad} V_k)] = g\\ &y|_{\partial\Omega_D} = 0, \qquad (-\operatorname{grad} y + y \operatorname{grad} V_k)\cdot \nu|_{\partial\Omega_N} = 0\end{aligned} \tag{3.6.17}$$

has to be solved. Employing the transformation $y = e^{V_k}\, du$, we obtain

$$\begin{aligned}&\text{a)}\quad -\operatorname{div}(\mu_n e^{V_k} \operatorname{grad} du) = g\\ &\text{b)}\quad du|_{\partial\Omega_D} = 0, \qquad \left.\frac{\partial du}{\partial \nu}\right|_{\partial\Omega_N} = 0.\end{aligned} \tag{3.6.18}$$

Application of the maximum principle (see [3.12]) gives that

$$\max_\Omega |du| \leqslant K e^{-\underline{V}} \max_\Omega |g|, \qquad \underline{V} := \min_\Omega V_k \tag{3.6.19}$$

holds, where the constant K depends only on the geometry of Ω and the minimum value of μ_n (see Problem 3.3). Transforming back into the y variable yields

$$\max_\Omega |y| \leqslant K e^{\overline{V}-\underline{V}} \max_\Omega |g|, \quad \overline{V} := \max_\Omega V_k. \tag{3.6.20}$$

So the stability bound for the linearized electron continuity equation depends exponentially on the potential difference. Remember that, at Ohmic contacts,

$$V|_{\partial\Omega_D} = U + \ln\left(\frac{C + \sqrt{C^2 + 4\delta^4}}{2\delta^2}\right)\Bigg|_{\partial\Omega_D} \tag{3.6.21}$$

holds. If we consider, for instance a device with a contacted n- and a contacted p- region and a piecewise constant doping profile ($C \equiv C_+ > 0$ in the n-region and $C \equiv -C_- < 0$ in the p- region), where a potential of U_n and U_p is applied, respectively, we obtain

$$e^{\overline{V}-\underline{V}} \geqslant \delta^{-4} e^{U_n - U_p} C_+ C_- . \tag{3.6.22}$$

If we use scaling factors corresponding to a 1μ device geometry and a maximal doping concentration of 10^{19} the right-hand side of (3.6.22) is of the order of 10^{18} in thermal equilibrium ($U_n = U_p$) and becomes even larger in the reverse bias case. The question which arises immediately is whether the bound (3.6.20) on the stability constant is sharp. Such high stability bounds are actually observed for complicated devices in numerical calculations (see [3.19]), but not for a simple P-N junction diode. Generally it can be said that the bound (3.6.20) is attained in the presence of so called floating regions. These are p- or n-regions without an Ohmic contact (see [3.34]). One can, at least in the one-dimensional case, show that a stability estimate of the form

$$\max_\Omega \left|\frac{y}{n}\right| \leqslant K \max_\Omega \left|\frac{g}{n}\right| \tag{3.6.23}$$

holds with a moderate constant K, if every p- and n- region is contacted.

We refer the reader to [3.3] for further details on the conditioning and stability of the drift diffusion equations and of its effect on the performance of iterative methods.

3.7 The Transient Problem

The remaining part of this Chapter is concerned with the analysis of the transient drift diffusion equations. So, we consider the problem

$$
\begin{aligned}
&\text{a)}\quad \operatorname{div}(\varepsilon \operatorname{grad} V) = q(n - p - C)\\
&\text{b)}\quad \operatorname{div} J_n = q(\partial_t n + R), \qquad \text{c)}\quad \operatorname{div} J_p = q(-\partial_t p - R)\\
&\text{d)}\quad J_n = q(D_n \operatorname{grad} n - \mu_n n \operatorname{grad} V)\\
&\text{e)}\quad J_p = q(-D_p \operatorname{grad} p - \mu_p p \operatorname{grad} V) \qquad \text{for} \quad x \in \Omega, t > 0\\
&\text{f)}\quad n(x,0) = n^I(x), \qquad p(x,0) = p^I(x)\\
&\text{g)}\quad V(x,t) = V_D(x,t), \qquad n(x,t) = n_D(x,t),\\
&\qquad p(x,t) = p_D(x,t) \qquad \text{for} \quad x \in \partial\Omega_D\\
&\text{h)}\quad \frac{\partial V}{\partial \nu}(x,t) = 0, \qquad J_n \cdot \nu = 0, \qquad J_p \cdot \nu = 0 \qquad \text{for} \qquad x \in \partial\Omega_N,
\end{aligned}
\tag{3.7.1}
$$

where the coefficients ε and q etc. have the same meaning as in the stationary case.

From a practical point of view the interest in the transient drift diffusion equations is of a different nature than that in the steady state problem. In the steady state problem one is interested primarily in the voltage current characteristics, that is in the current flow as a function of the applied potential difference. When considering the transient problem one is interested in the time required by the solution to reach a steady state. As far as numerical simulations are concerned, it is also of interest to calculate the transient response of several coupled devices constituing, say, a Flip-Flop. The structure of the transient solution is obviously more complex because of the additional dimension time. Numerical calculations are more expensive for the same reason. On the other hand, the transient problem is, in some sense, easier to deal with analytically. For instance, while the steady state problem admits multiple solutions, a feature which is exploited technically in c.f. a thyristor (see Chapter 4), the solution to the transient problem will, in general, be unique. Existence results for the steady state drift diffusion equations have to rely on the Schauder fixed point principle, while those for the transient problem can employ contraction arguments, at least locally in time.

Generally speaking, there are two types of transient analysis, one 'close' to a steady state and one 'far' from it. In the first case one is interested in what happens, if a small perturbation is introduced in a stationary initial state.

Thus, one considers the problem

$$\partial_t w = F(w), \qquad w(t = 0) = w^* + \varepsilon f, \tag{3.7.2}$$

where F, in our case, stands for the nonlinear differential operator corresponding to the steady state problem and $w = (n, p)$. w^* is the solution of the stationary problem, so $F(w^*) = 0$ holds. εf is a small perturbation, so f is some function and ε is a small parameter. If we set $w = w^* + \varepsilon z$, z satisfies

$$\partial_t z = F'(w^*)z + O(\varepsilon), \qquad z(t = 0) = f. \tag{3.7.3}$$

Thus, in first order, the behaviour of the solution w can be determined by analyzing the linearized problem

$$\partial_t \tilde{z} = F'(w^*)\tilde{z}, \qquad \tilde{z}(t = 0) = f. \tag{3.7.4}$$

In particular, it can be deduced that, if the solution $\tilde{z}$ of (3.7.4) decays to zero as $t \to \infty$, w^* is a dynamically stable steady state. We will treat the linearized problem (3.7.4) in the next Paragraph.
Of course, this approach is only valid if the perturbation εf is sufficiently small and, in this sense, is rather an extension of the steady state analysis. If the behaviour of transient solutions is of interest over longer periods of time, in particular if a *P-N* junction is switched from forward to reverse bias, the full, nonlinear problem (3.7.1) has to be dealt with. Here the analysis follows the same pattern as in the stationary case. Once again, the main instrument to understand the behaviour of the solution of (3.7.1) will be asymptotic analysis.

3.8 The Linearization of the Transient Problem

If we employ the forward bias scaling of Section 3.4 we obtain the system

$$\text{a)} \quad \lambda^2 \Delta V = n - p - C \tag{3.8.1}$$

$$\text{b)} \quad \frac{L^2}{U_T \tilde{\mu}} \partial_t n = \operatorname{div} J_n - R, \qquad \text{c)} \quad J_n = \mu_n(\operatorname{grad} n - n \operatorname{grad} V)$$

$$\text{d)} \quad \frac{L^2}{U_T \tilde{\mu}} \partial_t p = -\operatorname{div} J_p - R, \qquad \text{e)} \quad J_p = \mu_p(-\operatorname{grad} p - p \operatorname{grad} V).$$

Although the scaling is of no particular importance for the purpose of this Section we use the scaling of the stationary forward bias case for the time being. t in (3.8.1) is the unscaled time. (3.8.1) suggests the scaling

$$t = \frac{L^2}{U_T \tilde{\mu}} t_s, \tag{3.8.2}$$

where t_s is the scaled time variable. Since $\tilde{\mu}$ is the scaling factor of the electron and hole mobilities μ_n and μ_p, and $D_{n,p} = U_T \mu_{n,p}$ holds, the timescale given by (3.8.2) corresponds to the diffusion timescale of carriers. With this scaling

(3.8.1) reads

$$\begin{aligned} &\text{a)} \quad \lambda^2 \, \Delta V = n - p - C \\ &\text{b)} \quad \partial_{t_s} n = \operatorname{div} J_n - R, \qquad \text{c)} \quad J_n = \mu_n(\operatorname{grad} n - n \operatorname{grad} V) \\ &\text{d)} \quad \partial_{t_s} p = -\operatorname{div} J_p - R, \qquad \text{e)} \quad J_p = \mu_p(-\operatorname{grad} p - \operatorname{grad} V). \end{aligned} \tag{3.8.3}$$

From here on we will omit the subscript s.
If we linearize the operator F in (3.7.2), corresponding to the steady state problem, around any solution $w = (n, p)$ along the direction $z = (u, v)$ we obtain the linearized operator

$$\begin{aligned} &F'(w)z = \begin{cases} \operatorname{div} I_n - R_n u - R_p v \\ -\operatorname{div} I_p - R_n u - R_p v \end{cases} \\ &I_n = \mu_n(\operatorname{grad} u - u \operatorname{grad} V - n \operatorname{grad} \phi) \\ &I_p = \mu_p(-\operatorname{grad} v - v \operatorname{grad} V - p \operatorname{grad} \phi), \end{aligned} \tag{3.8.4}$$

where V and ϕ satisfy

$$\begin{aligned} &\lambda^2 \, \Delta V = n - p - C \\ &\lambda^2 \, \Delta \phi = u - v \end{aligned} \tag{3.8.5}$$

together with homogeneous boundary conditions for ϕ. There are four main questions which are of interest in connection with the transient linearized problem

1. Does the linearized problem (3.7.4) have a unique solution?
2. Is the linearization stable; i.e. if we consider the initial boundary value problem

$$\partial_t z = F'(w)z + g, \qquad z(t = 0) = f, \tag{3.8.6}$$

can z be bounded in terms of f and g?
3. Does the solution z of (3.8.6) with a homogeneous right-hand side $g \equiv 0$ tend to zero for $t \to \infty$?
4. Do the eigenvalues of F' have negative real parts?

Clearly questions 3 and 4 have to do with the dynamic stability of a steady state, if w is chosen as the stationary solution of $F(w) = 0$. The answers to questions 1 and 2 tell whether linearization is justified at all. Of course we cannot expect an unqualified 'yes' as the answer to questions 3 and 4 since we know that there exist dynamically unstable steady states to the drift diffusion equations (see Chapter 4). We will, however show that steady states, up to a certain amount of current flow and recombination, are stable. The answer to all four questions is based on an energy estimate; i.e. an estimate of the form

$$\langle z, F'(w)z \rangle \leqslant K \langle z, z \rangle, \tag{3.8.7}$$

where $\langle \cdot, \cdot \rangle$ denotes a suitable scalar product. This scalar product is of a problem specific form and has been used in [3.24] for the linearization around equilibrium ($J_n \equiv J_p \equiv 0$) and in [3.22] for the linearization around a general solution of the transient or steady state problem. Since, in order to deal with the eigenvalue problem, we will have to consider complex valued eigenfunctions we will define the scalar product for complex valued functions.

Definition 3.8.8: Let $z_1 = (u_1, v_1)$ and $z_2 = (u_2, v_2)$ be complex valued functions. Then we define the scalar product $\langle \cdot, \cdot \rangle$ and the norm $\| \cdot \|$ in the following way. First we solve

$$\lambda^2 \, \Delta\phi_2 = u_2 - v_2 \tag{3.8.9}$$

together with the boundary conditions

$$\phi_2(x) = 0 \quad \text{for} \quad x \in \partial\Omega_D, \quad \frac{\partial\phi_2}{\partial v}(x) = 0 \quad \text{for} \quad x \in \partial\Omega_N. \tag{3.8.10}$$

Then $\langle z_1, z_2 \rangle$ is given by

$$\langle z_1, z_2 \rangle = \int_\Omega \bar{u}_1 \left(\frac{u_2}{n} - \phi_2 \right) + \bar{v}_1 \left(\frac{v_2}{p} + \phi_2 \right) dx. \tag{3.8.11}$$

Note that the bars denote complex conjugation here.
We leave it to the reader to show that $\langle \cdot, \cdot \rangle$ really constitutes a scalar product (see Problem (3.9)). The energy estimate is now given by

Lemma 3.8.12: *Let $z = (u, v)$ be a complex valued function. Then*

$$\operatorname{Re}\langle z, F'(w)z \rangle \leqslant -\varepsilon \int_\Omega \mu_n n |\operatorname{grad} \alpha|^2 + \mu_p p |\operatorname{grad} \beta|^2 \, dx \tag{3.8.13}$$

$$+ \left(\frac{1}{1-\varepsilon} K_J + K_R \right) \langle z, z \rangle$$

holds for all $\varepsilon \in [0, 1)$ with α, β, K_J and K_R given by

$$\alpha = \frac{u}{n} - \phi, \qquad \beta = \frac{v}{p} + \phi, \tag{3.8.14}$$

$$K_J = \frac{1}{2}\left(\max_\Omega \left| \frac{J_n}{n} \right| + \max_\Omega \left| \frac{J_p}{p} \right| \right)$$

$$K_R = \frac{1}{2} \max_\Omega |np + 1| \max_\Omega \left| \frac{R_n}{p} - \frac{R_p}{n} \right|$$

and ϕ the solution of $\lambda^2 \, \Delta\phi = u - v$ together with the homogeneous boundary conditions (3.8.10).

The proof of this energy estimate is technically quite complicated and can be found in [3.22]. We will omit the proof here and instead discuss its implications on the questions 1–4 above.

The estimate (3.8.13) immediately implies the stability of (3.8.6). If we define the norm $\|z\|$ by

$$\|z(t)\| = \sqrt{\langle z(t), z(t)\rangle}, \tag{3.8.15}$$

and take $z = (u, v)$ to be real, we obtain

$$\partial_t \|z\| = \frac{1}{\|z\|} \operatorname{Re}\langle z, F'(w)z + g\rangle. \tag{3.8.16}$$

Using the energy estimate (3.8.13) and the Cauchy-Schwartz inequality, this gives

$$\partial_t \|z(t)\| \leqslant (K_J + K_R)\|z(t)\| + \|g(t)\| \tag{3.8.17}$$

and, by Gronwall's inequality

$$\|z(t)\| \leqslant e^{(K_J+K_R)t}\left(\|f\| + \int_0^t e^{-(K_J+K_R)s}\|g(s)\|\, ds\right).$$

To show that the solution z decays for sufficiently small current flow and recombination rate and homogeneous $g \equiv 0$ we estimate $\langle z, z\rangle$ in terms of $\int_\Omega (\mu_n n|\text{grad}\, \alpha|^2 + \mu_p p|\text{grad}\, \beta|^2)\, dx$. Rewriting $\langle z, z\rangle$, one obtains

$$\begin{aligned}\langle z, z\rangle &= \int_\Omega u\alpha + v\beta\, dx = \int_\Omega n|\alpha+\phi|^2 + p|\beta-\phi|^2 + (v-u)\phi\, dx \\ &= \int_\Omega n|\alpha+\phi|^2 + p|\beta-\phi|^2 + \lambda^2|\text{grad}\, \phi|\, dx \qquad (3.8.18)\\ &\leqslant \text{const}\left(\|\alpha\|^2_{L^2(\Omega)} + \|\beta\|^2_{L^2(\Omega)} + \int_\Omega (n+p)|\phi|^2 + \lambda^2|\text{grad}^2\, \phi|\, dx\right).\end{aligned}$$

Rewriting the Poisson equation in terms of α and β we obtain

$$\lambda^2\, \Delta\phi = (n + p)\phi + n\alpha - p\beta. \tag{3.8.19}$$

Multiplying both sides by ϕ, integrating by parts and again applying the Cauchy-Schwartz inequality yields

$$\int_\Omega (\lambda^2|\text{grad}\, \phi|^2 + (n + p)\phi^2)\, dx \leqslant \text{const}(\|\alpha\|_{L^2(\Omega)} + \|\beta\|_{L^2(\Omega)})\|\phi\|_{L^2(\Omega)}$$

$$\|\phi\|_{L^2(\Omega)} \leqslant \text{const}(\|\alpha\|_{L^2(\Omega)} + \|\beta\|_{L^2(\Omega)}). \tag{3.8.20}$$

Combining (3.8.18)–(3.8.20) gives

$$\langle z, z\rangle \leqslant \text{const}(\|\alpha\|_{L^2(\Omega)} + \|\beta\|_{L^2(\Omega)})^2. \tag{3.8.21}$$

Since the L^2-norm of a function can be bounded in terms of the L^2-norm of its gradient if the function is sufficiently smooth and vanishes on at least a part of the domain boundary $\partial\Omega$ (see [3.1]), there exists a constant K such that

$$\langle z, z\rangle \leqslant K \int_\Omega (\mu_n n|\text{grad}\, \alpha|^2 + \mu_p p|\text{grad}\, \beta|^2)\, dx \tag{3.8.22}$$

holds. Thus, it can be deduced from (3.8.13) that

$$\operatorname{Re}\langle z, F'(w)z\rangle \leqslant \left(\left(K_R + \frac{1}{1-\varepsilon}K_J\right)K - \varepsilon\right) \times \int_\Omega (\mu_n n|\operatorname{grad}\alpha|^2 + \mu_p p|\operatorname{grad}\beta|^2)\,dx \tag{3.8.23}$$

holds. If K_J and K_R are sufficiently small the right hand side can be made negative by an appropriate choice of ε and, using the same argument as above, $\lim_{t\to\infty}\|z(t)\| = 0$ holds for $g \equiv 0$.

To estimate the eigenvalues of $F'(w)$, (3.8.13) can be used immediately. If a complex eigenfunction z satisfies

$$\omega z = F'(w)z \tag{3.8.24}$$

multiplication by z yields

$$(\operatorname{Re}\omega)\langle z, z\rangle = \operatorname{Re}\langle z, F'(w)z\rangle. \tag{3.8.25}$$

Setting $\varepsilon = 0$ in (3.8.13), we obtain $\operatorname{Re}\omega \leqslant K_J + K_R$. Again, using the same argument as before, the real parts of all eigenvalues of $F'(w)$ are negative if K_J and K_R are sufficiently small. In [3.22] a version of Lemma 3.8.12 is used to show that the operator $F'(w)$ generates an analytic semigroup on $L^2(\Omega)$ equipped with the norm $\|\cdot\|$. Thus the existence of a solution to the problem (3.8.6) is guaranteed.

So the answer to all four questions asked in the beginning is a conditional 'yes'. There exists a solution to the linearized problem (3.7.4). This solution is stable in the sense of the estimate (3.8.17). The solution of the linearized problem with homogeneous boundary conditions tends to zero for $t \to \infty$ for sufficiently small current flow and recombination.

3.9 Existence for the Nonlinear Problem

In this Section we will show the existence of a solution to the transient drift diffusion equations (3.8.3). As mentioned previously, the existence of a solution can, at least for a sufficiently small time interval, be established by a contraction argument. There are various papers in the literature establishing existence for various models for the mobilities and the recombination rates (see c.f. [3.11], [3.32], [3.33]). It should be noted that the presence of velocity saturation effects complicates existence proofs considerably if the Einstein relations

$$\text{(a)}\quad D_n = U_T\mu_n, \qquad \text{(b)}\quad D_p = U_T\mu_p \tag{3.9.1}$$

are kept. In this case the differential operator loses its uniform parabolicity for large values of the electric field $-\operatorname{grad} V$ and the fixed point map used below will not necessarily be well defined.

The existence proof for the transient problem consists essentially of two

steps. First, short time existence is established by a contraction argument. Then a priori estimates are used to show that this solution can be continued arbitrarily in time. Other approaches, treating more sophisticated models (i.e. avalanche generation), use a Schauder type fixed point argument directly (see c.f. [3.32]).
We will follow the lines of [3.11] but, for the sake of simplicity, we will not consider velocity saturation effects. Let us first consider the existence of a solution for a sufficiently short time interval. We define the fixed point map F by $F(n, p) = (u, v)$ in the following way.

Step 1: Given n and p solve the Poisson equation

$$\lambda^2 \Delta V = n - p - C(x)$$
$$V|_{\partial\Omega_D} = V_D, \qquad \left.\frac{\partial V}{\partial \nu}\right|_{\partial\Omega_N} = 0 \tag{3.9.2}$$

for V.

Step 2: Solve

$$\text{a)} \quad \partial_t u = \operatorname{div}[\mu_n(\operatorname{grad} u - n \operatorname{grad} V)] - R(n, p) \qquad t \in (0, T)$$
$$\text{b)} \quad \partial_t v = \operatorname{div}[\mu_p(\operatorname{grad} v + p \operatorname{grad} V)] - R(n, p)$$
$$\text{c)} \quad u(t=0) = n^I, \qquad u|_{\partial\Omega_D} = n_D, \qquad \left.\frac{\partial u}{\partial \nu}\right|_{\partial\Omega_N} = 0 \tag{3.9.3}$$
$$\text{d)} \quad v(t=0) = p^I, \qquad v|_{\partial\Omega_D} = p_D, \qquad \left.\frac{\partial v}{\partial \nu}\right|_{\partial\Omega_N} = 0$$

for u and v.

For the fixed point argument we will use the following norm:

$$\|(n, p)\| = \left[\max_{0 \leqslant t \leqslant T} \{ \|n(t)\|^2_{L^2(\Omega)} + \|p(t)\|^2_{L^2(\Omega)} \} + \int_0^T \|n(s)\|^2_{H^1(\Omega)} + \|p(s)\|^2_{H^1(\Omega)}\, ds \right]^{1/2}. \tag{3.9.4}$$

If $\|(n, p)\|$ is finite then standard existence results (see [3.37]) for linear parabolic partial differential equations imply that (3.9.3) has a unique solution (u, v). We now show that the map F is contractive for a sufficiently small time interval $(0, T)$ on the set

$$M_a := \{(n, p) \colon \|(n, p)\| \leqslant a\}. \tag{3.9.5}$$

For the difference

$$(\delta u, \delta v) = F(n_1, p_1) - F(n_2, p_2) \tag{3.9.6}$$

we obtain the initial boundary value problem

$$
\begin{aligned}
&\text{a)}\quad \partial_t \delta u = \operatorname{div}[\mu_n(\operatorname{grad} \delta u - n_1 \operatorname{grad} V_1 + n_2 \operatorname{grad} V_2)] - \delta R \\
&\text{b)}\quad \partial_t \delta v = \operatorname{div}[\mu_p(\operatorname{grad} \delta v + p_1 \operatorname{grad} V_1 - p_2 \operatorname{grad} V_2)] - \delta R \\
&\text{c)}\quad \delta u(t=0) = 0, \qquad \delta u|_{\partial\Omega_D} = 0, \qquad \left.\frac{\partial \delta u}{\partial \nu}\right|_{\partial\Omega_N} = 0 \qquad (3.9.7)\\
&\text{d)}\quad \delta v(t=0) = 0, \qquad \delta v|_{\partial\Omega_D} = 0, \qquad \left.\frac{\partial \delta v}{\partial \nu}\right|_{\partial\Omega_N} = 0 \\
&\text{e)}\quad \delta R := R(n_1, p_1) - R(n_2, p_2).
\end{aligned}
$$

Multiplying (3.9.7)a) by δu, integrating by parts with respect to x and integrating with respect to t from $t = 0$ to $t = T$ yields

$$
\begin{aligned}
&\frac{1}{2}\|\delta u(t)\|_{L^2(\Omega)}^2 + \int_0^T \mu_n \|\operatorname{grad} \delta u(s)\|_{L^2(\Omega)}^2 \, ds \\
&\quad = \int_0^T \int_\Omega [\mu_n(n_1 \operatorname{grad} V_1 - n_2 \operatorname{grad} V_2)\cdot \operatorname{grad} \delta u - \delta R \delta u] \, dx \, ds.
\end{aligned} \tag{3.9.8}
$$

We estimate the right-hand side of (3.9.8) by

$$
\begin{aligned}
&\frac{1}{2}\|\delta u(t)\|_{L^2(\Omega)}^2 + \int_0^T \mu_n \|\operatorname{grad} \delta u(s)\|_{L^2(\Omega)}^2 \, ds \\
&\quad \leqslant \text{const} \int_0^T [(\|\delta n \operatorname{grad} V_2\|_{L^2(\Omega)} + \|n_1 \operatorname{grad} \delta V\|_{L^2(\Omega)}) \\
&\qquad \cdot \|\operatorname{grad} \delta u\|_{L^2(\Omega)} + (\|\delta n\|_{L^2(\Omega)} + \|\delta p\|_{L^2(\Omega)})\|\delta u\|_{L^2(\Omega)}] \, ds,
\end{aligned} \tag{3.9.9}
$$

where δn denotes $(n_1 - n_2)$ etc. The right-hand side of (3.9.9) can be estimated further using the definition of M_a, the Sobolev imbedding theorem and the Gagliardo-Nirenberg inequality (see c.f. [3.37])

$$
\begin{aligned}
\|g\|_{L^r(\Omega)} &\leqslant \text{const } \|g\|_{L^2(\Omega)}^{1-s} \|g\|_{H^1(\Omega)}^{s} \\
&\leqslant c(\varepsilon)\|g\|_{L^2(\Omega)} + \varepsilon \|g\|_{H^1(\Omega)}
\end{aligned} \tag{3.9.10}
$$

for any $\varepsilon > 0$, $r \geqslant 2$ and any $H^1(\Omega)$ function g with $s = d(\frac{1}{2} - \frac{1}{r})$ and d being the dimension of the problem. We obtain

$$
\begin{aligned}
&\|\delta n \operatorname{grad} V_2\|_{L^2(\Omega)} + \|n_1 \operatorname{grad} \delta V\|_{L^2(\Omega)} \\
&\quad \leqslant \text{const}[a(\|\delta n\|_{L^r(\Omega)} + \|\delta p\|_{L^r(\Omega)}) + (\|n_2\|_{L^2(\Omega)} + \|p_2\|_{L^2(\Omega)}) \\
&\qquad \cdot (c(\varepsilon)\|\delta n\|_{L^2(\Omega)} + \varepsilon \|\delta n\|_{H^1(\Omega)})] \|\operatorname{grad} \delta u\|_{L^2(\Omega)} \\
&\quad \leqslant c(a, \varepsilon)(\|\delta n\|_{L^2(\Omega)}^2 + \|\delta p\|_{L^2(\Omega)}^2) + \varepsilon(\|\delta n\|_{H^1(\Omega)}^2 \\
&\qquad + \|\delta p\|_{H^1(\Omega)}^2 + \|\delta u\|_{H^1(\Omega)}^2).
\end{aligned} \tag{3.9.11}
$$

Inserting (3.9.11) into (3.9.9) and using the inequality

$$
|ab| \leqslant \frac{1}{2\varepsilon} a^2 + \frac{\varepsilon}{2} b^2, \qquad a, b, \varepsilon \in \mathbb{R}, \qquad \varepsilon > 0 \tag{3.9.12}
$$

yields

$$\frac{1}{2}\|\delta u\|^2_{L^2(\Omega)} + \int_0^t \mu_n \|\text{grad}\, \delta u(s)\|^2_{L^2(\Omega)}\, ds \tag{3.9.13}$$

$$\leqslant \text{const} \int_0^t (c(a, \varepsilon)(\|\delta n\|^2_{L^2(\Omega)} + \|\delta p\|^2_{L^2(\Omega)})$$

$$+ \varepsilon(\|\delta n\|^2_{H^1(\Omega)} + \|\delta p\|^2_{H^1(\Omega)} + \|\delta u\|^2_{H^1(\Omega)}))\, ds$$

$$\leqslant \max\{Tc(a, \varepsilon), \varepsilon\}\, \|(\delta n, \delta p)\|^2 + \varepsilon \int_0^t \|\delta u(s)\|^2_{H^1(\Omega)}\, ds.$$

A similar inequality holds for δv and, by choosing ε and T sufficiently small, F can be made contractive in M_a; i.e.

$$\|(\delta u, \delta v)\| \leqslant \tfrac{1}{2}\|(\delta n, \delta p)\|, \qquad \forall (n_1, p_1), (n_2, p_2) \in M_a \tag{3.9.14}$$

holds. Because of the Banach fixed point theorem there exists exactly one fixed point (n^*, p^*) with $(n^*, p^*) = F(n^*, p^*)$ in M_a. This fixed point is the unique, weak solution of (3.8.3) in $(0, T)$.
In order to show the existence of a global solution for arbitrary time intervals $[0, T]$, it is necessary to show, in addition, an a priori estimate on the solution of the form

$$\|n(t)\|_{L^2(\Omega)} + \|p(t)\|_{L^2(\Omega)} + \int_0^t (\|n(s)\|_{H^1(\Omega)} + \|p(s)\|_{H^1(\Omega)})\, ds \leqslant c(t) \tag{3.9.15}$$

for the solution (n, p), where $c(t)$ is a bounded function. The proof of this a priori bound involves a quite complicated argument based on a problem specific Lyapunov function. We refer the reader to [3.11] for the details. Using (3.9.15), the existence of a solution can then be established for arbitrarily large time intervals.

Asymptotic Expansions for the Transient Drift Diffusion Equations

As in the steady state case, a great deal of insight into the structure of the solutions of the transient problem can be gained by asymptotic analysis. The structure of solutions of the transient drift diffusion equations is quite a bit more complicated than that of stationary equations. In addition to all the structural complexities already present in the steady state case, one is confronted with different time scales whose occurence depends on the form of the initial data. In this Section we will first derive an asymptotic approximation (for small λ^2) on the time scale given by the scaling (3.8.2). We will refer to this time scale as the 'slow' or diffusion time scale in the future. There we will observe the same spatial structure as in the stationary problem; i.e. layers around the P-N junctions etc. However, for this approximation to be valid, it will be necessary that the initial and boundary data satisfy certain conditions. If a steady state solution is taken as initial datum, which usually is the case, and if the applied boundary potential varies only on the diffusion time scale, then we can conclude from the results of Section 3.4 that these conditions are satisfied. So the fast time scale only occurs because of perturbations in steady state initial conditions or because of rapid changes in the externally applied bias. Perturbations in the initial conditions have to be considered as soon as the transient drift diffusion equations are solved numerically. It turns out that the structure of the temporal layer solutions depends strongly on the type of perturbation introduced.

3.10 Asymptotic Expansions on the Diffusion Time Scale

We will first derive the zero order term of the asymptotic expansion of the solution of the transient drift diffusion equations on the time scale given by (3.8.2). This time scale corresponds to the diffusion of carriers since the diffusion coefficients D_n and D_p in (3.7.1) have been scaled to order $O(1)$. As we will see, the structure of the zero order term on this time scale is not essentially different from that of the stationary problem in the sense that the reduced problem (3.4.37) now simply evolves in time. The spatial layers remain located at the *P*-*N* junctions and will only widen or narrow according to the applied bias.
Existence of a solution to the transient reduced problem could, in principle, be shown by the same methods as for the full problem in Section 3.9. However, we will sketch a different type of existence proof at the end of this Section in order to reflect on the structure of the solution of the transient reduced problem. This will express the fact that, for certain types of devices such as diodes or bipolar transistors, the free electrons will actually mainly populate the *N*-region and the holes will populate the *P*-region under moderate biasing conditions.
Setting the perturbation parameter $\lambda = 0$ in (3.8.3) we obtain the reduced equations

$$\begin{aligned}
&\text{a)}\quad 0 = \bar{n} - \bar{p} - C(x)\\
&\text{b)}\quad \partial_t \bar{n} = \operatorname{div} \bar{J}_n - \bar{R}\\
&\text{c)}\quad \partial_t \bar{p} = -\operatorname{div} \bar{J}_p - \bar{R}\\
&\text{d)}\quad \bar{J}_p = -\mu_p(\operatorname{grad} \bar{p} + \bar{p} \operatorname{grad} \bar{V})\\
&\text{e)}\quad \bar{J}_n = \mu_n(\operatorname{grad} \bar{n} - \bar{n} \operatorname{grad} \bar{V}),
\end{aligned} \tag{3.10.1}$$

which imply vanishing space charge everywhere. Again, as in the stationary case, (3.10.1)a) is consistent with the Dirichlet boundary conditions since, if only Ohmic contacts are present,

$$n_D - p_D - C(x) = 0 \qquad \text{for} \qquad x \in \partial\Omega_D \tag{3.10.2}$$

holds. The reduced equations (3.10.1) have the following interpretation. While the full transient drift diffusion equations state the conservation of the total current

$$J = J_n + J_p - \lambda^2 \operatorname{grad}(\partial_t V), \tag{3.10.3}$$

the displacement current $-\lambda^2 \operatorname{grad}(\partial_t V)$ can be neglected away from spatial and temporal layers, and only the drift diffusion current $J_n + J_p$ is conserved. The spatial layers on the slow time scale occur near the *P-N* junction where, because of steep gradients or discontinuities in the doping profile $C(x)$, the reduced concentrations $\bar{n}$ and $\bar{p}$ are discontinuous. The spatial layer terms, and the corresponding layer equations, are of the same form as for the stationary problem. For instance in two space dimensions the layer terms are functions of the form

$$\hat{w} = \hat{w}(\xi, s, t), \qquad \hat{w} = (\hat{V}, \hat{n}, \hat{p}, \hat{J}_n, \hat{J}_p)^T, \tag{3.10.4}$$

where (ξ, s) are again given by the same local coordinate transformation near the *P-N* junction as in the stationary case. s is the curve parameter of the *P-N* junction and ξ is the perpendicular distance of the point (x, y) to the *P-N* junction divided by λ. Carrying out this variable transformation and inserting the layer term (3.10.4) into the equations gives

$$\begin{aligned}
&\text{a)}\quad \partial_\xi^2 \hat{V} = \hat{n} - \hat{p} - C(s) + O(\lambda) \\
&\text{b)}\quad \partial_\xi \hat{n} - \hat{n}\partial_\xi \hat{V} = O(\lambda), \qquad \text{c)}\quad \partial_\xi(\hat{J}_n \cdot \vec{n}) = O(\lambda) \\
&\text{d)}\quad \partial_\xi \hat{p} + \hat{p}\partial_\xi \hat{V} = O(\lambda), \qquad \text{e)}\quad \partial_\xi(\hat{J}_p \cdot \vec{n}) = O(\lambda).
\end{aligned} \tag{3.10.5}$$

Letting λ go to zero, and after integration of the current relations and the continuity equations, one obtains, similar to the stationary case

$$\begin{aligned}
&\text{a)}\quad \hat{n}(\xi, s, t) = A_n(s, t)e^{\hat{V}(\xi, s, t)} \\
&\text{b)}\quad \hat{p}(\xi, s, t) = A_p(s, t)e^{-\hat{V}(\xi, s, t)} \\
&\text{c)}\quad \hat{J}_n(\xi, s, t) = \bar{J}_n e^{(\hat{V}(\xi, s, t) - \bar{V})} \\
&\text{d)}\quad \hat{J}_p(\xi, s, t) = \bar{J}_p e^{(-\hat{V}(\xi, s, t) + \bar{V})} \\
&\text{e)}\quad \partial_\xi^2 \hat{V} = \hat{n} - \hat{p} - C,
\end{aligned} \tag{3.10.6}$$

where A_n and A_p are given by

$$\begin{aligned}
&\text{a)}\quad A_n(s, t) = \bar{n}(X(s), Y(s), t)e^{-\bar{V}(X(s), Y(s), t)} \\
&\text{b)}\quad A_p(s, t) = \bar{p}(X(s), Y(s), t)e^{\bar{V}(X(s), Y(s), t)},
\end{aligned} \tag{3.10.7}$$

and the interface conditions at the *P-N* junction are given again by the requirement that

$$\bar{n}e^{-\bar{V}}, \quad \bar{p}e^{\bar{V}}, \quad \bar{J}_n \cdot \vec{n}, \quad \bar{J}_p \cdot \vec{n} \tag{3.10.8}$$

are continuous across the *P-N* junction. Equations (3.10.1) can be reformulated as a system of one elliptic coupled to one parabolic equation. Differentiating (3.10.1)a) with respect to time and inserting from

(3.10.1)b),c) gives

$$\operatorname{div}(\bar{J}_n + \bar{J}_p) = 0. \tag{3.10.9}$$

So the system (3.10.1) can be rewritten as

$$\begin{aligned} &\text{a)} \quad \operatorname{div}(\bar{J}_n + \bar{J}_p) = 0 \\ &\text{b)} \quad \partial_t \bar{p} = -\operatorname{div} \bar{J}_p - \bar{R} \\ &\text{c)} \quad \bar{J}_n = \mu_n[\operatorname{grad}(\bar{p} + C) - (\bar{p} + C) \operatorname{grad} \bar{V}] \\ &\text{d)} \quad \bar{J}_p = \mu_p[-\operatorname{grad} \bar{p} - \bar{p} \operatorname{grad} \bar{V}]. \end{aligned} \tag{3.10.10}$$

After inserting for $\bar{J}_n$ and $\bar{J}_p$ from (3.10.10)c),d) (3.10.10)a) is an elliptic equation for $\bar{V}$ while (3.10.10)b) is a parabolic equation for $\bar{p}$. Thus, the reduced problem is given by the equations (3.10.10) together with boundary conditions

$$\begin{aligned} &\text{a)} \quad \bar{p}(x,t) = p_D(x), \quad \bar{V}(x,t) = V_D(x,t) \quad \text{for} \quad x \in \partial\Omega_D, \quad t > 0 \\ &\text{b)} \quad \frac{\partial \bar{p}}{\partial \nu}(x,t) = 0, \quad \frac{\partial \bar{V}}{\partial \nu}(x,t) = 0, \quad \text{for} \quad x \in \partial\Omega_N, \quad t > 0, \end{aligned} \tag{3.10.11}$$

the interface conditions (3.10.8) and an initial condition for $\bar{p}$. $\bar{n}(x,t)$ is then given a posteriori by

$$\bar{n}(x,t) = \bar{p}(x,t) + C(x) \tag{3.10.12}$$

and satisfies automatically the boundary conditions

$$\bar{n}(x,t) = \bar{n}_D(x) \quad (= \bar{p}_D(x) + C(x)) \quad \text{for} \quad x \in \partial\Omega_D, \quad t > 0. \tag{3.10.13}$$

One feature of the solutions of the drift diffusion equations, that has not yet been expressed through our asymptotic analysis, is that, away from P-N junctions, the carrier densities n and p will be quite small in the P- and N-region, respectively (see [3.35]. In the case, when an N-region is adjacent only to P-regions, and when all the N- and P-regions are contacted this fact can be explained by an asymptotic analysis using the built in potential. This additional asymptotic analysis has been used in [3.26], [3.27] to show the existence of a solution to the reduced problem (3.10.10), (3.10.11), (3.10.8). We assume that the device domain Ω is the disjoint union of K P- and N-regions whose boundary contains exactly one contact each.

$$\begin{aligned} &\text{a)} \quad \Omega = \bigcup_{j=1}^{K} \Omega_j, \qquad \partial\Omega_j \cap \partial\Omega_D \neq \{\ \}, \qquad j = 1, \ldots, K \\ &\text{b)} \quad \operatorname{sign}(C(x)) \equiv \text{const} \quad \text{in } \Omega_j, \qquad j = 1, \ldots, K. \end{aligned} \tag{3.10.14}$$

Using the form of the boundary data V_D and p_D (see [3.19] or Section 3.1) in our scaling, we have

$$
\begin{aligned}
&\text{a)}\quad V_D(x,t) = V_j(t) + V_{bi}(x) \qquad \text{for} \qquad x \in \partial\Omega_j \cap \partial\Omega_D \\
&\text{b)}\quad V_{bi}(x) = \ln\left(\frac{C(x) + \sqrt{C(x)^2 + 4\delta^4}}{2\delta^2}\right) \\
&\text{c)}\quad p_D(x) = \delta^2 \exp(-V_{bi}(x)), \qquad x \in \partial\Omega_D \\
&\text{d)}\quad n_D(x) = p_D(x) + C(x).
\end{aligned}
\tag{3.10.15}
$$

Here V_j is the externally applied potential, so the difference between the V_j is the applied bias and V_{bi} is the so called built in potential, which is due only to the doping (see [3.19]). Note that $V_{bi} \equiv 0$ holds in the case of an undoped semiconductor. δ in (3.10.15) is given by

$$
\delta^2 = \frac{n_i}{\max\limits_{x\in\Omega} |C(x)|}
\tag{3.10.16}
$$

and will be quite small in practice. We have neglected the effect of δ, so far, since it appears only logarithmically in the boundary conditions (3.10.11) and V_{bi} will be of moderate size, even for small δ. In order to analyze the structure of solutions of the reduced problem, we employ the variable transformation $\bar{V} \to \phi$, $\bar{p} \to u$, $\bar{n} \to C(x) + u$, given by

$$
\begin{aligned}
&\text{a)}\quad \bar{V}(x,t) = \delta^4\phi(x,t) + V_{bi}(x) + V_j(t) \quad \text{for} \quad x \in \Omega_j \\
&\text{b)}\quad \bar{p}(x,t) = \delta^4 u(x,t) + \delta^2 \exp(-V_{bi}(x)) \quad \text{for} \quad x \in \Omega \\
&\text{c)}\quad \bar{J}_n = \delta^4 I_n, \qquad \text{d)}\quad \bar{J}_p = \delta^4 I_p.
\end{aligned}
\tag{3.10.17}
$$

Note, that this transformation implies automatically $\bar{n} - \bar{p} - C = 0$ everywhere. Inserting (3.10.17) into the equations (3.10.1) yields

$$
\begin{aligned}
&\text{a)}\quad \operatorname{div}(I_n + I_p) = 0 \\
&\text{b)}\quad \partial_t u = -\operatorname{div} I_p - S \\
&\text{c)}\quad I_n = \mu_n\Bigg[\operatorname{grad} u - u(\operatorname{grad} V_{bi} + \delta^4 \operatorname{grad}\phi) \\
&\qquad\qquad - \frac{-C + \sqrt{C^2 + 4\delta^4}}{2}\operatorname{grad}\phi\Bigg] \\
&\text{d)}\quad I_p = \mu_p\Bigg[-\operatorname{grad} u - u(\operatorname{grad} V_{bi} + \delta^4 \operatorname{grad}\phi) \\
&\qquad\qquad - \frac{C + \sqrt{C^2 + 4\delta^4}}{2}\operatorname{grad}\phi\Bigg].
\end{aligned}
\tag{3.10.18}
$$

S in (3.10.18)b) is the transformed recombination rate. If R is given by the Shockley Read Hall recombination term, we have

$$
\text{a)}\quad R = \frac{np - \delta^4}{n + p + 2\delta^2}, \qquad \text{b)}\quad S = \frac{u\sqrt{C^2 + 4\delta^4} + \delta^4 u^2}{2\delta^4 u + \sqrt{C^2 + 4\delta^4} + 2\delta^2}.
$$

The boundary conditions (3.10.11) become

$$\begin{aligned}
&\text{a)}\quad u(x,t)=0 \quad \text{for} \quad x\in\partial\Omega_D,\\
&\text{b)}\quad \frac{\partial u}{\partial \nu}(x,t)=0 \quad \text{for} \quad x\in\partial\Omega_N\\
&\text{c)}\quad \phi(x,t)=0 \quad \text{for} \quad x\in\partial\Omega_D,\\
&\text{d)}\quad \frac{\partial \phi}{\partial \nu}(x,t)=0 \quad \text{for} \quad x\in\partial\Omega_N,
\end{aligned} \tag{3.10.20}$$

and for the interface conditions we obtain that the functions

$$\begin{aligned}
&\text{a)}\quad [u(C+\sqrt{C^2+4\delta^4})+2]\exp(V_j+\delta^4\phi)\\
&\text{b)}\quad [u(-C+\sqrt{C^2+4\delta^4})+2]\exp(-V_j-\delta^4\phi)\\
&\text{c)}\quad I_n\cdot\vec{n} \qquad \text{and} \qquad \text{d)}\quad I_p\cdot\vec{n}
\end{aligned} \tag{3.10.21}$$

are continuous across *P-N* junctions. In (3.10.21) V_j takes on different values on different sides of the *P-N* junction (in the different subdomains Ω_j). If we now assume that (like c.f. in a MOSFET, see Chapter 4) an *N*-region is adjacent only to *P*-regions, and vice versa, the problem degenerates into a linear problem as $\delta\to 0$. This linear problem is given by

$$\begin{aligned}
&\text{a)}\quad \operatorname{div}(I_n^0+I_p^0)=0\\
&\text{b)}\quad \partial_t u^0=-\operatorname{div} I_p^0-S^0.\\
&\text{c)}\quad I_n^0=\mu_n[\operatorname{grad} u^0-u^0\operatorname{grad} V_{bi}-((-C+|C|)/2)\operatorname{grad}\phi^0]\\
&\text{d)}\quad I_p^0=\mu_p[-\operatorname{grad} u^0-u^0\operatorname{grad} V_{bi}-((C+|C|)/2)\operatorname{grad}\phi^0]
\end{aligned} \tag{3.10.22}$$

together with the boundary conditions (3.10.20). The interface conditions then become the conditions that

$$\text{a)}\quad I_n\cdot\vec{n}, \qquad \text{b)}\quad I_p\cdot\vec{n} \tag{3.10.23}$$

are continuous across *P-N* junctions and that

$$\begin{aligned}
&\text{a)}\quad u^0(x,t)C(x)+1=\exp(V_k(t)-V_j(t))\\
&\qquad\qquad \text{for} \quad x\in\partial\Omega_j\cap\partial\Omega_k \quad \text{if} \quad C(x)>0 \quad \text{in} \quad \Omega_j\\
&\text{or}\\
&\text{b)}\quad -u^0(x,t)C(x)+1=\exp(V_j(t)-V_k(t))\\
&\qquad\qquad \text{for} \quad x\in\partial\Omega_j\cap\partial\Omega_k \quad \text{if} \quad C(x)<0 \quad \text{in} \quad \Omega_j
\end{aligned} \tag{3.10.24}$$

holds. Here $u^0(x,t)$ and $C(x)$ are to be understood as the one sided limits of u^0 and C in Ω_j at the *P-N* junction $\partial\Omega_j\cap\partial\Omega_k$ (see Problem 3.10).
Note, that the problem (3.10.22), (3.10.23), (3.10.24) is linear. Standard arguments can be applied to establish the existence of a solution (see [3.26], [3.27]). If the appropriate spaces are chosen (see [3.26]) the existence of a

solution to the reduced problem (3.10.10)–(3.10.11) can be established by a perturbation argument for small δ.

3.11 Fast Time Scale Expansions

The slow time scale expansions, derived in the previous paragraph, imply that certain conditions for the inner and outer solution hold for all time. These are given by

$$\begin{aligned} &\text{a)} \quad \operatorname{div}(\bar{J}_n(x,t) + \bar{J}_p(x,t)) = 0, \qquad \forall t > 0 \\ &\text{b)} \quad \hat{n} = \bar{n}e^{\hat{\psi} - \bar{\psi}}, \qquad \text{c)} \quad \hat{p} = \bar{p}e^{\bar{\psi} - \hat{\psi}}. \end{aligned} \tag{3.11.1}$$

Therefore, in order for the approximation derived in the previous paragraph to hold, the initial data n^I and p^I have to be such that (3.11.1) is satisfied at $t = 0$. As we have seen in Section 3.4, this is the case when n^I and p^I are solutions of the stationary problem. In this and in the next paragraph we will analyze the structure of the solutions if this is not the case. The motivation for this analysis is twofold and we will distinguish between two principal cases. Firstly we are interested in steep, or in the extreme case discontinuous, changes in the applied bias. So n^I and p^I will be solutions of the steady state problem

$$\begin{aligned} &\text{a)} \quad \lambda^2 \Delta V^{SS} = n^I - p^I - C \\ &\text{b)} \quad \operatorname{div} J_n^{SS} = R \\ &\text{c)} \quad \operatorname{div} J_p^{SS} = -R \\ &\text{d)} \quad J_n^{SS} = \mu_n(\operatorname{grad} n^I - n^I \operatorname{grad} V^{SS}) \\ &\text{e)} \quad J_p^{SS} = \mu_p(-\operatorname{grad} p^I - p^I \operatorname{grad} V^{SS}) \\ &\text{f)} \quad V^{SS}(x) = V_D^{SS}(x), \qquad n^I(x) = n_D(x), \\ &\qquad p^I(x) = p_D(x) \qquad \text{for} \qquad x \in \partial\Omega_D \\ &\text{g)} \quad \frac{\partial V^{SS}}{\partial \nu} = 0, \qquad \frac{\partial n^I}{\partial \nu} = 0, \qquad \frac{\partial p^I}{\partial \nu} = 0 \qquad \text{for} \qquad x \in \partial\Omega_N . \end{aligned} \tag{3.11.2}$$

We then solve the transient initial value problem (3.7.1) together with the boundary conditions

$$V(x,t) = V_D(x,t) \qquad \text{for} \qquad x \in \partial\Omega_D, \tag{3.11.3}$$

where $V_D(x, 0) \neq V_D^{SS}(x)$ holds. Obviously, in this case, $V(x, 0)$ will not equal $V^{SS}(x)$ because of the different boundary values at the Dirichlet boundary $\partial\Omega_D$. Therefore

$$\operatorname{div}(\bar{J}_n + \bar{J}_p) = 0 \tag{3.11.4}$$

will also not hold at $t = 0$.

In the second case we will consider the solution of the transient problem for general initial functions n^I and p^I. The fundamental difference between the two cases is that in the first case V, J_n and J_p stay bounded at $t = 0$ for $\lambda \to 0$, while in the second case $V(x, t = 0) = \lambda^{-2}\Delta^{-1}(n^I - p^I - C)$ will be uniformly of order λ^{-2}, and so will be J_n and J_p at $t = 0$. This case can be interpreted as a random perturbation of the initial data caused by either external radiation or roundoff errors introduced by a numerical solution. An alternative interpretation would be to regard, after a rescaling of the potential V, the resulting problem as the transient drift diffusion equations under extreme reverse biasing conditions.

The Case of a Bounded Initial Potential

We will first consider the case when the initial data n^I and p^I are such that the potential V and the current densities J_n and J_p stay bounded for $\lambda \to 0$. So, again restricting ourselves to the two-dimensional case, we assume the same configuration as for the steady state problem in Section 3.4

$$\begin{aligned} &\text{a)}\quad n^I = \bar{n}^I(x, y) + \hat{n}^I(\xi, s) + O(\lambda) \\ &\text{b)}\quad p^I = p^I(x, y) + p^I(\xi, s) + O(\lambda), \end{aligned} \tag{3.11.5}$$

where $(x, y) \leftrightarrow (\xi, s)$ denotes again the local coordinate transformation around the P-N junction. Furthermore n^I and p^I shall be such that V^I, given by

$$\begin{aligned} &\text{a)}\quad \lambda^2\,\Delta V^I = n^I - p^I - C \\ &\text{b)}\quad V^I(x) = V_D(x, 0) \quad \text{for} \quad x \in \partial\Omega_D, \\ &\qquad \frac{\partial V^I}{\partial \nu}(x) = 0 \quad \text{for} \quad x \in \partial\Omega_N \end{aligned} \tag{3.11.6}$$

and J_n^I and J_p^I, given by

$$\begin{aligned} &\text{a)}\quad J_n^I = \mu_n(\operatorname{grad} n^I - n^I \operatorname{grad} V^I) \\ &\text{b)}\quad J_p^I = \mu_p(-\operatorname{grad} p^I - p^I \operatorname{grad} V^I) \end{aligned} \tag{3.11.7}$$

stay bounded as λ goes to zero. Now J_n^I and J_p^I will, in general, not satisfy (3.11.4) which makes a correction on a faster time scale necessary. In [3.26] it has been shown that the only possible time scale, in the absence of velocity saturation effects, is of the form

$$\tau = t/\lambda^2 \tag{3.11.8}$$

(see also Problem 3.11). Performing this change of variables, we obtain

$$\begin{aligned} &\text{a)}\quad \lambda^2\,\Delta\tilde{V} = \tilde{n} - \tilde{p} - C \\ &\text{b)}\quad \partial_\tau \tilde{n} = \lambda^2(\operatorname{div} \tilde{J}_n - \tilde{R}) \\ &\text{b)}\quad \partial_\tau \tilde{p} = \lambda^2(-\operatorname{div} \tilde{J}_p - \tilde{R}) \\ &\text{d)}\quad \tilde{J}_n = \mu_n(\operatorname{grad} \tilde{n} - \tilde{n} \operatorname{grad} \tilde{V}) \\ &\text{e)}\quad \tilde{J}_p = \mu_n(-\operatorname{grad} \tilde{p} - \tilde{p} \operatorname{grad} \tilde{V}), \end{aligned} \tag{3.11.9}$$

where '$\sim$' denotes the dependent variables on the fast time scale. Since $\tilde{J}_n$ and $\tilde{J}_p$ stay bounded as $\lambda \to 0$ we obtain

$$\tilde{n}(x, y, \tau) = n^I(x, y), \qquad \tilde{p}(x, y, \tau) = p^I(x, y). \tag{3.11.10}$$

The fast time scale equation for the potential $\tilde{V}$ is obtained by differentiating (3.11.9)a) with respect to τ. This yields

$$\partial_t \Delta \tilde{V} = \operatorname{div}(\tilde{J}_n + \tilde{J}_p), \tag{3.11.11}$$

where $\tilde{J}_n$ and $\tilde{J}_p$ are now given by (3.11.9) with the concentrations $\tilde{n}$ and $\tilde{p}$ replaced by n^I and p^I. The fast time scale equation is of the form

$$\partial_\tau \Delta \tilde{V} = -\operatorname{div}[(\mu_n n^I + \mu_p p^I) \operatorname{grad} \tilde{V}] + \operatorname{div}(\mu_n \operatorname{grad} n^I - u_p \operatorname{grad} p^I). \tag{3.11.12}$$

Thus, $\tilde{V}$ can satisfy the initial conditions and the boundary conditions

$$\begin{aligned} &\text{a)} \quad \tilde{V}(x, 0) = V^I(x) \\ &\text{b)} \quad \tilde{V}(x, \tau) = V_D(x, t) \quad \text{for} \quad x \in \partial\Omega_D, \\ &\qquad \frac{\partial \tilde{V}}{\partial \nu}(x, \tau) = 0 \quad \text{for} \quad x \in \partial\Omega_N. \end{aligned} \tag{3.11.13}$$

In [3.26] it is shown that the fast time scale problem (3.11.12) (3.11.13) has a unique solution $\tilde{V}$. A simple energy estimate (see again [3.26]) shows that the fast time scale potential decays towards a steady state solution $\tilde{V}^\infty$, $\tilde{J}_n^\infty$, $\tilde{J}_p^\infty$ satisfying

$$\begin{aligned} &\text{a)} \quad \operatorname{div}(\tilde{J}_n^\infty + \tilde{J}_p^\infty) = 0 \\ &\text{b)} \quad \tilde{J}_n^\infty = \mu_n(\operatorname{grad} n^I - n^I \operatorname{grad} \tilde{V}^\infty) \\ &\text{c)} \quad \tilde{J}_p^\infty = \mu_p(-\operatorname{grad} p^I - p^I \operatorname{grad} \tilde{V}^\infty). \end{aligned} \tag{3.11.14}$$

The corresponding current densities $\tilde{J}_n^\infty$ and $\tilde{J}_p^\infty$ now satisfy the condition (3.11.4) and so the fast time scale solution can be matched to the slow time solution, derived in (3.10.1)–(3.10.6), in the usual way (see [3.25]).
So, in the case of a bounded initial potential, the only correction on a faster time scale takes place in the potential V. The physical reason for this is that the timescale given by (3.11.8) corresponds to the dielectric relaxation time. By definition this is the time scale on which the electric potential V adjusts to a new charge distribution. The carriers move on a much slower time scale and therefore $\tilde{n} = n^I$ and $\tilde{p} = p^I$ holds in (3.11.10).
The above analysis can be used to describe the effect of rapid changes in the applied bias. If we start from a steady state solution, given by (3.11.2), and change the applied bias at Ohmic contacts on the dielectric relaxation time scale, i.e. if

$$V(x, t) = V_D\left(x, \frac{t}{\lambda^2}\right) \qquad \text{for} \qquad x \in \partial\Omega_D \tag{3.11.15}$$

holds, we can describe the behaviour of the solution on the fast time scale by corrections in the potential and in the drift current densities alone. We set

$$\begin{aligned} &\text{a)} \quad V(x,t) = V^{SS}(x) + \phi(x,\tau) \\ &\text{b)} \quad J_n(x,t) = J_n^{SS}(x) - \mu_n n^I(x) \operatorname{grad} \phi(x,\tau) \\ &\text{c)} \quad J_p(x,t) = J_p^{SS}(x) + \mu_p p^I(x) \operatorname{grad} \phi(x,\tau) \end{aligned} \tag{3.11.16}$$

with $\tau = t/\lambda^2$ and V^{SS}, J_n^{SS} and J_p^{SS} given by (3.11.2). $\phi(x, \tau)$ then satisfies the initial-boundary value problem

$$\begin{aligned} &\text{a)} \quad \partial_\tau \Delta\phi = -\operatorname{div}[(\mu_n n^I + \mu_p p^I) \operatorname{grad} \phi] \\ &\text{b)} \quad \phi(x,\tau) = V_D(x,\tau) - V_D(x,0) \qquad \text{for} \qquad x \in \partial\Omega_D \\ &\text{c)} \quad \frac{\partial\phi}{\partial\nu}(x,\tau) = 0 \qquad \text{for} \qquad x \in \partial\Omega_N \\ &\text{d)} \quad \phi(x,0) = 0. \end{aligned} \tag{3.11.17}$$

In this way the effect of, for instance, a steep ramp in the applied bias can be easily described (see Problem 3.13).
The situation is somewhat more complicated if velocity saturation effects are considered; i.e. if the mobilities μ_n and μ_p in (3.11.9) depend on the electric field $-\operatorname{grad} V$ as well. In this case an intermediate time scale of the form $\sigma = t/\lambda$ is present on which also the concentrations n and p can evolve inside the spatial layer regions. The presence of this intermediate time scale has, however, no effect on the over-all picture; that means on the solutions away from spatial layer regions. We refer the reader to [3.27] for a detailed analysis.

Fast Time Scale Solutions for General Initial Data

If arbitrary functions are prescribed as initial data for n^I and p^I in (3.1.1) severe complications are introduced into the preceeding asymptotic analysis. Most notably, the resulting potential at time $t = 0$, which satisfies

$$\begin{aligned} &\text{a)} \quad \lambda^2 \Delta V = n^I - p^I - C \\ &\text{b)} \quad V(x,0) = V_D(x,0) \quad \text{for} \quad x \in \partial\Omega_D, \\ &\qquad \frac{\partial V}{\partial\nu}(x,0) = 0 \quad \text{for} \quad x \in \partial\Omega_N \end{aligned} \tag{3.11.18}$$

will become of order $O(\lambda^{-2})$, uniformly in Ω, as λ tends to zero and so will the current densities J_n and J_p at $t = 0$. One would hope that, after initially being very large, V, J_n and J_p would settle down to $O(1)$, so that they can be matched to the diffusion time scale solution derived before. Unfortunately, no general results, which are valid in more than on space dimension, are available to this end. As we will see, the resulting fast time scale equations are of mixed elliptic-hyperbolic type and this makes the analysis of their behaviour for large τ highly complicated. It could be argued that this complication is somewhat artificial. It arises from the fact that n^I and p^I in

(3.1.1) are regarded as initial data and $V(x, 0)$ is given a posteriori by the solution of the Poisson equation. If one would regard the initial potential and one carrier density as initial datum instead, the fast time scale expansion derived in the previous paragraph would completely suffice. The results of this paragraph actually suggest that one should proceed in this manner, when solving the transient drift diffusion equations numerically since otherwise the resulting equation represent a differential algebraic system of index 2 (see [3.13] for the definition of the index of a differential algebraic system). The case of large ($O(\lambda^{-2})$) initial potential is, however, of additional interest since it will also describe the transient behaviour of the drift diffusion equations under extremely large bias.

Inserting the potential V, given by (3.11.18), and the initial functions n^I and p^I into the continuity equations and current relations (3.8.3)b)–e) at time $t = 0$ shows that the time derivatives of n and p are of order $O(\lambda^{-2})$ at $t = 0$. This implies again the fast time scale variable to be of the form

$$\tau = \frac{t}{\lambda^2}.$$

Also, because of the size of the potential at time $t = 0$, V, J_n and J_p have to be rescaled. We set

$$V(x, t) = \frac{\phi(x, \tau)}{\lambda^2}, \tag{3.11.19}$$

$$J_p(x, \tau) = \frac{I_p(x, \tau)}{\lambda^2}, \qquad J_n(x, \tau) = \frac{I_n(x, \tau)}{\lambda^2}.$$

This yields the equations

$$\begin{aligned} &\text{a)} \quad \Delta\phi = \tilde{n} - \tilde{p} - C \\ &\text{b)} \quad \partial_\tau \tilde{n} = \operatorname{div}(I_n) - \lambda^2 R, \qquad \text{c)} \quad \partial_\tau \tilde{p} = -\operatorname{div}(I_p) - \lambda^2 R \\ &\text{c)} \quad I_n = \mu_n(\lambda^2 \operatorname{grad} \tilde{n} - \tilde{n} \operatorname{grad} \phi) \\ &\text{d)} \quad I_p = \mu_p(-\lambda^2 \operatorname{grad} \tilde{p} - \tilde{p} \operatorname{grad} \phi) \end{aligned} \tag{3.11.20}$$

subject to the initial and boundary conditions

$$\begin{aligned} &\text{a)} \quad \tilde{n}(x, 0) = n^I(x), \qquad \text{b)} \quad \tilde{p}(x, 0) = p^I(x) \\ &\text{c)} \quad \phi(x, \tau) = \lambda^2 V_D(x, \lambda\tau) \qquad \text{for} \qquad x \in \partial\Omega_D \\ &\text{d)} \quad \frac{\partial\phi}{\partial\nu}(x, \tau) = 0 \qquad \text{for} \qquad x \in \partial\Omega_N. \end{aligned} \tag{3.11.21}$$

Because of the rescaling the small parameter λ^2 does not appear in the Poisson equation anymore, it appears as small diffusion coefficient in the current relations. Also, by (3.11.20)a) the effect of recombination-generation is $O(\lambda^2)$ on the dielectric relaxation time scale.

Note, that the problem (3.11.20)–(3.11.21) is the transient equivalent of the steady state problem (3.5.5) under extreme reverse biasing conditions. Thus, the problem (3.11.20)–(3.11.21) has two interpretations. It can either be interpreted as the transient problem under moderate biasing conditions with general initial data n^I and p^I. In this case $\phi(x, \tau)$ will be of order $O(\lambda^2)$ at the Dirichlet boundary $\partial\Omega_D$. Or it can be interpreted as the transient problem under extreme reverse biasing conditions, in which case $\phi(x, \tau)$ will be of order $O(1)$ at $\partial\Omega_D$. For $\lambda \to 0$ (zero diffusion limit in (3.11.20)) we obtain the mixed elliptic-hyperbolic system

$$\begin{aligned} &\text{a)} \quad \Delta\phi^0 = \tilde{n}^0 - \tilde{p}^0 - C \\ &\text{b)} \quad \partial_\tau \tilde{n}^0 = -\operatorname{div}(\tilde{n}^0 \operatorname{grad} \phi^0), \\ &\text{c)} \quad \partial_\tau \tilde{p}^0 = \operatorname{div}(\tilde{p}^0 \operatorname{grad} \phi^0) \end{aligned} \tag{3.11.22}$$

subject to the initial and boundary conditions

$$\begin{aligned} &\text{a)} \quad \tilde{n}^0(x, 0) = n^I(x), \qquad \text{b)} \quad \tilde{p}^0(x, 0) = p^I(x) \\ &\text{c)} \quad \phi^0(x, \tau) = \phi_D^0(x) \qquad \text{for} \qquad x \in \partial\Omega_D \\ &\text{d)} \quad \frac{\partial\phi^0}{\partial\nu}(x, \tau) = 0 \qquad \text{for} \qquad x \in \partial\Omega_N \end{aligned} \tag{3.11.23}$$

with $\phi_D^0(x) = \lim_{\lambda \to 0} \lambda^2 V_D(x, \lambda\tau)$, i.e. $\phi_D^0(x) = 0$ holds for moderate biasing conditions and $\phi_D^0(x) = O(1)$ holds for extreme biasing conditions. An additional boundary condition has to be imposed on $\tilde{n}^0$ whenever the characteristic directions of (3.11.22)b) point inward from the boundary, i.e. wherever

$$\operatorname{grad} \phi^0 \cdot \nu < 0 \tag{3.11.24}$$

holds. (Here, as always, ν denotes the unit outward normal vector on the boundary $\partial\Omega$.) Similarly, a boundary condition has to be used for $\tilde{p}^0$ wherever $\operatorname{grad} \phi^0 \cdot \nu > 0$ holds. For the one dimensional case, a proof that the solution $(\phi, \tilde{n}, \tilde{p})$ of the fast time scale problem actually converges to the solution of (3.11.22)–(3.11.23) in the weak sense as $\lambda \to 0$, can be found in [3.20].

To analyze the solution of (3.11.22)–(3.11.23) is extremely tricky since the parts of the domain boundary $\partial\Omega$, where a boundary condition on the concentrations $\tilde{n}^0$ and $\tilde{p}^0$ is required, will depend on the solution ϕ^0 itself. Several special cases have been investigated in [3.36] and we will present here only the simplest one of them in order to give the reader some idea of the situation.

A standard device to reduce the drift diffusion equations to the simplest possible case is to assume a one-dimensional semiconductor with a piecewise constant and antisymmetric doping concentration (see [3.21]). In this case (3.11.20) reduces to

$$\begin{aligned} &\text{a)} \quad \partial_x^2 \phi = \tilde{n} - \tilde{p} - C \\ &\text{b)} \quad \partial_\tau \tilde{n} = \partial_x I_n - \lambda^2 R, \qquad \text{c)} \quad \partial_\tau \tilde{p} = -\partial_x I_p - \lambda^2 R \\ &\text{d)} \quad I_n = \lambda^2 \partial_x \tilde{n} - \tilde{n}\partial_x \phi, \qquad \text{e)} \quad I_p = -\lambda^2 \partial_x \tilde{p} - \tilde{p}\partial_x \phi. \end{aligned} \tag{3.11.25}$$

(For the sake of simplicity we have taken the mobilities to be constant, and therefore scaled to 1, here.) The semiconductor occupies the region $\Omega = (-1, 1)$. The piecewise constant doping concentration $C(x)$ satisfies

$$C(x) = \begin{cases} -1 & \text{for} \quad -1 \leqslant x < 0, \\ 1 & \text{for} \quad 0 \leqslant x \leqslant 1. \end{cases} \tag{3.11.26}$$

Thus, the P-N junction is located at $x = 0$. This setting allows the reduction of the drift diffusion equations via the symmetry (see [3.21])

$$\begin{aligned} &\text{a)} \quad n(-x, \tau) = p(x, \tau), \qquad \text{b)} \quad I_n(-x, \tau) = I_p(x, \tau), \\ &\text{c)} \quad \phi(-x, \tau) = -\phi(x, \tau) \qquad \text{for} \qquad x \in (0, 1) \end{aligned} \tag{3.11.27}$$

(see also Problem 3.14). To simplify matters even further Szmolyan in [3.36] assumes that the hole concentration $\tilde{p}$ is neglegibly small in the N-region; i.e. he assumes that

$$\tilde{p}(x, \tau) = 0 \qquad \text{for} \qquad x \in (0, 1) \tag{3.11.28}$$

holds. For $\lambda \to 0$, this leads to the unipolar problem

$$\begin{aligned} &\left.\begin{aligned} &\text{a)} \quad \partial_x^2 \phi^0 = \tilde{n}^0 - 1 \\ &\text{b)} \quad \partial_\tau \tilde{n}^0 = -\partial_x(\tilde{n}^0 \partial_x \phi^0) \end{aligned}\right\} \quad \text{for} \quad x \in (0, 1) \\ &\text{c)} \quad \tilde{n}^0(x, 0) = n^I(x), \\ &\text{d)} \quad \tilde{n}^0(0, \tau) = 0, \qquad \text{e)} \quad \tilde{n}^0(1, \tau) = 1 \\ &\text{f)} \quad \phi^0(0, \tau) = 0, \qquad \text{g)} \quad \phi^0(1, \tau) = \phi_D^0. \end{aligned} \tag{3.11.29}$$

The boundary conditions (3.11.29)d) and f) arise from the symmetry assumption (3.11.27). $\phi_D^0 = 0$ would hold if the transient problem is considered under 'moderate' biasing conditions. If an extreme bias is applied in the sense of Section 3.5 ($V_D = O(\lambda^{-2})$ in (3.11.21)c)) a value $\phi_D^0 \neq 0$ is possible. We introduce the electric field

$$E(x, \tau) = -\partial_x \phi^0(x, \tau) \tag{3.11.30}$$

as a new variable. Differentiating (3.11.29)a) with respect to τ and integration gives

$$\begin{aligned} &\text{a)} \quad \partial_\tau E - E \partial_x E = -E + A(\tau) \\ &\text{b)} \quad \partial_\tau \tilde{n}^0 - E \partial_x \tilde{n}^0 = \tilde{n}^0(1 - \tilde{n}^0) \\ &\text{c)} \quad A(\tau) = E(1, \tau) + \partial_\tau E(1, \tau). \end{aligned} \tag{3.11.31}$$

Both equations (3.11.31)a) and (3.11.31)b), have now the same characteristics $\hat{x}(x, s, \tau)$ satisfying

$$\begin{aligned} &\text{a)} \quad \partial_s \hat{x}(x, s, \tau) = -E(\hat{x}, s) \\ &\text{b)} \quad \hat{x}(x, \tau, \tau) = x. \end{aligned} \tag{3.11.32}$$

If we denote by $\hat{n}$ and $\hat{E}$ the values of $\tilde{n}^0$ and E along these characteristics; i.e.

$$\begin{aligned} &\text{a)} \quad \hat{E}(x, s, \tau) = E(\hat{x}(x, s, \tau), s) \\ &\text{b)} \quad \hat{n}(x, s, \tau) = \tilde{n}^0(\hat{x}(x, s, \tau), s), \end{aligned} \tag{3.11.33}$$

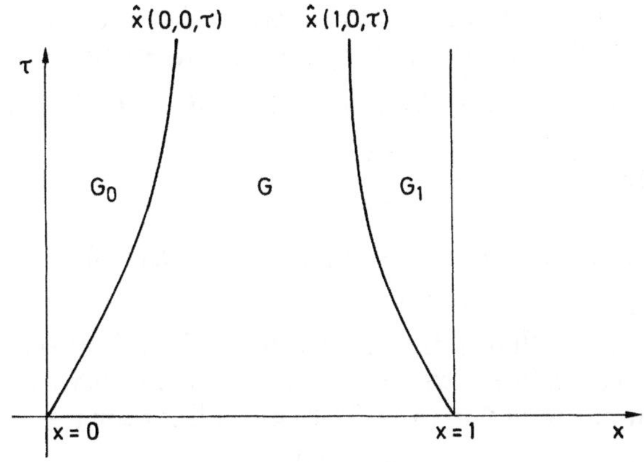

Fig. 3.11.1

we obtain

$$\text{a)}\quad \partial_s \hat{E} = -\hat{E} + A(s)$$
$$\text{b)}\quad \partial_s \hat{n} = \hat{n}(1 - \hat{n}). \tag{3.11.34}$$

The values of $E(x, \tau)$ and $\tilde{n}^0(x, \tau)$ at the starting point of the characteristic curve are the initial values for the ordinary differential equations (3.11.34). In particular, the values of $\tilde{n}^0$ and E at $\tau = 0$ and the values of $\tilde{n}^0$ at inflow boundaries are known. If we assume that there is no flux of holes from the P-region into the N-region at the initial time $\tau = 0$, this means (see [3.36])

$$\text{a)}\quad E(0, 0) > 0, \qquad \text{b)}\quad E(1, 0) < 0. \tag{3.11.35}$$

The assumption (3.11.35) implies that the characteristic curves $\hat{x}(0, s, 0)$ and $\hat{x}(1, s, 0)$, starting at time $\tau = 0$ at the boundaries $x = 0$ and $x = 1$, point inward. Consequently boundary values for $\tilde{n}^0(0, \tau)$ and $\tilde{n}^0(1, \tau)$ are prescribed initially. So, the situation is as depicted in Fig. 3.11.1.
The characteristics $\hat{x}(0, s, 0)$ and $\hat{x}(1, s, 0)$ split the domain $(0, 1) \times (0, \tau)$ in the three subdomains G_0, G_1 and G. It follows from the equation (3.11.34)b) that the boundary values are transported into the domain in G_0 and G_1. Therefore $\tilde{n}^0 \equiv 0$ holds in G_0 and $\tilde{n}^0 \equiv 1$ holds in G_1. This fact, and equation (3.11.29)a) imply

$$E(x, \tau) = \begin{cases} (E(0, \tau) + x & \text{for} \quad (x, \tau) \in G_0, \\ E(1, \tau) & \text{for} \quad (x, \tau) \in G_1. \end{cases} \tag{3.11.36}$$

Using the differentiated version of Poisson's equation once more, one easily obtains

$$\text{a)}\quad E(0, \tau) = E(1, \tau) + y(\tau) + \alpha$$
$$\text{b)}\quad y(\tau) = \int_0^\tau E(1, s)\, ds, \qquad \text{c)}\quad \alpha = E(0, 0) - E(0, 1). \tag{3.11.37}$$

The solution E and $\tilde{n}^0$ in the middle subdomain G can now be determined by straightforward integration along the characteristics. One obtains

$$\text{a)}\quad \hat{x}(x, s, 0) = x + (E(1, 0) - E(x, 0))(1 - e^{-s}) - y(s)$$

$$\text{b)}\quad \hat{n}(x, s, 0) = \frac{n_I(x)}{n_I(x)(1 - e^{-s}) + e^{-s}} \tag{3.11.38}$$

$$\text{c)}\quad \hat{E}(x, s, 0) = (E(x, 0) - E(1, 0))e^{-s} + E(1, s)$$

with $y(s)$ defined as in (3.11.37)b).
In [3.36] it is shown that, if the applied potential ϕ_D^0 satisfies $\phi_D^0 \leqslant \alpha^2/2$, the characteristics $\hat{x}(0, s, 0)$ and $\hat{x}(1, s, 0)$ remain in the interval $(0, 1)$ for all $s > 0$, and that the function $y(\tau)$ in (3.11.37)b) has a limit for $\tau \to \infty$.

$$\lim_{\tau \to \infty} y(\tau) = y(\infty) = -\alpha - \sqrt{2\phi_D^0} \tag{3.11.39}$$

holds. Thus, using (3.11.38)a), we obtain

$$\text{a)}\quad \lim_{s \to \infty} \hat{x}(0, s, 0) = \sqrt{2\phi_D^0}$$

$$\text{b)}\quad \lim_{s \to \infty} \hat{x}(1, s, 0) = 1 + \alpha + \sqrt{2\phi_D^0}. \tag{3.11.40}$$

Using (3.11.38)b)c), we obtain the steady state solution for E and $\tilde{n}^0$.

$$\text{a)}\quad \tilde{n}^0(x, \infty) = \begin{cases} 0 & \text{for } 0 \leqslant x \leqslant \sqrt{2\phi_1}, \\ 1 & \text{for } \sqrt{2\phi_1} \leqslant x \leqslant 1, \end{cases}$$

$$\text{b)}\quad E(x, \infty) = \begin{cases} x - \sqrt{2\phi_1} & \text{for } 0 \leqslant x \leqslant \sqrt{2\phi_1}, \\ 0 & \text{for } \sqrt{2\phi_1} \leqslant x \leqslant 1. \end{cases} \tag{3.11.41}$$

For a moderate value of the applied bias, that means for $\phi_D^0 = 0$, $\tilde{n}^0$ has a jump discontinuity at $x = 0$, the P-N junction, and the boundary $x = 0$ becomes a characteristic. In any event $\tilde{n}^0$ and E (and therefore also ϕ) converge to a steady state solution satisfying

$$\text{a)}\quad \tilde{n}^0(x, \infty) - C(x) = 0, \qquad \text{b)}\quad \partial_x I_n = 0. \tag{3.11.42}$$

This steady state solution can then be matched to the slow (diffusion-) time scale solution derived before. For a large value of the applied bias ($V_D = O(\lambda^{-2})$ in (3.11.18)b)), which implies $\phi_D^0 \neq 0$, the layer moves away from the P-N junction towards the edge of the depletion region at $x = \sqrt{2\phi_D^0}$. At least for this simple model problem, the corresponding limit solution for $\tau \to \infty$ is identical to the stationary solution derived for extremely large reverse bias values in Section 3.5 (see Problem 3.16).
Unfortunately, this type of analysis breaks down if more than one carrier is present since it heavily relies on the fact that the equations for the field E and the concentration $\tilde{n}^0$ have the same characteristics (see Problem 3.15). The above example, and the other special cases analyzed in [3.36], lead one to expect that the solutions of the fast time scale problem under extreme biasing conditions in general tends to a solution of the corresponding

stationary problem for the fast time variable $\tau \to \infty$, as it is the case for moderate biasing conditions. Note, that, if (3.11.29) is regarded as the transient problem with general initial data but under moderate biasing conditions (i.e. $\phi_D^0 = 0$), the limiting solution $\phi(x, \tau = \infty)$ vanishes identically. This suggests that, for large τ, the scaling (3.11.19) has to be reversed again and the fast time scale solution could be matched to the slow time scale solution derived before. The precise mechanism for this transition from the fast to the slow time scale is, however, not known yet.

So, as for the stationary problem, the asymptotic analysis of the drift diffusion equations in the reverse bias case is much more complicated than in the forward bias case. The reason is that in the reverse bias case the reduced problem is of mixed elliptic-hyperbolic type, leading to a free boundary value problem in the stationary case. In general, it can be said that a great deal of additional analysis is needed for the drift diffusion equations in the extreme reverse bias scaling. It should be pointed out, however, that the value of singular perturbation analysis lies in the general understanding of the qualitative behaviour of solutions of the drift diffusion equations and that singular perturbation analysis cannot replace numerical calculations techniques. For particular applications the asymptotics has to be adjusted to the device geometry, the models for the mobilities and the recombination rate etc., all of which will influence the actual solution significantly. Therefore it does not seem reasonable to analyze the general drift diffusion equations in too much detail without specifying the geometry and the model parameters. For this reason we will, in the next Chapter, turn to the study of particular devices where we apply the tools and results developed in this Chapter for the general problem.

Problems

3.1 Derive the remainder term $G(x, y)$ in (3.4.27) and show that it is of order $O(\lambda)$ uniformly in Ω.

3.2 Suppose $F(w) = 0$ is solved by Newton's method. Show that if the variable transformation $w = G(s)$ is used and Newton's method is applied to the transformed equation the linearization of the new updating strategy coincides with the old strategy. What does this mean for the local convergence rates?

3.3 Use the maximum principle to show (3.6.19).

3.4 Show that V_0^c in (3.4.24) satisfies the Neumann boundary conditions at $\partial\Omega_N$ if (3.4.12) holds.

3.5 Derive the boundary value problems defining the first-order approximation in (3.4.46).

3.6 Show that the boundary value problem consisting of the equation (3.5.28)a) and the boundary conditions (3.5.28)e) has a solution.
Hint: Use the same approach as used when showing the existence of a solution to the drift diffusion equations in Section 3.3; i.e. find an appropriate fixed point map.

3.7 Derive the $O(\lambda)$ and the $O(\lambda^2)$ terms in the asymptotic expansion for the strongly reverse biased one dimensional case (3.5.31).

3.8 Derive the boundary value problem for the smooth variable w, given by the transformation (3.5.20) in the two dimensional case. Under what conditions is this boundary value problem uniformly elliptic?

3.9 Show that the product defined in (3.8.11) satisfies all the requirements of a scalar product.

3.10 How do the terms $(\psi_k(1) - \psi_j(t))$ in (3.10.24) behave for $\delta \to 0$. What does this mean for forward and reverse biased P-N junctions?

3.11 Show that away from spatial layers (that means when the spatial derivatives stay bounded) the only fast time scale, which does not yield a trivial solution, is of the form (3.11.8).

3.12 Derive the $O(\lambda)$ and $O(\lambda^2)$ terms in the asymptotic expansion of the fast time scale equation (3.11.9).

3.13 Assume a one dimensional P-N diode model with a piecewise constant antisymmetric doping concentration, where after scaling the diode is located in the interval $[-1, 1]$ (like in (3.11.25)–(3.11.27)). Assume the applied bias is varied linearly from 0 V to -0.5 V forward bias in the scaled time interval from $t = 0$ to $t = \lambda^2$. Use the approach in (3.11.16)–(3.11.17) to compute the current at the end of the switching period approximately.

3.14 Verify that the symmetry 'ansatz' (3.11.27) is consistent with the boundary conditions (3.11.21).

3.15 Derive the characteristic equations corresponding to (3.11.32) and (3.11.34) in the bipolar case. That means when $p \neq 0$ in (3.11.29)a).

3.16 Verify that the solution (3.5.18), (3.5.25) of the stationary problem under extreme reverse biasing conditions reduces to the limit solution (3.11.41), for $\tau \to \infty$, of the transient problem if a unipolar one dimensional device with antisymmetric doping is considered.

References

[3.1] R. A. Adams: Sobolev Spaces. Academic Press, New York (1975).

[3.2] F. Alabeau: A Singular Perturbation Analysis of the Semiconductor Device and the Electrochemistry Equations. Report, Institut National de Recherche en Informatique et en Automatique, Paris (1984).

[3.3] U. Ascher, P. A. Markowich, C. Schmeiser, H. Steinrück, R. Weiß: Conditioning of the Steady State Semiconductor Device Problem. SIAM J. Appl. Math. *49*, 165–185 (1989).

[3.4] R. E. Bank, et al.: Analytical and Numerical Aspects of Semiconductor Device Modelling. Report 82-11274-2, Bell Laboratories, Murray Hill (1982).

[3.5] R. E. Bank, D. J. Rose: Global Approximate Newton Methods. Numerische Mathematik *37*, 279–295 (1981).

[3.6] F. Brezzi: Theoretical and Numerical Problems in Reverse Biased Semiconductor Devices. Proc. 7th Intern. Conf. on Comput. Meth. in Appl. Sci. and Engng., INRIA, Paris (1985).

[3.7] F. Brezzi, A. C. S. Capelo, L. Gastaldi: A Singular Perturbation Analysis of Reverse Biased Semiconductor Diodes. SIAM J. Math. Anal. *20*, 372–387 (1989).

[3.8] F. Brezzi, L. Gastaldi: Mathematical Properties of One-Dimensional Semiconductors. Mat. Apl. Comp. *5*, 123–137 (1986).

[3.9] L. Caffarelli, A. Friedman: A Singular Perturbation Problem for Semiconductors. Bolletino U.M.I. *1–B* (7), 409–421 (1987).

[3.10] H. Gajewski: On the Existence of Steady-State Carrier Distributions in Semiconductors. ZAMM (to appear).

[3.11] H. Gajewski: On Existence, Uniqueness and Asymptotic Behavior of Solutions of the Basic Equations for Carrier Transport in Semiconductors. ZAMM *65*, 101–108 (1985).

[3.12] D. Gilbarg, N. S. Trudinger: Elliptic Partial Differential Equations of Second Order, 2nd edn. Springer, Berlin (1984).

[3.13] E. Griepentrog, R. Maerz: Differential Algebraic Equations and Their Numerical Treatment. Teubner, Leipzig (1986).

[3.14] H. K. Gummel: A Self-Consistent Iterative Scheme for One-Dimensional Steady State Transistor Calculations. IEEE Trans. Electron. Devices *11*, 455–465 (1964).

[3.15] J. Henri, B. Louro: Singular Perturbation Theory Applied to the Electrochemistry Equations in the Case of Electroneutrality. Nonlinear Analysis TMA *13*, 787–801 (1989).

[3.16] J. W. Jerome: Consistency of Semiconductor Modelling: An Existence/Stability Analysis for the Stationary Van Roosbroeck System. SIAM J. Appl. Math. *45*, 565–590 (1985).

[3.17] T. Kerkhoven: A Proof of the Convergence of Gummel's Method for Realistic Device Geometries. SIAM J. Num. Anal. *23*, 1121–1137 (1986).

[3.18] P. Markowich: A Singular Perturbation Analysis of the Fundamental Semiconductor Device Equations. SIAM J. Appl. Math. *44*, 896–928 (1984).

[3.19] P. Markowich: The Stationary Semiconductor Device Equations. Springer, Wien-New York (1986).

[3.20] P. Markowich, P. Szmolyan: A System of Convection-Diffusion Equations with Small Diffusion Coefficient Arising in Semiconductor Physics. J. Diff. Equ. *81*, 234–254 (1989).

[3.21] P. Markowich, C. Ringhofer: A Singularly Perturbed Boundary Value Problem Modelling a Semiconductor Device. SIAM J. Appl. Math. *44*, 231–256 (1984).

[3.22] P. Markowich, C. Ringhofer: Stability of the Linearized Transient Semiconductor Device Equations. ZAMM *67*, 319–332 (1987).

[3.23] P. Markowich, C. Schmeiser: Uniform Asymptotic Representations of the Basic Semiconductor Device Equations. IMA J. Appl Math. *36*, 43–57 (1986).

[3.24] M. S. Mock: Analysis of Mathematical Models of Semiconductor Devices. Boole Press, Dublin (1983).

[3.25] R. E. O'Malley jr.: Introduction to Singular Perturbations. Academic Press, New York (1974).

[3.26] C. Ringhofer: An Asymptotic Analysis of a Transient *P-N* Junction Model. SIAM J. Appl. Math. *47*, 624–642 (1987).

[3.27] C. Ringhofer: A Singular Perturbation Analysis for the Transient Semiconductor Device Equations in One Space Dimension. IMA J. Appl. Math. *39*, 17–32 (1987).

[3.28] C. Ringhofer, C. Schmeiser: An Approximate Newton Method for the Solution of the Basic Semiconductor Device Equations. SIAM J. Num. Anal. *26*, 507–516 (1989).

[3.29] C. Ringhofer, C. Schmeiser: A Modified Gummel Method for the Basic Semiconductor Device Equations. IEEE Trans. CAD 7, 251–253 (1988).

[3.30] C. Schmeiser: On Strongly Reverse Biased Semiconductor Diodes. SIAM J. Appl. Math. (1989) (to appear).

[3.31] C. Schmeiser, R. Weiss: Asymptotic Analysis of Singular Singularly Perturbed Boundary Value Problems. SIAM J. Math. Anal. *17*, 560–579 (1986).

[3.32] T. Seidman, G. Troianiello: Time Dependent Solution of a Nonlinear System Arising in Semiconductor Theory. Nonlinear Analysis T.M.A. *9*, 1137–1157 (1985).

[3.33] T. Seidman, G. Troianiello: Time Dependent Solution of a Nonlinear System Arising in Semiconductor Theory, II: Boundedness and Periodicity. Nonlinear Analysis T.M.A. *10*, 491–502 (1986).

[3.34] S. Selberherr: Analysis and Simulation of Semiconductor Devices. Springer-Verlag, Wien New York (1984).

[3.35] S. M. Sze: Physics of Semiconductor Devices, 2nd edn. John Wiley & Sons, New York (1981).

[3.36] P. Szmolyan: Initial Transients of Solutions of the Semiconductor Device Equations. Nonlinear Analysis TMA (to appear).
[3.37] H. Triebel: Interpolation Theory, Function Spaces, Differential Operators. Verlag der Wissenschaften, Berlin (1973).
[3.38] J. H. Wilkinson: Rounding Errors in Algebraic Processes. Prentice Hall, New Jersey (1963).

Devices 4

4.1 Introduction

Depending on the semiconductor material, the doping profile and on the geometry, semiconductor devices show a variety of different kinds of electrical behaviour. This Chapter is concerned with an analysis of the performance of the practically most important devices.

The starting point is a mathematical model which is general enough to include the considered physical effects. Obviously there is not only one choice, and it is tempting to take a model which is as simple as possible. However, this involves the danger of neglecting important detail. On the other hand, a very complex model which represents a detailed picture of the physical world, possibly contains a good deal of superfluous information. Examples of such complicated models are the transport equations in Chapter 1 if the mean free path is very small compared to the length of the active region of a device. Methods for the systematic simplification of these models belong to the basic parts of the tool kit of the applied mathematician [4.7]. When the model has been chosen, an appropriate scaling is carried out and the relevant dimensionless parameters are identified (such as the scaled mean free path). After these preparatory steps, the formal machinery of perturbation theory is a tool for the systematic simplification of the problem. For the examples given in this Chapter the simplification goes far enough such that information about the relations between a few observable quantities can be obtained by analytical methods.

This systematic approach leads to mathematically justified results which in many cases confirm established text book formulas [4.22]. In the original derivation of these formulas the necessary simplifications are justified by physical arguments. The power of the formal approach is demonstrated by the fact that some of the results go significantly beyond the classical analysis. Striking examples can be found in the Sections 4.3 and 4.5 on the bipolar transistor and the thyristor.

For all the devices considered here, the smallness of the scaled mean free path is a valid assumption. The analysis will therefore be based on the drift diffusion equations with the appropriate models for the mobilities and the recombination-generation rate. A second assumption which holds throughout this Chapter is that the scaled minimal Debye length λ and the scaled intrinsic number δ^2 are small. These parameters have been defined in Chapter 3, where the role of the singular perturbation parameter λ is analyzed in detail. This analysis sheds light on the solution structure but does not simplify the problem sufficiently to obtain explicit information. Since we still have the small parameter δ^2 at our disposal, additional simplifications are possible.

In a complete mathematical model for a semiconductor device, boundary conditions reflect the interaction of the device with the circuit which it is imbedded in. In the following we shall mostly be interested in *dc operating conditions.* This means that the times between two switching events are long enough for quasi steady states to be reached. In this case most of the important information is contained in the *static voltage-current characteristics*, i.e. the relation between contact voltages and currents through the contacts under steady conditions. Only in one case (the Gunn diode, Section 4.8) a time dependent problem will be considered. This is due to a lack of maturity of the theory of those semiconductor devices, whose performance is based on dynamic effects. A unified account of a variety of the possible dynamic behaviour of semiconductors can be found in [4.16].

Static Voltage-Current Characteristics

Suppose that $\Omega \subseteq \mathbb{R}^k$, $k = 1, 2$ or 3, represents the geometry of a semiconductor device and that a stationary operating point is described by the drift-diffusion equations

$$\begin{aligned} &\lambda^2 \, \Delta V = n - p - C, \\ &J_n = \mu_n(\operatorname{grad} n - n \operatorname{grad} V), \qquad \operatorname{div} J_n = R, \\ &J_p = -\mu_p(\operatorname{grad} p + p \operatorname{grad} V), \qquad \operatorname{div} J_p = -R \end{aligned} \tag{4.1.1}$$

for $x \in \Omega$. Note that (4.1.1) is already in scaled form. For the scaling see Section 3.4.

We assume that the device has m Ohmic contacts $\Gamma_0, \dots, \Gamma_{m-1}$ and $l + 1 - m$ oxide regions $\Gamma_m, \dots, \Gamma_l$ attached to its boundary $\partial\Omega$ ($\Gamma_i \subseteq \partial\Omega$, $i = 0, \dots, l$). At the Ohmic contacts we have the boundary conditions

$$\begin{aligned} &n - p - C = 0, \qquad np = \delta^4, \qquad V = V_{bi} - U_i \\ &\text{at} \quad \Gamma_i, \qquad i = 0, \dots, m-1, \end{aligned} \tag{4.1.2}$$

where the built-in potential V_{bi} is given by

$$V_{bi} = \operatorname{areasinh}(C/2\delta^2).$$

Since we are free to choose a reference point for the potential we take $U_0 = 0$. Without going into the details of the description of the oxide regions at the moment we only mention that they lead to the additional applied voltages $U_m, \dots, U_l$ and that current flow into the oxide is not possible. Thus, the solutions of (4.1.1) depend on l applied voltages $U_1, \ldots, U_l$. The current I_i leaving the device through the contact Γ_i is given by

$$I_i = \int_{\Gamma_i} (J_n + J_p) \cdot \nu \, ds, \qquad i = 0, \ldots, m - 1,$$

where ν denotes the unit outward normal and s the $(k-1)$-dimensional Lebesgue measure. A static voltage-current characteristic is given by an equation of the form

$$F(I_i, U_1, \ldots, U_l) = 0 \tag{4.1.3}$$

representing an l-dimensional surface in the $(l+1)$-dimensional $(I_i, U_1, \ldots, U_l)$-space. Since the solution of the voltage driven problem (prescribed U_j, $j = 1, \ldots, l$) cannot be expected to be unique in general (see e.g. Section 4.5), it might be impossible to solve (4.1.3) for I_i. A complete description of the stationary behaviour is given by an l-dimensional surface in the $(l + m - 1)$-dimensional $(I_1, \ldots, I_{m-1}, U_1, \ldots, U_l)$-space. Since the total current density $J_n + J_p$ is divergence free no information is lost by not considering I_0 which can be computed from

$$\sum_{i=0}^{m-1} I_i = 0.$$

As a first step in the analysis of voltage-current characteristics we consider the situation close to *thermal equilibrium*. This means that the applied voltages and, therefore, also the currents through the contacts are small in absolute value. Our approach is a perturbation analysis which takes into account the smallness of both dimensionless parameters λ and δ. As in Section 3.4 we introduce the Slotboom variables [4.19] by the transformation

$$n = \delta^2 e^V u, \qquad p = \delta^2 e^{-V} v.$$

The variables u and v are related to the (scaled) quasi Fermi levels φ_n and φ_p by

$$u = e^{-\varphi_n}, \qquad v = e^{\varphi_p}.$$

The Ohmic contact boundary conditions (4.1.2) in terms of u and v read

$$u = e^{U_i}, \quad v = e^{-U_i}, \quad V = V_{bi} - U_i \quad \text{at } \Gamma_i, \quad i = 0, \ldots, m-1.$$

Since in thermal equilibrium $u \equiv v \equiv 1$ holds (see Section 3.4), u and v are well scaled as long as the applied voltages are not large compared to the thermal voltage (which is our reference quantity for voltages).
Close to thermal equilibrium large electric fields are only expected in small parts of the device. Thus, position dependent mobility models and the Shockley-Read-Hall recombination-generation term

$$R = \frac{np - \delta^4}{\tau_p(n + \delta^2) + \tau_n(p + \delta^2)}$$

are certainly sufficient for describing the relevant effects (see Chapter 2).

From the formula for the built-in potential we see that V is of the order of magnitude of $\ln \delta^{-2}$. The (in practice not very restrictive) assumption $\lambda^2 \ln \delta^{-2} \ll 1$ justifies the use of the zero space charge approximation which amounts to replacing λ^2 by zero in (4.1.1). If the zero space charge assumption is written in terms of the Slotboom variables it can be used to compute the potential in terms of u and v. Thus, it remains to consider the equations

$$\begin{aligned} J_n &= \mu_n \frac{C + \sqrt{C^2 + 4\delta^4 uv}}{2u} \operatorname{grad} u, &\quad \operatorname{div} J_n &= R, \\ J_p &= -\mu_p \frac{-C + \sqrt{C^2 + 4\delta^4 uv}}{2v} \operatorname{grad} v, &\quad \operatorname{div} J_p &= -R. \end{aligned} \tag{4.1.4}$$

As discussed in Chapter 3 there is an important difference between the Slotboom variables and the carrier densities. In the terminology of singular perturbation theory, u and v are *slow variables* whereas n and p are *fast variables*. This means that the derivatives of n and p are locally unbounded as $\lambda \to 0$ and that they converge pointwise to discontinuous functions. On the other hand, the derivatives of u and v are uniformly bounded as $\lambda \to 0$. This implies that their limits are continuous.

The above expressions contain the approximations

$$n = \frac{C + \sqrt{C^2 + 4\delta^4 uv}}{2}, \qquad p = \frac{-C + \sqrt{C^2 + 4\delta^4 uv}}{2} \tag{4.1.5}$$

for the carrier densities. For small values of δ they imply

$$n = C + O(\delta^4), \qquad p = O(\delta^4) \qquad \text{in } n\text{-regions } (C > 0)$$

and (4.1.6)

$$p = -C + O(\delta^4), \qquad n = O(\delta^4) \qquad \text{in } p\text{-regions } (C < 0).$$

Remark: Strictly speaking, the Landau order symbols O and o (see e.g. [4.1] for a definition) in (4.1.6) require the supplement "as $\delta \to 0$". Here and in the following the meaning of the order symbols is determined by phrases like "for small values of δ" instead. Note that the conclusion (4.1.6) relies on the assumption that the scaling of u and v is correct (i.e. $u, v = O(1)$), which is not justified a priorily. Nevertheless, this type of argument will be used repeatedly in this Chapter. If it leads to consistent approximations of solutions, this is considered as an—at least heuristic—justification.

(4.1.4) and (4.1.6) imply that J_n is $O(\delta^4)$ in p-regions and J_p is $O(\delta^4)$ in n-regions. Disregarding situations where an n- or p-region has more than one contact we conclude that the current densities are $O(\delta^4)$ throughout Ω. This is motivated by the following argument: Consider an n-region with one Ohmic contact. Then the electron current through the adjacent PN-junctions is $O(\delta^4)$. The same holds for the recombination-generation rate and, by the divergence theorem, for the current through the Ohmic contact. This implies that the electron current density is $O(\delta^4)$ throughout the n-

region. The same argument with the obvious changes holds for the hole current density in p-regions.
Appropriately scaled current densities are introduced by

$$J_n = \delta^4 J_{ns}, \qquad J_p = \delta^4 J_{ps}. \tag{4.1.7}$$

With these assumptions we are left with the problem, now expressed in terms of the rescaled current densities

$$\begin{aligned}
\delta^4 J_n &= \mu_n \frac{C + \sqrt{C^2 + 4\delta^4 uv}}{2u} \operatorname{grad} u, \\
\operatorname{div} J_n &= \frac{uv - 1}{\tau_p(n + \delta^2) + \tau_n(p + \delta^2)}, \\
\delta^4 J_p &= -\mu_p \frac{-C + \sqrt{C^2 + 4\delta^4 uv}}{2v} \operatorname{grad} v, \\
\operatorname{div} J_p &= -\frac{uv - 1}{\tau_p(n + \delta^2) + \tau_n(p + \delta^2)}
\end{aligned} \tag{4.1.8}$$

subject to the boundary conditions

$$\begin{aligned}
&u = e^{U_i}, \qquad v = e^{-U_i} \qquad \text{at} \qquad \Gamma_i, \quad i = 0, \ldots, m - 1, \\
&\frac{\partial u}{\partial v} = \frac{\partial v}{\partial v} = 0 \qquad \text{at} \qquad \partial\Omega_N = \partial\Omega \backslash \bigcup_{i=0}^{m-1} \Gamma_i.
\end{aligned} \tag{4.1.9}$$

In (4.1.8) we dropped the index s of the rescaled current densities.
Now we make use of the smallness of δ. Solutions of (4.1.8), (4.1.9) are approximated by letting δ tend to zero which gives

$$\operatorname{grad} u = 0, \qquad \operatorname{div}\left(\frac{\mu_p u}{C} \operatorname{grad} v\right) = \frac{uv - 1}{\tau_p C} \qquad \text{in } n\text{-regions}$$

and (4.1.10)

$$\operatorname{grad} v = 0, \qquad \operatorname{div}\left(\frac{\mu_n v}{|C|} \operatorname{grad} u\right) = \frac{uv - 1}{\tau_n |C|} \qquad \text{in } p\text{-regions.}$$

Our approximation amounts to replacing the quasi Fermi level of the majority carriers by a constant. Then the other quasi Fermi level can be computed by solving a linear elliptic equation.
Consider the problem of determining v in an n-region Ω_+. Assuming the constant value of u as given we observe that $v - 1/u$ solves a homogeneous linear equation. The boundary of Ω_+ splits into Neumann segments $\partial\Omega_+ \cap \partial\Omega_N$ and Dirichlet segments $S_1, \ldots, S_p$ which consist of PN-junctions adjacent to Ω_+ and (possibly) a contact. The boundary conditions at Ohmic contacts and the fact that v is constant in p-regions imply that v takes constant values $v_1, \ldots, v_p$ along $S_1, \ldots, S_p$, respectively. Introducing the functions $\varphi_1, \ldots, \varphi_p$ by solving

$$\operatorname{div}\left(\frac{\mu_p}{C}\operatorname{grad}\varphi_j\right) = \frac{\varphi_j}{\tau_p C}, \tag{4.1.11}$$
$$\varphi_j|_{S_i} = \delta_{ij}, \qquad i = 1, \ldots, p, \qquad \left.\frac{\partial \varphi_j}{\partial \nu}\right|_{\partial\Omega_+ \cap \partial\Omega_N} = 0$$

we can determine v in terms of the φ_j's:

$$v = \frac{1}{u} + \sum_{j=1}^{p}\left(v_j - \frac{1}{u}\right)\varphi_j .$$

As a consequence we obtain

$$J_p = -\sum_{j=1}^{p}\frac{\mu_p(uv_j - 1)}{C}\operatorname{grad}\varphi_j .$$

Since the φ_j do not depend on the applied voltages they only have to be computed once for a given device. In the same way formulas for u and J_n in p-regions can be obtained.
These ideas will be used in the following Sections for the computation of voltage-current characteristics. They will lead to explicit formulas if the functions φ_j for all n- and p-regions of the device are assumed to be known.

4.2 P-N Diode

The importance of understanding the electrical behaviour of P-N junctions is twofold. On one hand the P-N junction diode itself is a device with a wide variety of applications and on the other hand the performance of more complicated devices (such as the bipolar transistor and the thyristor) is based on the interaction of P-N junctions.
In this Section we consider steady states of a P-N diode with Ohmic contacts. The mathematical model consists of the steady state drift diffusion equations (4.1.1). The choice of models for the mobilities and the recombination-generation rate R will depend on the effects we want to take into account. We consider (4.1.1) in the domain $\Omega \subseteq \mathbb{R}^k$, $k = 1, 2$ or 3 which is the disjoint union of the p-region Ω_- $(C < 0)$, the n-region Ω_+ $(C > 0)$, and the P-N junction Γ (see Fig. 4.2.1). For simplification we assume an abrupt P-N junction, i.e. the doping profile C has jump discontinuities along Γ. At the contacts $\Gamma_0 \subseteq \partial\Omega \cap \partial\Omega_-$ and $\Gamma_1 \subseteq \partial\Omega \cap \partial\Omega_+$ the boundary conditions (4.1.2) are satisfied with $U_0 = 0$ and $U_1 = U$ where U is the applied voltage. Along the insulating (or artificial) part of the boundary $\partial\Omega_N = \partial\Omega \backslash (\Gamma_0 \cup \Gamma_1)$ homogeneous Neumann boundary conditions for V, n and p are prescribed. In this Section the singular perturbation analysis of the Sections 3.4 and 3.5 is extended by further simplifications of the approximating reduced and layer problems. These additional simplifications are carried out by exploiting the smallness of the scaled intrinsic number $\delta^2 = n_i/\tilde{C}$. The solution of the simplified problems allows to obtain explicit information on the geome-

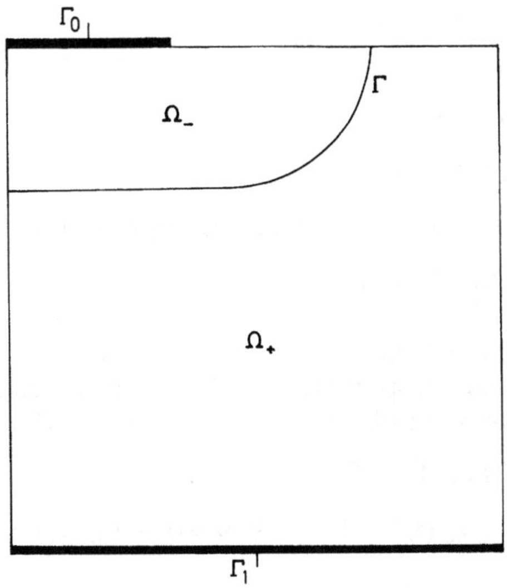

Fig. 4.2.1 Two-dimensional cross section of a P-N diode

try of the *depletion region* and on the *voltage-current characteristics* close to thermal equilibrium.
Additional information on the voltage-current characteristics in *high injection* and *large reverse bias* situations (in particular the critical voltages causing *punch-through* and *avalanche breakdown*) will be obtained by considering the simplified one-dimensional model

$$\Omega = (0, 1), \qquad \Gamma_0 = \{0\}, \qquad \Gamma_1 = \{1\}, \qquad \Gamma = \{x_0\} \tag{4.2.1}$$

with constant mobilities and a piecewise constant doping profile

$$C(x) = \begin{cases} -C_- < 0 & \text{for} \quad x < x_0, \\ C_+ > 0 & \text{for} \quad x > x_0. \end{cases} \tag{4.2.2}$$

The Depletion Region in Thermal Equilibrium

If the potential drop across a P-N junction is large compared to the thermal voltage a certain region around the P-N junction is depleted of charge carriers. As discussed in Section 3.5 this effect is especially important for reverse biased P-N junctions. In thermal equilibrium a depletion region occurs if the built-in potential is large enough which always is the case if a significantly large doping concentration is assumed.
Recalling the results from Section 3.4 for the thermal equilibrium problem

$$\begin{gathered} \lambda^2 \Delta V = 2\delta^2 \sinh V - C, \\ (V - V_{bi})|_{\Gamma_0 \cup \Gamma_1} = 0, \qquad \left.\frac{\partial V}{\partial v}\right|_{\partial\Omega_N} = 0, \end{gathered} \tag{4.2.3}$$

the solution is approximated by the solution

$$\bar{V}(x) = V_{bi}(x)$$

of the reduced equation

$$0 = 2\delta^2 \sinh \bar{V} - C$$

away from Γ whereas close to Γ it can be approximated by a layer solution

$$\hat{V}(s, \xi) = \lim_{\lambda \to 0} V(s, \lambda\xi). \tag{4.2.4}$$

In (4.2.4), V is written in terms of the local coordinates $(s, r) \in \mathbb{R}^{k-1} \times \mathbb{R}$ along Γ with the tangential components s and the normal component r ($r > 0$ in Ω_+). $\xi = r/\lambda$ is a fast variable and $\hat{V}$ is a solution of the layer problem

$$\begin{aligned} &\partial_\xi^2 \hat{V} = 2\delta^2 \sinh \hat{V} - \hat{C} \\ &\hat{V}(s, -\infty) = V_{bi}(s, 0-), \qquad \hat{V}(s, \infty) = V_{bi}(s, 0+), \end{aligned} \tag{4.2.5}$$

where the notation

$$\hat{C}(s, \xi) := \begin{cases} -C_-(s) = C(s, 0-) & \text{for } \xi < 0, \\ C_+(s) = C(s, 0+) & \text{for } \xi > 0 \end{cases}$$

was introduced. A detailed discussion of the formalism leading to (4.2.5) is given in Section 3.4.

Since the tangential component s only appears as a parameter, (4.2.5) is essentially a one-dimensional problem. An approximate solution can be obtained by exploiting the smallness of δ^2 for significantly large doping concentrations. We rewrite the boundary conditions at $\xi = \pm\infty$ as

$$\begin{aligned} \hat{V}(s, \infty) &= \ln \delta^{-2} + \ln C_+ + O(\delta^4), \\ \hat{V}(s, -\infty) &= -\ln \delta^{-2} - \ln C_- + O(\delta^4). \end{aligned} \tag{4.2.6}$$

This shows that $\hat{V}$ is badly scaled because its boundary values at $\pm\infty$ become unbounded as $\delta^2 \to 0$. Instead of δ^2 we introduce the small parameter $\gamma = (\ln \delta^{-2})^{-1}$ and the scaling

$$W = \gamma \hat{V}$$

of the dependent variable. The new reference value

$$U_T/\gamma = U_T \ln(\tilde{C}/n_i)$$

for voltages is of the order of magnitude of the built-in voltage. For small values of γ it is large compared to the thermal voltage. After substitution of W in (4.2.5) the factor $1/\gamma$ appears in front of the second derivative. It can be eliminated by the rescaling

$$\eta = \xi\sqrt{\gamma}$$

of the independent variable. The resulting problem reads

$$\partial_\eta^2 W = \exp\left(\frac{W-1}{\gamma}\right) - \exp\left(\frac{-W-1}{\gamma}\right) - \hat{C}, \tag{4.2.7}$$

$$W(\infty) = 1 + \gamma \ln C_+ + \mathrm{TST}, \tag{4.2.8}$$

$$W(-\infty) = -1 - \gamma \ln C_- + \mathrm{TST}, \tag{4.2.9}$$

where TST stands for *t*ranscendentally *s*mall *t*erms, i.e. terms which are $O(e^{-c/\gamma})$ with a positive constant c. In general, the symbol "TST" denotes terms which are small compared to any power of the small parameter considered.

Since the derivative of the right-hand side of (4.2.7) with respect to W is positive, the maximum principle (see [4.13]) can be applied. Lemma 3.3.14 implies the estimates

$$-1 + O(\gamma) \leqslant W \leqslant 1 + O(\gamma)$$

which in turn imply the boundedness of the carrier densities $\exp[(W-1)/\gamma]$ and $\exp[(-W-1)/\gamma]$ uniformly in γ. By the differential equation (4.2.7) $\partial_\eta^2 W$ is also uniformly bounded, which can be used for a justification of the limiting process $\gamma \to 0$ in (4.2.7), (4.2.8), (4.2.9) (see [4.2]). Going to the limit we obviously have

$$\begin{aligned} &-1 \leqslant W \leqslant 1, \qquad W(-\infty) = -1, \qquad W(\infty) = 1, \\ &\partial_\eta^2 W + \hat{C} \leqslant 0 \quad \text{for} \quad W < 1, \\ &\partial_\eta^2 W + \hat{C} \geqslant 0 \quad \text{for} \quad W > -1, \end{aligned} \tag{4.2.10}$$

and W is continuously differentiable by the boundedness of $\partial_\eta^2 W$. The problem (4.2.10) is a standard *double obstacle problem.* This term originates from an interpretation of W as the displacement of a string under the action of a lateral load $\hat{C}$ which lies between two obstacles represented by the bounds 1 and -1 in our situation. It can be shown that (4.2.10) has a unique solution by rewriting it as a variational inequality [4.2]. The above treatment outlines the proof of a convergence result which is well known, because (4.2.7), (4.2.8), (4.2.9) is a standard penalisation of (4.2.10) (see [4.2]).

The solution of (4.2.10) splits Ω into three subdomains: the *coincidence sets* where $W = \pm 1$ holds and the *noncoincidence set* where $-1 < W < 1$ and $\partial_\eta^2 W + \hat{C} = 0$ holds. Leaving the solution of (4.2.10) to the reader (see Problem 4.2) we only state the result that the noncoincidence set is given by the η-interval

$$\left(-2/\sqrt{C_-(C_-/C_+ + 1)},\ 2/\sqrt{C_+(C_+/C_- + 1)}\right) \tag{4.2.11}$$

with the length $2\sqrt{1/C_+ + 1/C_-}$.

Note that the limiting carrier densities

$$\lim_{\gamma\to 0} \exp\left(\frac{W-1}{\gamma}\right), \qquad \lim_{\gamma\to 0} \exp\left(\frac{-W-1}{\gamma}\right)$$

vanish in the noncoincidence set of (4.2.10) which means that the noncoin-

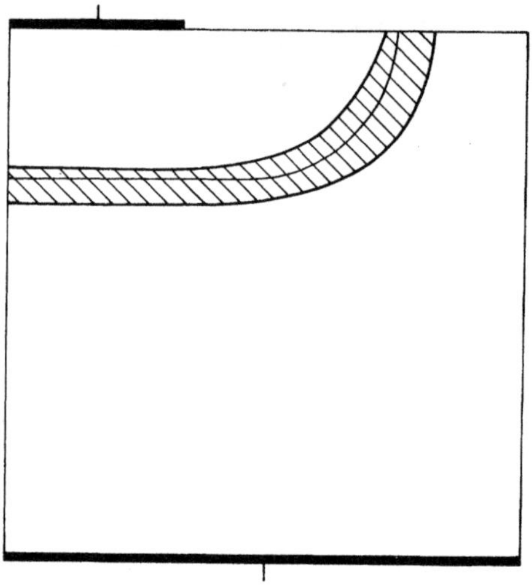

Fig. 4.2.2 The depletion region in thermal equilibrium

cidence set is depleted of charge carriers. An interval of the form (4.2.11) is obtained for every point along the P-N junction Γ. The depletion region is, thus, given by the union of the η-intervals (4.2.11) taken along the values of the tangential coordinate s along Γ (see Fig. 4.2.2).

With the approximation $U_{bi} = 2U_T/\gamma$ of the unscaled built-in voltage, the unscaled width of the depletion region is given by

$$\sqrt{\frac{2\varepsilon_s}{q} U_{bi}(1/C_+ + 1/C_-)}\,.$$

Remark: Here and in the sequel was denote scaled and unscaled quantities by the same symbols. Formulas in terms of unscaled quantities are explicitely announced in the text in order to avoid misinterpretations.

The above expression for the width of the depletion region can be found in textbooks on semiconductor device physics (see e.g. [4.22]). There it is obtained by the a priori assumption that a depleted region exists, and that zero space charge prevails in the rest of the device.

We conclude this Paragraph with a mathematical remark. The results have been obtained by the limiting process $\lambda \to 0$ followed by $\delta \to 0$ applied to the thermal equilibrium problem (4.2.3). We point out that the limits commute. By introducing W in (4.2.3) and letting $\delta \to 0$ we arrive at a multi-dimensional double obstacle problem which contains the singular perturbation parameter $\lambda/\sqrt{\gamma}$. An analysis of this singularly perturbed obstacle problem can be found in [4.15]. There it is shown that its solution can be approximated by solving the one-dimensional obstacle problem (4.2.10) in the junction layer. Apart from these two approaches an analysis where λ and δ tend to zero simultaneously is also possible (see [4.11]).

Strongly Asymmetric Junctions

The performance of most semiconductor devices relies on the interaction of strongly asymmetric P-N junctions, i.e. junctions with doping levels of different orders of magnitude in the *n*- and *p*-regions. In this Paragraph an analysis of the junction layer problem (4.2.7), (4.2.8), (4.2.9) for strongly asymmetric junctions is given. The presentation is based on the work of Ward [4.25] who also pointed out the similarity of the problem to a MIS diode (see Section 4.6) in strong inversion. In this paragraph, advanced concepts of singular perturbation theory are introduced. They are also important for the analysis in the Sections 4.6 and 4.7 on the MIS diode and the MOSFET.

In the case of an N^+P junction $C_+ \gg C_-$ holds. The depletion region (4.2.11) obtained in the preceding Paragraph spreads mostly into the *p*-region. Concentrating on the potential in the *p*-region, we assume that the doping concentration in the *p*-region has been chosen as reference value in the scaling. This implies $C_- = 1$. Note that this is in contrast to the scaling philosophy adopted in the biggest part of this book where the doping profile is scaled to the maximal value 1. It is now justified since the main interest lies in an analysis of the region with lower doping. The problem (4.2.7), (4.2.8), (4.2.9) will again be analyzed by letting γ tend to zero. The asymmetry of the junction is taken into account by keeping the quantity

$$A := \gamma \ln C_+$$

fixed as $\gamma \to 0$ (A doping concentration of $10^{18}/\text{cm}^3$ in the *n*-region and 10^{14} cm^3 in the *p*-region for Si at room temperature leads to $A \sim 1$).

The first step is the computation of the value of the potential at the junction $\eta = 0$. Multiplication of (4.2.7) by $\partial_\eta W$ and integration gives

$$(\partial_\eta W)^2/2 = \gamma \exp\left(\frac{W-1}{\gamma}\right) + \gamma \exp\left(\frac{-W-1}{\gamma}\right) - \hat{C}W + k \tag{4.2.12}$$

with

$$k = \begin{cases} e^{A/\gamma}(A + 1 - \gamma) + \text{TST} & \text{for} \quad \eta > 0, \\ 1 - \gamma + \text{TST} & \text{for} \quad \eta < 0. \end{cases}$$

The values of k have been determined from (4.2.12) evaluated at $\eta = \pm\infty$ and (4.2.8), (4.2.9). Requiring continuity of the derivative of W at $\eta = 0$, the value of W at the junction

$$W(0) = A + 1 - \gamma + \text{TST} \tag{4.2.13}$$

can be computed. Since this value is equal to $W(\infty)$ up to $O(\gamma)$, most of the variation of the potential takes place in the *p*-region as expected. There, (4.2.12) can be rewritten as

$$\partial_\eta W/\sqrt{2} = \sqrt{\gamma \exp\left(\frac{W-1}{\gamma}\right) + \gamma \exp\left(\frac{-W-1}{\gamma}\right) + W + 1 - \gamma + \mathrm{TST}}\,. \tag{4.2.14}$$

Since the initial value (4.2.13) is larger than 1, the derivative of W at $\eta = 0$ becomes unbounded as $\gamma \to 0$, and (4.2.14) is singularly perturbed. The asymptotic analysis below will show that the solution has multiple layers. A very thin initial layer is followed by a transition layer describing the behaviour of the solution close to $W = 1$. This transition layer connects the initial layer to a depletion region where W is between -1 and 1. The edge of the depletion region is a free boundary where W takes the value -1 which is asymptotically equal to its value at $\eta = -\infty$.
Initially, the solution is expected to vary fast. An approximation depends on a layer variable

$$\sigma = \frac{\eta}{\delta(\gamma)}$$

where $\delta(\gamma) = o(1)$ holds. A differential equation for the approximation is obtained by letting $\gamma \to 0$ in the transformed differential equation (4.2.14). However, for nonlinear equations the limit in general depends on the values which the solution takes. For (4.2.14) in particular, the relative orders of magnitude of the terms under the square root depend on the value of W. It is, for example, possible to generate different situations by choosing different values W_0 in

$$W = W_0 + \gamma y \tag{4.2.15}$$

where the new variable y has been introduced. The choices of $\delta(\gamma)$ and of W_0 are governed by an heuristic principle: *The limiting equation should contain as many terms as possible.* A limiting equation which satisfies this requirement is called a *significant degeneration* (see [4.1], sometimes also *distinguished limit* [4.5]).
The initial condition (4.2.13) requires the choice $W_0 = A + 1$ and, thus,

$$W_{in}(\sigma) = A + 1 + \gamma y(\sigma)$$

for the initial approximation W_{in}. With this choice, the heuristic principle implies $\delta(\gamma) = \sqrt{\gamma}\exp(-A/2\gamma)$. With the fast variable

$$\sigma = \frac{\eta}{\sqrt{\gamma}} e^{A/2\gamma},$$

we obtain the degeneration

$$\partial_\sigma y = \sqrt{2} e^{y/2}, \qquad y(0) = -1 \tag{4.2.16}$$

with the solution

$$y = -2 \ln(\sqrt{e} - \sigma/\sqrt{2}).$$

The approximation (4.2.16) has been obtained by dropping all the terms under the square root in (4.2.14) except the first. Obviously, this term only dominates as long as W is larger than one. The transition of W through values close to 1 is governed by a significant degeneration of (4.2.14), obtained by setting

$$W_{tr}(\tau) = 1 + \gamma \ln \gamma^{-1} + \gamma z(\tau), \qquad \tau = \eta/\gamma$$

and letting γ tend to zero:

$$\partial_\tau z = \sqrt{2(e^z + 2)}\,. \tag{4.2.17}$$

The independent variable τ is fast compared to the original variable η but slow compared to the initial layer variable σ. The solutions of (4.2.17) are given by

$$z = -\ln\left(\tfrac{1}{2}\sinh^2(\tau + c)\right), \tag{4.2.18}$$

where c is a constant of integration. This constant is determined by a procedure called *matching*, which has already been introduced in Section 3.4. The underlying idea is that the domains of validity of the approximations W_{in} and W_{tr} overlap. In its simplest form, matching would amount to equating the limit of W_{in} as $\sigma \to -\infty$ to the value of W_{tr} at $\tau = 0$. However, the limit of W_{in} does not exist and we have to proceed differently. Rewriting W_{in} in terms of the slower variable τ gives

$$W_{\text{in}}(\tau\sqrt{\gamma}e^{A/2\gamma}) \sim 1 + \gamma \ln \gamma^{-1} - \gamma \ln(\tau^2/2). \tag{4.2.19}$$

Obviously, this coincides with W_{tr} for small values of τ if we set $c = 0$ in (4.2.18).

The transistion layer approximation W_{tr} takes W from values larger than 1 to values smaller than 1. For W between 1 and -1, the carrier densities vanish as $\gamma \to 0$. The approximation $W_{\text{depl}}(\eta)$ satisfies the limiting equation

$$\partial_\eta W_{\text{depl}} = \sqrt{2(W_{\text{depl}} + 1)} \tag{4.2.20}$$

with the general solution $W_{\text{depl}} = -1 + (\eta/\sqrt{2} + c)^2$. The free constant c can again be computed by matching. A straightforward computation gives

$$\lim_{\gamma \to 0} W_{\text{tr}}(\eta/\gamma) = 1 + 2\eta\,. \tag{4.2.21}$$

This agrees for small η with W_{depl} if we set $c = \sqrt{2}$. Thus, W_{depl} is given by

$$W_{\text{depl}} = -1 + (\eta + 2)^2/2\,.$$

Since the solution of (4.2.14) is obviously monotone, W_{depl} looses its validity at the point $\eta = -2$, which represents the edge of the depletion region. The monotonicity also implies that for $\eta < -2$ the potential can be approximated by the constant value -1. This is a continuously differentiable extension of the depletion layer solution W_{depl} and a singular solution of the differential equation (4.2.20).

For the carrier densities we obtain that in a thin region adjacent to the P-N junction the electron density is much larger than the hole density. This thin layer is followed by a depletion region with a sharp edge. Outside the depletion region the hole density is approximately equal to the doping concentration and the electron density is much smaller (compare to (4.1.6)).

The Voltage-Current Characteristic Close to Thermal Equilibrium

The ideal diode operates like a perfect valve. The total current through the device is zero in reverse bias (i.e. for negative values of applied voltage U) and grows with the applied voltage in forward bias (see Fig. 4.2.3). In this Paragraph we shall demonstrate that—at least close to thermal equilibrium—the P-N diode approximately exhibits this behaviour.
For the P-N diode problem, (4.1.10) implies

$$u|_{\Omega_+} = e^U, \qquad v|_{\Omega_-} = 1, \tag{4.2.22}$$

$$\operatorname{div}\left(\frac{\mu_n}{|C|}\operatorname{grad} u\right) = \frac{u-1}{\tau_n|C|} \quad \text{in } \Omega_-, \quad u|_{\Gamma_0} = 1, \quad u|_\Gamma = e^U, \tag{4.2.23}$$

$$e^U \operatorname{div}\left(\frac{\mu_p}{C}\operatorname{grad} v\right) = \frac{e^U v - 1}{\tau_p C} \quad \text{in } \Omega_+, \quad v|_\Gamma = 1, \quad v|_{\Gamma_1} = e^{-U}. \tag{4.2.24}$$

The solutions of (4.2.23) and (4.2.24) are given by

$$u|_{\Omega_-} = (e^U - 1)\varphi_1 + 1, \qquad J_n|_{\Omega_-} = (e^U - 1)\frac{\mu_n}{|C|}\operatorname{grad}\varphi_1,$$

$$v|_{\Omega_+} = (1 - e^{-U})\varphi_2 + e^{-U}, \qquad J_p|_{\Omega_+} = (1 - e^U)\frac{\mu_p}{C}\operatorname{grad}\varphi_2,$$

where the functions φ_1 and φ_2 are the unique solutions of the problems

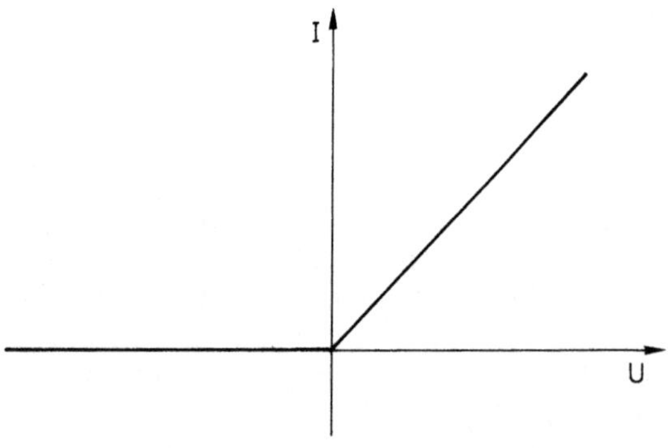

Fig. 4.2.3 Ideal diode characteristic

$$\operatorname{div}\left(\frac{\mu_n}{|C|}\operatorname{grad}\varphi_1\right)=\frac{\varphi_1}{\tau_n|C|}\qquad \text{in}\quad \Omega_-,$$

$$\varphi_1|_{\Gamma_0}=0,\qquad \varphi_1|_{\Gamma}=1,\qquad \left.\frac{\partial\varphi_1}{\partial\nu}\right|_{\partial\Omega_N\cap\partial\Omega_-}=0$$

and

$$\operatorname{div}\left(\frac{\mu_p}{C}\operatorname{grad}\varphi_2\right)=\frac{\varphi_2}{\tau_p C}\qquad \text{in}\quad \Omega_+,$$

$$\varphi_2|_{\Gamma}=1,\qquad \varphi_2|_{\Gamma_1}=0,\qquad \left.\frac{\partial\varphi_2}{\partial\nu}\right|_{\partial\Omega_N\cap\partial\Omega_+}=0.$$

The current I through the device in the direction from Γ_0 to Γ_1 is obtained by computing the current across the P-N junction Γ. If ν denotes the unit normal vector along Γ oriented into Ω_+, we have

$$\begin{aligned}I&=\int_\Gamma (J_n+J_p)\cdot\nu\,ds\\&=(e^U-1)\int_\Gamma\left(\frac{\mu_n}{|C|}\operatorname{grad}\varphi_1-\frac{\mu_p}{C}\operatorname{grad}\varphi_2\right)\cdot\nu\,ds.\end{aligned}$$

The first term on the right-hand side can be rewritten as

$$\int_{\partial\Omega_-}\varphi_1\frac{\mu_n}{|C|}\operatorname{grad}\varphi_1\cdot\nu\,ds=\int_{\Omega_-}\frac{1}{|C|}\left(\mu_n|\operatorname{grad}\varphi_1|^2+\frac{\varphi_1^2}{\tau_n}\right)dx$$

using integration by parts. An analogous computation for the second term finally leads to a voltage-current characteristic of the form

$$I=I_s(e^U-1),\tag{4.2.25}$$

which approximately exhibits the behaviour described at the beginning of this Paragraph. Under reverse bias ($U<0$) the current does not vanish but it is bounded below by the leakage current

$$\begin{aligned}I_s=&\int_{\Omega_-}\frac{1}{|C|}\left(\mu_n|\operatorname{grad}\varphi_1|^2+\frac{\varphi_1^2}{\tau_n}\right)dx\\&+\int_{\Omega_+}\frac{1}{|C|}\left(\mu_p|\operatorname{grad}\varphi_2|^2+\frac{\varphi_2^2}{\tau_p}\right)dx\end{aligned}$$

(see Fig. 4.2.4).

Our result (4.2.25) is a scaled version of the famous *Shockley equation* [4.18] which in terms of unscaled variables reads

$$I=I_s(e^{U/U_T}-1).$$

In the simplified one-dimensional case (4.2.1), (4.2.2) the problems for φ_1 and φ_2 can be solved explicitely (see Problem 4.4) and the unscaled saturation

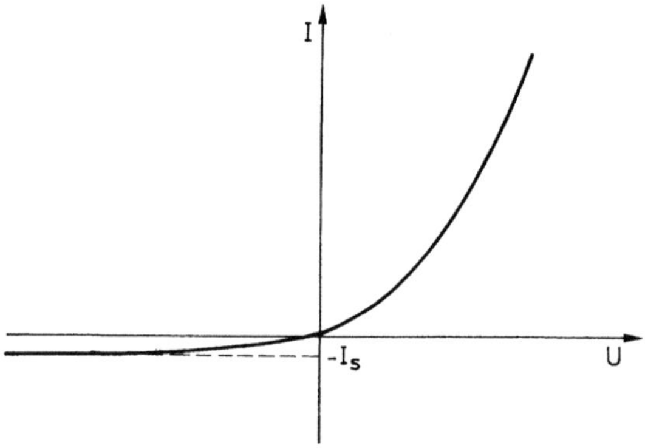

Fig. 4.2.4 Voltage-current characteristic

current is then given by

$$I_s = q\sqrt{U_T}n_i^2\left(\frac{1}{C_-}\sqrt{\frac{\mu_n}{\tau_n}}\coth\left(\frac{x_0}{\sqrt{\mu_n\tau_nU_T}}\right)\right.$$

$$\left.+\frac{1}{C_+}\sqrt{\frac{\mu_p}{\tau_p}}\coth\left(\frac{L-x_0}{\sqrt{\mu_p\tau_pU_T}}\right)\right), \tag{4.2.26}$$

where L denotes the unscaled length of the device. For devices, where the lengths x_0 and $L - x_0$ of the p- and n-regions are large compared to the diffusion lengths $\sqrt{\mu_n\tau_nU_T}$ and $\sqrt{\mu_p\tau_pU_T}$, the coth-terms can be replaced by 1 which leads to the saturation current of the classical Shockley equation. In the case of a very short device (i.e. the arguments of the coth-terms are small) (4.2.26) reduces to

$$I_s = qU_Tn_i^2\left(\frac{\mu_n}{x_0C_-}+\frac{\mu_p}{(L-x_0)C_+}\right)$$

for the unscaled saturation current. This formula shows that for very short devices the characteristic close to thermal equilibrium is essentially independent of recombination-generation effects.

In the derivation of the Shockley equation we used the assumptions that the zero space charge approximation holds (with the exception of a thin layer) and that the error made in the approximation (4.2.22), (4.2.23), (4.2.24) remains small. The first assumption is violated in strong reverse bias when the width of the depletion region is not small compared to the length of the device. The second assumption is satisfied as long as the (scaled) applied voltage U is small enough for e^U to be small compared to δ^{-4}, i.e. if U is small compared to the built-in potential. If this is not the case the assumption that the charge carrier densities are close to their equilibrium values does not hold any more. This situation is called *high injection.*

High Injection—A Model Problem

In [4.10] the solution of the zero space charge problem for a simple model device, the one-dimensional *symmetric diode*, was computed explicitely. As no further approximations are necessary in this case the resulting current-voltage characteristic remains valid in high injection. The necessary computations are outlined below.

Suppose that the center of the device is represented by the point $x = 0$, that the doping profile is piecewise constant and an odd function. We take constant mobilities and neglect the effects of recombination-generation. Symmetry arguments allow the ansatz

$$V(-x) = V(x), \qquad n(-x) = p(x),$$
$$J_n/\mu_n = J_p/\mu_p = J/(\mu_n + \mu_p) =: I,$$

which in turn allows to compute the solution by considering only half of the device (say, the n-region). Thus, the problem is posed on the scaled n-region (0, 1) and the boundary conditions

$$V(0) = 0, \qquad n(0) = p(0)$$

at $x = 0$ follow from the symmetry ansatz. As expected the solution of the resulting problem has a boundary layer at zero and the solution of the zero space charge approximation satisfies

$$\begin{aligned} 0 &= n - p - 1, \\ I &= n' - nV', \\ I &= -p' - pV' \end{aligned} \tag{4.2.27}$$

subject to the boundary conditions

$$\begin{aligned} &V(0) - \ln n(0) = -V(0) - \ln p(0), \\ &V(1) = V_{bi}(1) - U/2, \qquad p(1) = \tfrac{1}{2}(-1 + \sqrt{1 + 4\delta^4}), \end{aligned} \tag{4.2.28}$$

where the constant 1 in the reduced Poisson equation is the scaled doping concentration and

$$V_{bi}(1) = \operatorname{areasinh}(1/2\delta^2)$$

holds. The boundary condition at $x = 0$ can be interpreted as the symmetry condition $\varphi_n(0) = -\varphi_p(0)$ for the reduced quasi Fermi potentials.

After integration of (4.2.27) subject to the boundary conditions at $x = 1$ the carrier densities and the potential can be expressed in terms of I:

$$\begin{aligned} n &= p + 1 = \tfrac{1}{2}(1 + \sqrt{1 + 4\delta^4 + 4I(1 - x)}\,), \\ V &= V_{bi} - U/2 + \sqrt{1 + 4\delta^4 + 4I(1 - x)} - \sqrt{1 + 4\delta^4}\,. \end{aligned} \tag{4.2.29}$$

The voltage-current characteristic in implicit form is obtained by substitution of (4.2.29) into the boundary condition at $x = 0$:

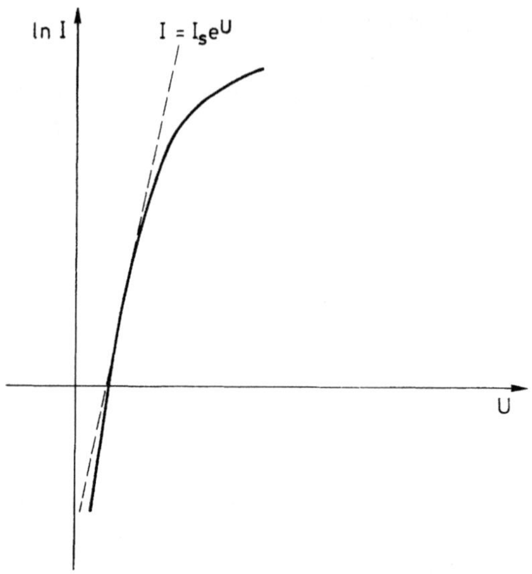

Fig. 4.2.5 Forward bias characteristic (logarithmic scale)

$$U/2 = V_{bi} - \ln\frac{1+\sqrt{1+4\delta^4+4I}}{\sqrt{4\delta^4+4I}} + \sqrt{1+4\delta^4+4I} - \sqrt{1+4\delta^4}\,. \tag{4.2.30}$$

With $I = O(\delta^4)$ this equation includes the Shockley equation as the limiting case $\delta^4 \to 0$. In a high injection situation it predicts a quadratic dependence of the current on the applied voltage which, in unscaled variables, takes the form

$$J \sim \frac{q\tilde{C}(\mu_n + \mu_p)}{8LU_T} U^2 .$$

This equation illustrates the result that the growth of the voltage-current characteristic slows down in high injection, which is well know from experiments and numerical computations (see Fig. 4.2.5). The solution (4.2.29) also shows that the charge carrier densities grow linearly with the applied bias and that a significant potential drop takes place throughout the device as opposed to low injection.

Large Reverse Bias

The Shockley equation looses its validity as soon as the depletion region covers a significant part of the device. This is the case if the scaled reverse bias $-U$ is $O(\lambda^{-2})$. As discussed in Chapter 3 a scaling different from that in (4.1.1) has to be used in this situation. In order to be able to obtain

explicit information on the voltage-current characteristic we restrict ourselves to the one-dimensional situation (4.2.1), (4.2.2).
With the transformation

$$\varphi = \lambda^2(V - V_{bi}(0))$$

(4.1.1) becomes

$$\begin{aligned} \varphi'' &= n - p - C, \\ \lambda^2 J_n &= \mu_n(\lambda^2 n' - n\varphi'), \qquad J_n' = R, \\ \lambda^2 J_p &= -\mu_p(\lambda^2 p' + p\varphi'), \qquad J_p' = -R \end{aligned} \tag{4.2.31}$$

subject to the boundary conditions

$$\begin{aligned} &\varphi(0) = 0, \qquad \varphi(1) = \varphi_1, \\ &n(0) - p(0) + C_- = n(1) - p(1) - C_+ = 0, \\ &n(0)p(0) = n(1)p(1) = \delta^4, \end{aligned} \tag{4.2.32}$$

where $\varphi_1 = \lambda^2(V_{bi}(1) - V_{bi}(0) - U) > 0$ holds. For R we take the Shockley-Read-Hall term

$$R = \frac{np - \delta^4}{\tau_p(n + \delta^2) + \tau_n(p + \delta^2)}.$$

As in Section 3.5 an approximate solution is obtained by going to the limit $\lambda \to 0$ in (4.2.31), (4.2.32). The reduced equations

$$\begin{aligned} \varphi'' &= n - p - C, \\ n\varphi' &= p\varphi' = 0 \end{aligned}$$

imply that the device splits into a depletion region ($n = p = 0$) and a zero space charge region ($n - p - C = 0$). The reduced potential φ is a solution of the double obstacle problem

$$\begin{aligned} &0 \leqslant \varphi \leqslant \varphi_1, \qquad \varphi(0) = 0, \qquad \varphi(1) = \varphi_1, \\ &\varphi'' + C \leqslant 0 \quad \text{for} \quad \varphi < \varphi_1, \\ &\varphi'' + C \geqslant 0 \quad \text{for} \quad \varphi > 0 \end{aligned} \tag{4.2.33}$$

with the noncoincidence set (see Problem 4.6)

$$(x_l, x_r) = \left(x_0 - \sqrt{\frac{2\varphi_1}{C_-(1 + C_-/C_+)}},\ x_0 + \sqrt{\frac{2\varphi_1}{C_+(1 + C_+/C_-)}} \right), \tag{4.2.34}$$

which represents the depletion region. It has the length

$$x_r - x_l = \sqrt{2\varphi_1(1/C_+ + 1/C_-)}\,. \tag{4.2.35}$$

The limiting problem in $(0, x_l) \cup (x_r, 1)$ is completed by (see Chapter 3)

$$0 = n - p - C,$$

$$(np)' = \frac{J_n}{\mu_n} p - \frac{J_p}{\mu_p} n, \tag{4.2.36}$$

$$J_n' = -J_p' = \frac{np - \delta^4}{\tau_p(n + \delta^2) + \tau_n(p + \delta^2)}$$

subject to the auxiliary conditions

$$(np)(0) = (np)(1) = \delta^4, \qquad (np)(x_l) = (np)(x_r) = 0,$$

$$J_n(x_r) - J_n(x_l) = -\frac{\delta^2}{\tau_n + \tau_p} \sqrt{2\varphi_1(1/C_+ + 1/C_-)}, \tag{4.2.37}$$

$$J_p(x_r) - J_p(x_l) = \frac{\delta^2}{\tau_n + \tau_p} \sqrt{2\varphi_1(1/C_+ + 1/C_-)}.$$

The fact that np is a slow variable, which does not have a jump across the edges of the depletion region, explains the equations for np at x_l and x_r. The right hand sides of the last two equations are caused by generation effects in the depletion region and can be obtained by integrating the differential equations of the current densities.

For $\delta^2 = 0$ (4.2.36), (4.2.37) obviously has the solution $np = J_n = J_p = 0$. This motivates the rescaling $\delta^2 w = np$, $\delta^2 I_n = J_n$, $\delta^2 I_p = J_p$. The transformed problem is approximated by going to the limit $\delta^2 \to 0$ which implies for the charge carrier densities:

$$n = \max(0, C), \qquad p = \max(0, -C) \qquad \text{in} \quad (0, x_l) \cup (x_r, 1).$$

The variables w, I_n, I_p solve the problem

$$w' = \begin{cases} I_n C_-/\mu_n & \text{in} \quad (0, x_l), \\ -I_p C_+/\mu_p & \text{in} \quad (x_r, 1), \end{cases}$$

$$I_n' = -I_p' = \begin{cases} w/\tau_n C_- & \text{in} \quad (0, x_l), \\ w/\tau_p C_+ & \text{in} \quad (x_r, 1), \end{cases}$$

subject to the auxiliary conditions

$$w(0) = w(x_l) = w(x_r) = w(1) = 0,$$

$$I_n(x_r) - I_n(x_l) = -\frac{1}{\tau_n + \tau_p} \sqrt{2\varphi_1(1/C_+ + 1/C_-)},$$

$$I_p(x_r) - I_p(x_l) = \frac{1}{\tau_n + \tau_p} \sqrt{2\varphi_1(1/C_+ + 1/C_-)}.$$

The solution of this problem only involves the integration of linear differential equations with constant coefficients and is left to the reader. A typical charge carrier distribution is depicted in Fig. 4.2.6.

The only source of current flow is generation in the depletion region and

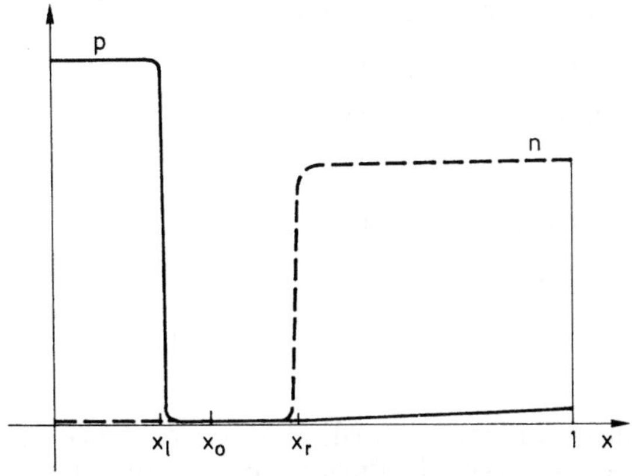

Fig. 4.2.6 Charge carrier distribution in strong reverse bias

the approximate current-voltage characteristic reads

$$J = \frac{\delta^2}{\tau_n + \tau_p} \sqrt{2\varphi_1(1/C_+ + 1/C_-)}\,. \tag{4.2.38}$$

In terms of unscaled quantities this gives

$$J = \frac{n_i}{\tau_n + \tau_p} \sqrt{2\varepsilon_s q(1/C_+ + 1/C_-)(-U)}\,,$$

which is a standard textbook formula [4.22]. It states that the leakage current caused by generation in the depletion region is proportional to the width of the depletion region and, thus, to the square root of the applied bias.

Avalanche Breakdown

In large reverse bias impact ionization might cause junction breakdown. This phenomenon can be explained by analysing (4.2.31), (4.2.32) with the generation term

$$R = -\alpha_n \exp\left(-\frac{\lambda E_n}{|\varphi'|}\right)|J_n| - \alpha_p \exp\left(-\frac{\lambda E_p}{|\varphi'|}\right)|J_p|\,.$$

This is a scaled version of the impact ionization model (2.6.4), (2.6.5). The factor λ in the scaled critical field strengths λE_n, λE_p has been introduced because λE_n and λE_p are of the order of magnitude of the scaled Debye length λ in typical cases. The dimensionless parameters $E_{n/p}$, $\alpha_{n/p}$ are assumed to take moderate values.

For $\lambda \to 0$ the reduced potential is that of the preceding Paragraph and, obviously, the depletion region is the same, too. Introducing

$$w = \lim_{\delta^4 \to 0} (np\delta^{-4}), \quad I_n = \lim_{\delta^4 \to 0} (J_n\delta^{-4}), \qquad I_p = \lim_{\delta^4 \to 0} (J_p\delta^{-4})$$

the problem in the zero space charge regions $(0, x_l)$ and $(x_r, 1)$ becomes

$$
\begin{aligned}
w' &= \begin{cases} I_n C_-/\mu_n & \text{in} \quad (0, x_l), \\ -I_p C_+/\mu_p & \text{in} \quad (x_r, 1), \end{cases} \\
I_n' &= I_p' = 0, \\
w(0) &= w(1) = 1, \qquad w(x_l) = w(x_r) = 0.
\end{aligned} \tag{4.2.39}
$$

From (4.2.39) we obtain

$$
I_n|_{(0,x_l)} = -\frac{\mu_n}{x_l C_-}, \qquad I_p|_{(x_r,1)} = -\frac{\mu_p}{(1-x_r)C_+}.
$$

Whereas the ionization rates vanish in the zero space charge regions as $\lambda \to 0$, they take their maximal values α_n and α_p within the depletion region. As long as the current densities I_n and I_p remain negative the differential equations

$$
I_n' = -I_p' = \alpha_n I_n + \alpha_p I_p
$$

hold in the depletion region (x_l, x_r). When subjected to the boundary conditions

$$
I_n(x_l) = -\frac{\mu_n}{x_l C_-}, \qquad I_p(x_r) = -\frac{\mu_p}{(1-x_r)C_+}
$$

this system can be solved explicitely giving the following equation for the current $I = I_n + I_p$:

$$
\begin{aligned}
&(\alpha_p e^{x_l(\alpha_p - \alpha_n)} - \alpha_n e^{x_r(\alpha_p - \alpha_n)}) \frac{I}{\alpha_p - \alpha_n} \\
&= -\frac{\mu_n}{x_l C_-} e^{x_l(\alpha_p - \alpha_n)} - \frac{\mu_p}{(1-x_r)C_+} e^{x_r(\alpha_p - \alpha_n)}.
\end{aligned} \tag{4.2.40}
$$

The current blows up when the term in the brackets on the left hand side of (4.2.40) vanishes. This occurs for

$$
x_r - x_l = \frac{1}{\alpha_p - \alpha_n} \ln(\alpha_p/\alpha_n). \tag{4.2.41}
$$

From (4.2.41) and the formula (4.2.35) for the length of the depletion region the breakdown voltage

$$
\varphi_{1b} = \frac{1}{2(1/C_+ + 1/C_-)} \left(\frac{1}{\alpha_p - \alpha_n} \ln(\alpha_p/\alpha_n) \right)^2 \tag{4.2.42}
$$

can be computed. In terms of unscaled variables this gives

$$
U_b = \frac{q}{2\varepsilon_s(1/C_+ + 1/C_-)} \left(\frac{1}{\alpha_p^\infty - \alpha_n^\infty} \ln(\alpha_p^\infty/\alpha_n^\infty) \right)^2.
$$

Note that in the case $\alpha_n = \alpha_p = \alpha$ the condition (4.2.41) becomes

$$\alpha(x_r - x_l) = 1,$$

which is a special case of the classical condition [4.22]: Breakdown occurs as soon as the integral of the ionization rate over the depletion region takes the value 1.

Equation (4.2.40) shows that the current also becomes unbounded if one of the boundaries of the depletion region reaches a contact, i.e. if $x_l = 0$ or $x_r = 1$ holds. The results of the preceding Paragraph show that this happens for the critical voltages

$$\tfrac{1}{2}x_0^2 C_-(1 + C_-/C_+) \qquad \text{or} \qquad \tfrac{1}{2}(1 - x_0)^2 C_+(1 + C_+/C_-)$$

respectively. This effect is called *punch through*.

There are three possible reasons for junction breakdown: Avalanche generation and punch through at the left or right contact. Breakdown occurs as soon as one of the two critical voltages given above or the avalanche breakdown voltage (4.2.42) is exceeded. Naturally, only the smallest of the three values is of practical interest.

An existence result by Markowich [4.8] for (4.2.31), (4.2.32) with a simplified version of the ionization rates shows that for every value of the applied potential a stationary solution with finite current exists. Thus, the fact that the current obtained by our analysis tends to infinity at a critical voltage does not necessarily mean that the same holds for the solution of the full problem (4.2.31), (4.2.32). It only tells us that we reached the limit of the validity of our asymptotic analysis. The following Paragraph will show that, by rescaling and carrying out a different perturbation procedure, the continuation of the current-voltage characteristic after punch through can be computed.

In reality, however, the sharp increase in the current causes thermal effects which are not taken into account by our model and which might very well be the actual reason for breakdown.

Punch Through

For simplicity we consider the model problem which we dealt with in the Paragraph on high injection. As we are interested in the situation after punch through, we introduce the rescaled current density

$$I = \lambda^2 J/(\mu_n + \mu_p)$$

as suggested above. With the scaling for large reverse bias the problem now reads

$$\begin{aligned} \varphi'' &= n - p - 1, \\ I &= \lambda^2 n' - n\varphi', \\ I &= -\lambda^2 p' - p\varphi' \end{aligned} \tag{4.2.43}$$

subject to the boundary conditions

$$\varphi(0) = 0, \qquad n(0) = p(0), \tag{4.2.44}$$
$$\varphi(1) = \varphi_1, \qquad n(1) = p(1) + 1 = \tfrac{1}{2}(1 + \sqrt{1 + 4\delta^4}).$$

The analysis of the Paragraph "Large reverse bias" implies that the part of the depletion region on the n-side is given by the interval $(0, \sqrt{2\varphi_1})$. Thus, punch through occurs at the applied bias $\varphi_1 = \frac{1}{2}$ and, therefore, we assume $\varphi_1 > \frac{1}{2}$ for the following. By multiplying the Poisson equation by φ' and by substituting the reduced current relations into (4.2.43), the differential equation

$$\varphi'(\varphi'' + 1) = 0$$

for the reduced potential follows. Since φ is a slow variable it is expected to satisfy the original boundary conditions after going to the limit $\lambda \to 0$. Thus, we obtain

$$\varphi = \varphi_1 x + \frac{x(1 - x)}{2}.$$

The reduced charge carrier densities are given by

$$n = p = -I/\varphi' = -I/(\varphi_1 + \tfrac{1}{2} - x).$$

The stability of boundary layer equations at $x = 1$ for n and p depend on the sign of $\varphi'(1)$. Since $\varphi'(1) > 0$ holds, the layer equation for n is stable while that for p is unstable. This means that p cannot have a boundary layer at $x = 1$. The requirement that p takes the prescribed boundary value at $x = 1$ gives the approximate voltage-current characteristic

$$I = -(\varphi_1 - \tfrac{1}{2})\tfrac{1}{2}(-1 + \sqrt{1 + 4\delta^4}) \sim -\delta^4(\varphi_1 - \tfrac{1}{2}).$$

After punch through, the characteristic can be extended linearly. The order of magnitude of the currents is larger than before by the factor λ^{-2}. In terms of unscaled quantities the above equation reads

$$J = \frac{n_i^2 q}{2L\tilde{C}}(\mu_n + \mu_p)\left(U + \frac{qL^2\tilde{C}}{\varepsilon_s}\right).$$

The punch through voltage is given by $U_{pt} = -qL^2\tilde{C}/\varepsilon_s$ and the above approximation for the characteristic is valid for $U < U_{pt}$.

4.3 Bipolar Transistor

The bipolar transistor is a device whose performance is based on the interaction of two P-N junctions. Thus, there are two possibilities: A PNP- and an NPN-configuration. We restrict our discussion to PNP-transistors because the resulting theory carries over to the NPN-case with the obvious changes.

Each of the three differently doped regions has an Ohmic contact. This means that an appropriate model has to be at least two-dimensional. Thus,

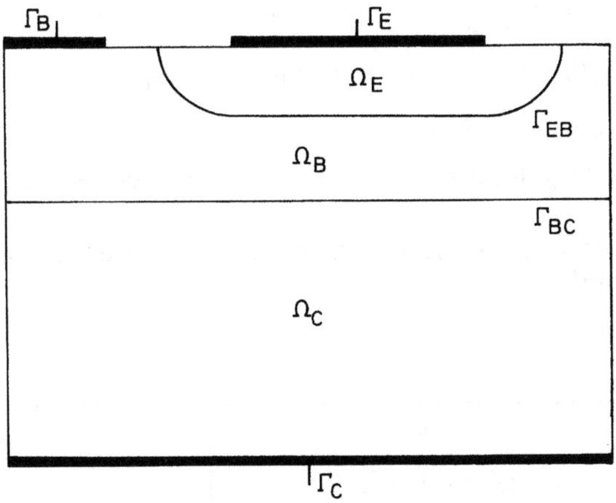

Fig. 4.3.1 Two-dimensional cross section of a bipolar transistor

the device is represented by a domain $\Omega \subseteq \mathbb{R}^k$ with $k = 2$ or 3. The middle (in our case n-) region is usually called *base* region (Ω_B) whereas the two outer (p-) regions are the *emitter* and resp. *collector* regions (Ω_E and Ω_C). The contacts corresponding to these regions are denoted by $\Gamma_q \subseteq \partial\Omega_q \cap \partial\Omega$, $q = B, E, C$, the emitter junction by Γ_{EB} and the collector junction by Γ_{BC} (see Fig. 4.3.1).

For a three terminal device the possible steady states constitute a two parameter manifold (see Section 4.1). An appropriate choice of parameters which depends on the circuit configuration permits a convenient interpretation of the results. Here we concentrate on the so called *common-emitter configuration* where usually the current through the base contact and the collector-emitter voltage are prescribed. The relevant output quantity in this situation is the collector current.

In the common-emitter configuration the transistor acts as an amplifier. For significant values of the collector-emitter voltage the *common-emitter current gain*

$$\beta = \frac{\partial I_C}{\partial I_B}$$

is large. Here I_C is the collector current and I_B the base current leaving the device.

Current Gain Close to Thermal Equilibrium

We consider the drift-diffusion equations (4.1.1) subject to the boundary conditions

$$(n - p - C)|_{\Gamma_E \cup \Gamma_B \cup \Gamma_C} = 0, \qquad np|_{\Gamma_E \cup \Gamma_B \cup \Gamma_C} = \delta^4,$$
$$(V - V_{bi})|_{\Gamma_E} = 0, \qquad (V - V_{bi})|_{\Gamma_B} = -U_{BE},$$
$$(V - V_{bi})|_{\Gamma_C} = -U_{CE}$$

where U_{BE} and U_{CE} denote the base-emitter and collector-emitter voltages, respectively.
The simplified equations (4.1.10) imply the following representations for the Slotboom variables:

	u	v
emitter	$(e^{U_{BE}} - 1)\varphi_1 + 1$	1
base	$e^{U_{BE}}$	$(1 - e^{-U_{BE}})\varphi_2 + (e^{-U_{CE}} - e^{-U_{BE}})\varphi_3 + e^{-U_{BE}}$
collector	$(e^{U_{BE}} - e^{U_{CE}})\varphi_4 + e^{U_{CE}}$	$e^{-U_{CE}}$

where the functions $\varphi_1, \ldots, \varphi_4$ are the solutions of boundary value problems of the form (4.1.11) with boundary conditions which are depicted in Fig. 4.3.2.
Now we are able to compute the currents across the emitter and collector junctions which are obviously equal to the currents through the emitter and collector contacts:

$$\begin{aligned} I_E &= \int_{\Gamma_{EB}} (J_n + J_p) \cdot v \, ds \\ &= (e^{U_{BE}} - 1) \int_{\Gamma_{EB}} \left(\frac{\mu_n}{|C_E|} \operatorname{grad} \varphi_1 - \frac{\mu_p}{C_B} \operatorname{grad} \varphi_2 \right) \cdot v \, ds \\ &\quad - (e^{U_{BE} - U_{CE}} - 1) \int_{\Gamma_{EB}} \frac{\mu_p}{C_B} \operatorname{grad} \varphi_3 \cdot v \, ds \end{aligned}$$

for the emitter current and

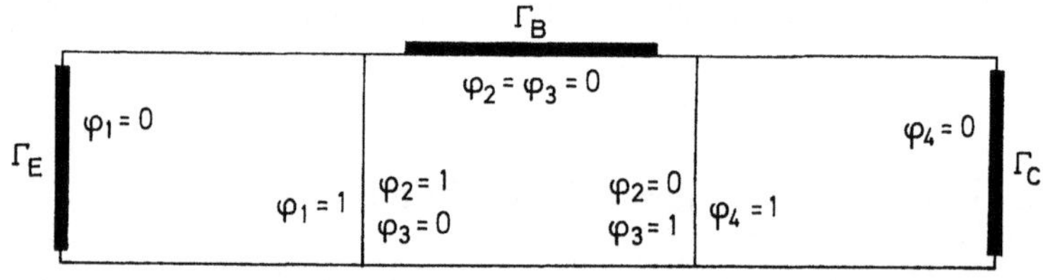

Fig. 4.3.2 Boundary conditions for the φ_j

$$I_C = \int_{\Gamma_{BC}} (J_n + J_p)\cdot v\, ds$$
$$= -(e^{U_{BE}} - 1)\int_{\Gamma_{BC}} \frac{\mu_p}{C_B}\operatorname{grad}\varphi_2\cdot v\, ds$$
$$-(e^{U_{BE}-U_{CE}} - 1)\int_{\Gamma_{BC}} \left(\frac{\mu_p}{C_B}\operatorname{grad}\varphi_3 - \frac{\mu_n}{|C_C|}\operatorname{grad}\varphi_4\right)\cdot v\, ds$$

for the collector current. In the above equations $C_E = C|_{\Omega_E}$ and similar definitions for C_B and C_C hold. v denotes the unit normal vector along the junctions pointing in the direction from the emitter to the collector. The base current is given as the difference $I_B = I_E - I_C$.

In the common emitter configuration we are interested in prescribing U_{CE} and I_B instead of U_{BE}. Fortunately the above formula for I_B turns out to be a one-to-one relation between I_B and U_{BE}. The output characteristic is obtained by computing U_{BE} in terms of I_B and substitution in the equation for I_C:

$$I_C = I_B\left(\frac{a_1}{a_3 + a_4\exp(-U_{CE})} - \frac{a_2}{a_3\exp(U_{CE}) + a_4}\right)$$
$$- a_1 + a_2 + (a_1 - a_2\exp(-U_{CE}))\frac{a_3 + a_4}{a_3 + a_4\exp(-U_{CE})} \tag{4.3.1}$$

and

$$U_{BE} = \ln\left(\frac{I_B + a_3 + a_4}{a_3 + a_4\exp(-U_{CE})}\right) \tag{4.3.2}$$

with the parameters

$$a_1 = -\int_{\Gamma_{BC}} \frac{\mu_p}{C_B}\operatorname{grad}\varphi_2\cdot v\, ds,$$
$$a_2 = \int_{\Gamma_{BC}} \left(\frac{\mu_p}{C_B}\operatorname{grad}\varphi_3 - \frac{\mu_n}{|C_C|}\operatorname{grad}\varphi_4\right)\cdot v\, ds,$$
$$a_3 = \int_{\Gamma_{EB}} \frac{\mu_n}{|C_E|}\operatorname{grad}\varphi_1\cdot v\, ds - \int_{\Gamma_B} \frac{\mu_p}{C_B}\operatorname{grad}\varphi_2\cdot v\, ds,$$
$$a_4 = -\int_{\Gamma_{BC}} \frac{\mu_n}{|C_C|}\operatorname{grad}\varphi_4\cdot v\, ds - \int_{\Gamma_B} \frac{\mu_p}{C_B}\operatorname{grad}\varphi_3\cdot v\, ds.$$

Since the dependence of I_C on I_B is linear, the common-emitter current gain only depends on the collector-emitter voltage, i.e. $\beta = \beta(U_{CE})$. For collector-emitter voltages which are significantly larger than the thermal voltage (which is our reference quantity for voltages), it can be approximated by

$$\beta(\infty) = a_1/a_3. \tag{4.3.3}$$

An application of the maximum principle implies that $0 < \varphi_j < 1$ holds in the interior of the respective n- and p-regions where the φ_j are defined. This

property is sufficient for determining the signs of the normal derivatives along the contacts and PN-junctions. If this knowledge is taken into account, it turns out that all the parameters a_i are sums of positive quantities. Thus, $\beta(\infty)$ is large as long as both terms which sum up to a_3 are small compared to a_1. For the first term this certainly holds if the doping in the emitter region is large compared to that in the base region.
The second requirement

$$-\int_{\Gamma_B} \frac{\mu_p}{C_B} \operatorname{grad} \varphi_2 \cdot v \, ds \ll -\int_{\Gamma_{BC}} \frac{\mu_p}{C_B} \operatorname{grad} \varphi_2 \cdot v \, ds \tag{4.3.4}$$

essentially refers to the flow of minority carriers in the base region of the device. In the simplified case of a constant doping profile in Ω_B, the above condition only depends on the geometry of the base region. The function φ_2 describes a situation where a certain current carried by minority carriers (in our case holes) is injected into the base region through the emitter junction. It is required that this current leaves the base region essentially through the collector junction without causing a significant minority carrier current through the base contact.
The classical analysis of bipolar transistors uses a one-dimensional model. This obviously raises the question of how to model the influence of the base contact. Since the usual Ohmic contact boundary conditions cannot be applied in this situation, simplifying assumptions have to be made. The major assumption turns out to be that the current through the base contact is entirely caused be majority carriers. The one-dimensional analysis can therefore only produce reliable results for devices which satisfy the requirement (4.3.4).
The formula (4.3.2) shows that U_{BE} tends to a positive limiting value as $U_{CE} \to \infty$. This shows that the emitter junction is forward biased with a voltage which remains bounded as $U_{CE} \to \infty$ whereas the collector junction is reverse biased with a voltage growing with U_{CE}. As in the case of the PN-diode we expect a depletion region of significant width around Γ_{BC} which has not been taken into account by the above analysis. The influence of the widening of the depletion region on the output characteristics of bipolar transistors is usually referred to as the *Early effect* in the literature (see e.g. [4.22]). Similarly to the case of a reverse biased diode it can be shown that the collector current does not saturate for large collector-emitter voltages as opposed to the above result (4.3.1).
A second limitation for the above theory are high injection situations where the collector current becomes large enough for our scaling to be invalid. This leads to a decrease of the current gain which is commonly referred to as the *Webster effect* (see [4.22]).

4.4 PIN-Diode

The PIN-diode is a device where an *n*- and a *p*-region are separated by an intrinsic region (*i*-region) with a very low concentration of ionized impurities. It behaves essentially like a P-N diode but has some peculiar additional features. Its practical importance is due to a higher conductivity in the forward bias regime and a larger breakdown voltage compared to PN-diodes. The first of these properties will discussed below, for the second we refer to the literature (see e.g. [4.12]).

Thermal Equilibrium

If the PIN-diode is represented by the domain $\Omega \subseteq \mathbb{R}^k$, $k = 1$, 2 or 3, the doping profile satisfies

$$C(x) \begin{cases} <0 & \text{in} \quad \Omega_-, \\ =0 & \text{in} \quad \Omega_0, \\ >0 & \text{in} \quad \Omega_+, \end{cases}$$

where $\Omega_-, \Omega_0, \Omega_+$ are disjoint, simply connected subdomains of Ω with

$$\bar{\Omega}_- \cup \bar{\Omega}_0 \cup \bar{\Omega}_+ = \bar{\Omega}, \qquad \bar{\Omega}_- \cap \bar{\Omega}_+ = \{\ \}.$$

We consider the idealized situation of a vanishing doping concentration in the *i*-region Ω_0. The doping profile is supposed to have jumps along the *pi*- and *ni*-junctions and the *p*- and *n*-regions have Ohmic contacts

$$\Gamma_- \subseteq \partial\Omega \cap \partial\Omega_-, \qquad \Gamma_+ \subseteq \partial\Omega \cap \partial\Omega_+$$

(see Fig. 4.4.1).

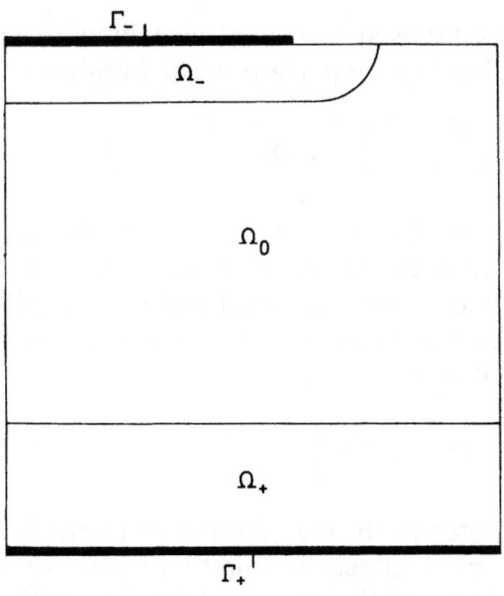

Fig. 4.4.1 Cross section of the PIN-diode

We consider the thermal equilibrium problem

$$\lambda^2 \Delta V = 2\delta^2 \sinh V - C \qquad \text{in} \quad \Omega,$$

$$(V - V_{bi})|_{\Gamma_- \cup \Gamma_+} = 0, \qquad \left.\frac{\partial V}{\partial \nu}\right|_{\partial\Omega_N} = 0,$$

where $\partial\Omega_N = \partial\Omega \backslash (\Gamma_- \cup \Gamma_+)$ and $V_{bi} = \operatorname{areasinh}(C/2\delta^2)$ holds.
As in the case of the P-N diode approximations of the solution can be obtained by exploiting the smallness of λ and δ. Introducing the small parameter $\gamma = (\ln \delta^{-2})^{-1}$ and the rescaled variable $W = \gamma V$ we expect W to be uniformly bounded in λ and δ. The new problem reads (compare to (4.2.7), (4.2.8), 4.2.9))

$$\tilde{\lambda}^2 \Delta W = \exp\left(\frac{W-1}{\gamma}\right) - \exp\left(\frac{-W-1}{\gamma}\right) - C, \tag{4.4.1}$$

$$W|_{\Gamma_-} = -1 + O(\gamma), \qquad W|_{\Gamma_+} = 1 + O(\gamma), \qquad \left.\frac{\partial W}{\partial \nu}\right|_{\partial\Omega_N} = 0$$

with the new parameter $\tilde{\lambda} = \lambda/\sqrt{\gamma}$ which we also assume to be small. For P-N diodes we noted that the limits $\tilde{\lambda} \to 0$ and $\gamma \to 0$ in the problem corresponding to (4.4.1) commute. This statement is wrong for the PIN-diode. Letting $\tilde{\lambda} \to 0$ in (4.4.1) we obtain the reduced solution

$$\overline{W} = \gamma V_{bi} = \begin{cases} -1 + O(\gamma) & \text{in} \quad \Omega_-, \\ 0 & \text{in} \quad \Omega_0, \\ 1 + O(\gamma) & \text{in} \quad \Omega_+. \end{cases}$$

It has jump discontinuities at the junctions which are smoothed by appropriate layer terms. The layer equation in the i-region reads

$$\partial_\xi^2 \hat{W} = \exp\left(\frac{\hat{W}-1}{\gamma}\right) - \exp\left(\frac{-\hat{W}-1}{\gamma}\right),$$

where ξ is the fast variable. An estimate for the thickness of the layer can be obtained from the linearization of this equation at the stationary point $\hat{W} = 0$. The result is the characteristic length $\sqrt{\gamma}\, \exp(1/2\gamma)$ in terms of the variable ξ. If we return to the original length scale the layer thickness is of the order of magnitude of

$$\tilde{\lambda}\sqrt{\gamma}\, \exp(1/2\gamma) = \lambda/\delta = \frac{1}{L}\sqrt{\frac{\varepsilon U_T}{q n_i}},$$

which can be interpreted as the scaled intrinsic Debye length. Our approach is justified if the layer thickness is small which is equivalent to smallness of the intrinsic Debye length compared to the width of the i-region or, equivalently, to the relation

$$\lambda \ll \delta$$

in terms of our original parameters.
If on the other hand $\delta \ll \lambda$ holds (PIN-diode with short i-region) we approximate the solution of (4.4.1) by first letting γ tend to zero. Arguments as in the case of the PN-diode show that the result is the double obstacle problem

$$\begin{aligned} &-1 \leqslant W \leqslant 1, \qquad W|_{\Gamma_-} = -1, \qquad W|_{\Gamma_+} = 1, \\ &\tilde{\lambda}^2 \, \Delta W + C \leqslant 0 \qquad \text{for} \qquad W < 1, \\ &\tilde{\lambda}^2 \, \Delta W + C \geqslant 0 \qquad \text{for} \qquad W > -1 \end{aligned} \tag{4.4.2}$$

which can be rewritten as the following singularly perturbed variational inequality: Find $W \in \mathbb{K}$ with

$$\tilde{\lambda}^2 \int_\Omega \operatorname{grad} W \cdot \operatorname{grad}(\varphi - W)\, dx \geqslant \int_\Omega C(x)(\varphi - W)\, dx, \qquad \forall \varphi \in \mathbb{K}, \tag{4.4.3}$$

where

$$\mathbb{K} = \{\varphi \in H^1(\Omega)\colon \varphi|_{\Gamma_-} = -1, \varphi|_{\Gamma_+} = 1, -1 \leqslant \varphi \leqslant 1\}$$

is a closed, convex subset of the space $H^1(\Omega)$ of square integrable functions with a square integrable gradient. Since the doping profile vanishes in the i-region, the formal limit of the variational inequality as $\tilde{\lambda} \to 0$ only consists of contributions from the p- and n-regions. The formal limiting problem does not have a unique solution. The set of solutions is given by

$$\chi = \{\varphi \in H^1(\Omega)\colon \varphi|_{\Omega_-} = -1, \varphi|_{\Omega_+} = 1, -1 \leqslant \varphi \leqslant 1\}.$$

We expect the limit of W as $\tilde{\lambda} \to 0$ to be an element of χ.
For $\psi \in \chi$ the limiting variational inequality

$$0 \geqslant \int_\Omega C(x)(\varphi - \psi)\, dx, \qquad \forall \varphi \in \mathbb{K}, \tag{4.4.4}$$

holds. Replacing φ by ψ in (4.4.3) and by W in (4.4.4) and adding up the resulting inequalities gives

$$\int_\Omega \operatorname{grad} W \cdot \operatorname{grad}(\psi - W)\, dx \geqslant 0, \qquad \forall \psi \in \chi. \tag{4.4.5}$$

Since we require W to be an element of χ in the limit $\tilde{\lambda} \to 0$ the limiting solution can be determined uniquely from the variational inequality (4.4.5). A justification of this formal procedure can be found [4.6].
The solution of (4.4.5) satisfies

$$W|_{\Omega_-} = -1, \qquad W|_{\Omega_+} = 1, \qquad \Delta W = 0 \quad \text{in} \quad \Omega_0 \tag{4.4.6}$$

which differs strongly from the limiting solution $\overline{W}$ for the case $\lambda \ll \delta$ above. For a short PIN-diode the zero space charge approximation does not hold within the i-region.

Behaviour Close to Thermal Equilibrium

We cannot apply the considerations from the beginning of this Chapter to the PIN-diode because the *i*-region requires special care. For simplicity we restrict our attention to a one-dimensional PIN-diode with long *i*-region ($\lambda \ll \delta$) and constant mobilities. The geometry is given by

$$\Omega = (0, 1), \qquad \Omega_- = (0, x_1), \qquad \Omega_+ = (x_2, 1) \tag{4.4.7}$$

$$\text{with} \qquad 0 < x_1 < x_2 < 1 .$$

The zero space charge approximation ($\lambda = 0$ in (4.1.1)) leads to the following relations between the carrier densities and the Slotboom variables u and v in the *i*-region:

$$n = p = \delta^2 \sqrt{uv} \qquad \text{in} \quad (x_1, x_2).$$

For the Shockley-Read-Hall recombination-generation term this implies

$$R = \frac{\delta^2(uv - 1)}{\tau_p(\sqrt{uv} + 1) + \tau_n(\sqrt{uv} + 1)} = \frac{\delta^2(\sqrt{uv} - 1)}{\tau_n + \tau_p} \qquad \text{in} \quad (x_1, x_2).$$

Within the *i*-region R is $O(\delta^2)$ which is in contrast to *n*- or *p*-regions where R is $O(\delta^4)$. This implies that properly rescaled current densities are introduced by

$$J_n = \delta^2 J_{ns}, \qquad J_p = \delta^2 J_{ps}$$

instead of (4.1.7) where a factor δ^4 was used instead of δ^2. This fact demonstrates the higher conductivity of the PIN-diode mentioned in the introduction of this Section. After substitution of the rescaled current densities in (4.1.4) we let $\delta \to 0$ and obtain

$$v' = 0, \qquad J_n = 0 \qquad \text{in the } p\text{-region } (0, x_1),$$

$$u' = 0, \qquad J_p = 0 \qquad \text{in the } n\text{-region } (x_2, 1),$$

$$J_n = \mu_n \sqrt{v/u}\, u', \qquad J_p = -\mu_p \sqrt{u/v}\, v',$$

$$J_n' = -J_p' = \frac{\sqrt{uv} - 1}{\tau_n + \tau_p} \qquad \text{in the } i\text{-region } (x_1, x_2).$$

With U denoting the applied voltage we have the boundary conditions

$$u(0) = v(0) = 1, \qquad u(1) = e^U, \qquad v(1) = e^{-U},$$

which imply

$$v = 1 \qquad \text{in} \quad (0, x_1)$$

and

$$u = e^U \qquad \text{in} \quad (x_2, 1).$$

With $w := \sqrt{uv}$ a straightforward computation gives

$$w' = J_n/\mu_n - J_p/\mu_p \qquad \text{in} \quad (x_1, x_2)$$

and, thus,

$$w'' = \frac{w-1}{L_a^2} \qquad \text{in} \quad (x_1, x_2) \qquad \text{with} \qquad L_a = \sqrt{\frac{\tau_n + \tau_p}{1/\mu_n + 1/\mu_p}}, \tag{4.4.8}$$

where L_a denotes the *ambipolar diffusion length* (see [4.22]). The solution of (4.4.8) is

$$w = 1 + (w(x_1) - 1)\frac{\sinh(x_2 - x)/L_a}{\sinh(x_2 - x_1)/L_a} + (w(x_2) - 1)\frac{\sinh(x - x_1)/L_a}{\sinh(x_2 - x_1)/L_a} \tag{4.4.9}$$

and the values of the Slotboom-variables at the junctions are

$$u(x_1) = w(x_1)^2, \qquad v(x_2) = w(x_2)^2 e^{-U}.$$

A system of algebraic equations for $w(x_1)$ and $w(x_2)$ is obtained by noting that

$$\frac{u'}{u} = \frac{J_n}{\mu_n w}, \qquad \frac{v'}{v} = -\frac{J_p}{\mu_p w}$$

holds. Integrating these equations gives

$$2\ln w(x_1) - U = -\int_{x_2}^{x_1} \frac{1}{\mu_n w} \int_{x_1}^{x} \frac{w-1}{\tau_n + \tau_p}\, ds\, dx,$$

$$2\ln w(x_2) - U = -\int_{x_2}^{x_1} \frac{1}{\mu_p w} \int_{x}^{x_2} \frac{w-1}{\tau_n + \tau_p}\, ds\, dx.$$

Although the integrations on the right hand sides can be carried out explicitely the above system does not allow for an explicit solution for $w(x_1)$ and $w(x_2)$. However, it is amenable to an asymptotic analysis as $U \to \pm\infty$ which shows that

$$\lim_{U \to -\infty} w(x_1) = \lim_{U \to -\infty} w(x_2) = 0$$

holds and that $w(x_1)$ and $w(x_2)$ are $O(e^{U/2})$ as $U \to \infty$. Since the total current density is given by

$$J = J_n(x_2) = \int_{x_1}^{x_2} \frac{w-1}{\tau_n + \tau_p}\, dx$$

we arrive at the conclusion that the voltage-current characteristic of the PIN-diode has the same qualitative behaviour as that of the P-N-diode. The reverse bias saturation current can be computed by substituting $w(x_1) = w(x_2) = 0$ in (4.4.9) and evaluating the above integral. In the forward bias situation we have the result

$$J = O(e^{U/2}).$$

Remark: Compare this result to the P-N-diode where the current grows like e^U.

4.5 Thyristor

Thyristors are devices with four differently doped regions. Their practical importance is due to the existence of multiple steady state solutions under certain biasing conditions. In particular it is possible to switch between an OFF state with very low current and a conducting ON state.

In this Section we discuss the static voltage-current characteristic of a two terminal device, the so called *Shockley-diode* (Fig. 4.5.1a). The S-shaped forward bias characteristic is depicted qualitatively in Figure 4.5.1b. It consists of three parts which are separated by the critical points corresponding to the *holding voltage* U_h and the *break over voltage* U_{br}. The currents on the lowest (*forward blocking*) branch are small leakage currents. The points on the middle branch between U_{br} and U_h correspond to solutions which are dynamically unstable in the voltage driven case. Solutions on the upper (conduction) branch are characterized by significant current flow. Note, that for all applied voltages between U_h and U_{br} two dynamically stable steady state solutions exist.

The qualitative behaviour of the characteristic has traditionally been analyzed by replacing the thyristor by two bipolar transistors where the base region of one transistor is identified with the collector region of the other, and vice versa. Only quite recently Rubinstein [4.14] and Steinrück [4.20], [4.21] explained the structure of the characteristic by applying perturbation methods and bifurcation theory to the drift-diffusion model.

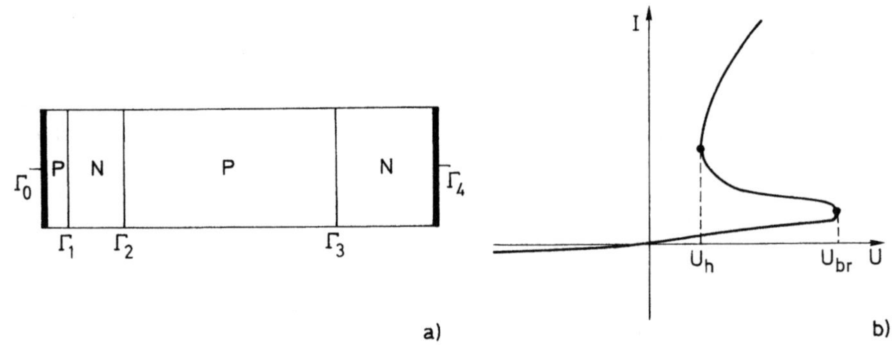

Fig. 4.5.1 (a) Cross section of the Shockley diode, (b) voltage-current characteristic

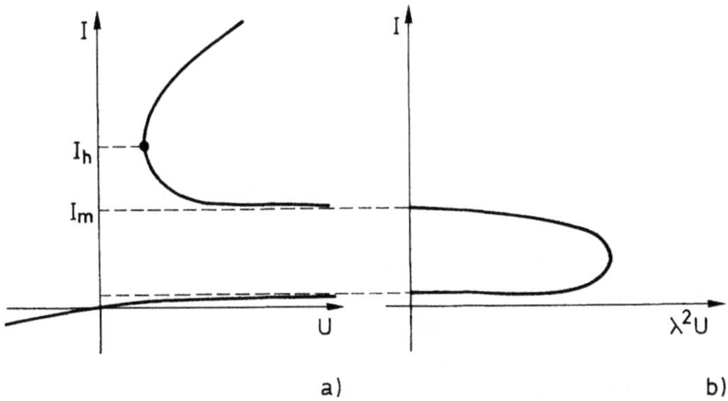

Fig. 4.5.2 Thyristor characteristic for (a) small voltage scaling (b) large voltage scaling

The key to the analysis is the observation that different scalings have to be used for obtaining different parts of the characteristic in order to single out the dominant effects. Two observations are important in this context. Firstly, the currents on the blocking branch are several orders smaller than on the middle branch. Second, the break over voltage usually assumes very large values compared to the holding voltage which is of the order of magnitude of the built-in voltage. This leads to the conclusion that a small-voltage-scaling might result in an approximation of the characteristic which has two saturation currents corresponding to the blocking and the middle branch (Fig. 4.5.2 a). Since the currents on these two branches have different orders of magnitude, they are obtained separately by considering different scalings. Finally, the situation close to the break over voltage can be analyzed by considering the large-voltage-scaling which was already used for P-N-diodes under large reverse bias. This should provide a piece of the characteristic which connects the two parts discussed above (Fig. 4.5.2 b).

In the following Paragraph the lower branch for small voltages is considered which can be analyzed for a multi-dimensional model. Here the same ideas which have been used in the Sections on diodes and bipolar transistors for the situation close to thermal equilibrium are applied. Since a four layer device can be obtained by a small perturbation of the doping profile of a PIN-diode, it is clear that not every four layer device has a characteristic like that depicted in Fig. 4.5.1 b. It will be demonstrated that there are two possibilities. In one case the behaviour expected of a thyristor is obtained: An approximation for the lower branch of the characteristic saturates for large voltages. In the second case the characteristic grows exponentially like that of a diode. Only in the former situation the device has a chance to show the performance which we expect of a thyristor. The distinction between the two cases is determined by the sign of a parameter which depends on the geometry, the doping profile and the recombination-generation rate.

For discussing the situation along the middle and conducting branches we rescale the currents and restrict ourselves to a one-dimensional model. If the

above mentioned parameter takes small values and has the appropriate sign the existence of a saturation current corresponding to the middle branch can be shown if an additional condition on the device parameters is satified. It will also be demonstrated how the critical point at the holding voltage can be computed in this case.
The part of the characteristic connecting the lower and middle branches is essentially governed by two physical effects. One is the widening of the depleton region around the middle P-N-junction on the characteristic, the second is impact ionization. Only the first effect is taken into account in the final Paragraph of this Section where an approximation for the break over voltage is computed. This result can in general only be expected to be qualitatively correct because there is strong evidence that the influence of impact ionization on the value of the break over voltage cannot be neglected (see [4.22]).

Characteristic Close to Thermal Equilibrium

The aim of this Paragraph is to derive an approximation of the lower branch of the characteristic of a thyristor close to thermal equilibrium. The equations (4.1.10) for the Slotboom variables will be used with the simplifying assumption of vanishing recombination-generation effects. It does not cause mathematical problems to include Shockley-Read-Hall recombination, only the resulting calculations would be much more involved (see Problem 4.9).
We denote the differently doped regions of the device by $\Omega_1, \ldots, \Omega_4$ and assume Ω_1 and Ω_3 to be p-regions and Ω_2 and Ω_4 to be n-regions. The junction separating Ω_i and Ω_{i+1} is denoted by Γ_i. The regions Ω_1 and Ω_4 have the Ohmic contacts Γ_0 and Γ_4, respectively (see Fig. 4.5.1 a).
By the approximate equations (4.1.10) the Slotboom variable u corresponding to electrons is constant in Ω_2 and Ω_4 whereas v is constant in Ω_1 and Ω_3:

$$v|_{\Omega_1} = 1, \qquad u|_{\Omega_2} = e^V, \qquad v|_{\Omega_3} = e^{-W}, \qquad u|_{\Omega_4} = e^U \tag{4.5.1}$$

holds, where U denotes the applied voltage and the values of the constants V and W are as yet unknown. The Slotboom variables corresponding to the minority carriers can be expressed in terms of the functions φ_i, $i = 1, \ldots, 4$ defined on Ω_i as solutions of problems of the form (4.1.11) with

$$\varphi_i|_{\Gamma_{i-1}} = 0, \qquad \varphi_i|_{\Gamma_i} = 1.$$

If V and W are considered as given u and v are completely determined by (4.5.1) and

$$\begin{aligned} u|_{\Omega_1} &= 1 + (e^V - 1)\varphi_1, & v|_{\Omega_2} &= 1 + (e^{-W} - 1)\varphi_2, \\ u|_{\Omega_3} &= e^V + (e^U - e^V)\varphi_3, & v|_{\Omega_4} &= e^{-W} + (e^{-U} - e^{-W})\varphi_4. \end{aligned} \tag{4.5.2}$$

The simplicity of these formulas is due to the assumption of vanishing recombination-generation.

It remains to determine the values of V and W. Note that the current through the device is equal to the current across any of the surfaces Γ_i. By the zero recombination assumption the same holds for the electron and hole currents taken separately. The conditions that the electron current across Γ_1 is equal to that across Γ_3 and that the hole current across Γ_2 is equal to that across Γ_4 lead to the equations

$$\begin{aligned} (e^V - 1)\kappa_1 &= e^{-W}(e^U - e^V)\kappa_3, \\ e^V(e^{-W} - 1)\kappa_2 &= e^U(e^{-U} - e^{-W})\kappa_4, \end{aligned} \tag{4.5.3}$$

where the κ_i are given by

$$\kappa_i = \int_{\Gamma_i} \mu_{n(p)}/|C| \operatorname{grad} \varphi_i \cdot v \, ds, \qquad i = 1, \ldots, 4$$

with v being the unit normal vector along Γ_i pointing outward of Ω_i. The maximum principle implies that φ_i takes its maximal value 1 along Γ_i which in turn implies that κ_i is positive. κ_i can be interpreted as a measure for the conductivity of Ω_i for minority carriers. The total current through the device is given by

$$I = (e^V - 1)\kappa_1 + (e^{U-W} - 1)\kappa_4 .$$

Elimination of e^V from (4.5.3) leads to a quadratic equation for e^{-W} which has a unique positive solution. The asymptotic behaviour as U tends to infinity depends on the sign of the parameter

$$A := \kappa_1\kappa_4 - \kappa_2\kappa_3 . \tag{4.5.4}$$

For negative values of A, W tends to a positive limiting value as $U \to \infty$ whereas V grows like U. Thus, the current grows as e^U in this situation which is not what we expect from a thyristor. A possible reason for $A < 0$ is low doping in the middle regions compared to the outer regions which corresponds to a device which is close to a PIN-diode. If A is positive W grows as U and V tends to a limiting value as $U \to \infty$. The current saturates in this case and we are led to the conclusion that $A > 0$ is a necessary condition for a four layer device having a thyristor characteristic. Our results also show that the potential drops across the (forward biased) outer junctions Γ_1 and Γ_3 remain bounded whereas that across the (reversed biased) juncton Γ_2 grows as U.

The reverse bias characteristic is analyzed by considering the asymptotic behaviour of the solution as $U \to -\infty$. It turns out that the current saturates at a negative value. Again the voltage drop is concentrated to one P-N-junction. If the parameter

$$B := \kappa_1\kappa_2 - \kappa_3\kappa_4$$

is positive V and W tend to negative limiting values as $U \to -\infty$. This implies that the voltage drop across the junction Γ_3 grows with U. For negative values of B both V and W tend to $-\infty$ linearly in U. In this case the biggest part of the voltage drop takes place at Γ_1.

Forward Conduction

In this Paragraph an approximation of the voltage current characteristic of a thyristor in a neighbourhood of the holding current will be derived. A one-dimensional model with vanishing recombination-generation rate is considered for simplicity, but recombination rates of the Shockley-Read-Hall as well as Auger type could be easily included (see [4.20]).
Our approach is based on an observation concerning the current controlled problem. From Fig. 4.5.2 we conclude that this problem is quite ill conditioned. However, the analysis below will show that only the determination of the potential is critical, whereas the carrier densities depend on the current in a smooth way. It is possible to take advantage of this fact by decoupling the current controlled problem and using Taylor expansions in terms of the current for approximating the carrier densities. In a second step the potential will be computed leading to an approximation of the voltage-current characteristic in the form $U = U(I)$. Under the assumption that the parameter A, which has been defined in the preceding Paragraph, is positive and an additional condition on the device is satisfied, $U(I)$ has the expected behaviour. It is defined for $I > I_m$, which is the saturation current of the middle branch of the characteristic, and takes its minimum at $I = I_h$ which is the holding current.
In order to keep the calculations as simple as possible, the doping profile is assumed to be piecewise constant and the mobilities are assumed to be constant. The domain Ω of the preceding Paragraph corresponds to the interval (0, 1), and the surfaces $\Gamma_0, \ldots, \Gamma_4$ to the points

$$0 = x_0 < \cdots < x_4 = 1. \tag{4.5.5}$$

The doping profile is given by

$$C(x) = \begin{cases} -C_1 & \text{in} \quad (x_0, x_1), \\ C_2 & \text{in} \quad (x_1, x_2), \\ -C_3 & \text{in} \quad (x_2, x_3), \\ C_4 & \text{in} \quad (x_3, x_4). \end{cases} \tag{4.5.6}$$

The zero space charge approximation reads

$$\begin{aligned} 0 &= n - p - C, \\ J_n &= \mu_n(n' - nV'), \\ J_p &= -\mu_p(p' + pV') \end{aligned} \tag{4.5.7}$$

with the jump conditions

$$[V]_{x_i} = [\ln n]_{x_i} = -[\ln p]_{x_i}, \qquad i = 1, 2, 3$$

which can be interpreted as the condition that the quasi Fermi levels do not have jumps across the junctions. Replacing δ^4 in the Ohmic contact boundary conditions by zero the minority carrier densities vanish in thermal equilibrium. This motivates the introduction of a new variable w by setting

$$np = Iw$$

where $I = J_n + J_p$ is the total current. This leads to the representation

$$n = \tfrac{1}{2}(C + \sqrt{C^2 + 4Iw}), \qquad p = \tfrac{1}{2}(-C + \sqrt{C^2 + 4Iw})$$

for the carrier densities. The current densities are rescaled by replacing J_n and J_p by IJ_n and IJ_p, respectively. By elimination of the potential from (4.5.7) the problem

$$w' = \frac{J_n}{2\mu_n}(-C + \sqrt{C^2 + 4Iw}) - \frac{J_p}{2\mu_p}(C + \sqrt{C^2 + 4Iw}), \tag{4.5.8}$$

$$w(0) = w(1) = 0, \qquad J_n + J_p = 1$$

for w is obtained. Note that the trivial solution corresponding to $I = 0$ has been eliminated by our choice of scaling. The problem for the determination of the potential reads

$$V' = -I\frac{J_n/\mu_n + J_p/\mu_p}{\sqrt{C^2 + 4Iw}} \quad \text{for} \quad x \neq x_i, \quad i = 1, 2, 3,$$

$$V(0) = V_{bi}(0), \qquad [V]_{x_i} = [\ln(C + \sqrt{C^2 + 4Iw})]_{x_i}, \qquad i = 1, 2, 3. \tag{4.5.9}$$

The fact that w and V can be computed consecutively from (4.5.8), (4.5.9) is the decoupling mentioned above. The solution of (4.5.8) can be approximated by Taylor expansion in terms of I. For the leading terms w_0, J_{n0}, J_{p0} the differential equation in (4.5.8) reduces to

$$w_0' = \begin{cases} -J_{n0}C/\mu_n & \text{for} \quad C < 0, \\ -J_{p0}C/\mu_p & \text{for} \quad C > 0. \end{cases}$$

The solution can be written in terms of the quantities κ_i defined in the preceding Paragraph which are given by

$$\kappa_1 = \frac{\mu_n}{C_1(x_1 - x_0)}, \qquad \kappa_2 = \frac{\mu_p}{C_2(x_2 - x_1)},$$

$$\kappa_3 = \frac{\mu_n}{C_3(x_3 - x_2)}, \qquad \kappa_4 = \frac{\mu_p}{C_4(x_4 - x_3)}.$$

in the one-dimensional case. w_0 is piecewise linear and takes the values

$$\begin{aligned} w_0(x_1) &= \kappa_3(\kappa_2 + \kappa_4)/\Sigma, \\ w_0(x_2) &= -A/\Sigma, \\ w_0(x_3) &= \kappa_2(\kappa_1 + \kappa_3)/\Sigma \end{aligned} \tag{4.5.10}$$

at the junctions. A is defined in the preceding Paragraph and

$$\Sigma = \kappa_1\kappa_2\kappa_3 + \kappa_1\kappa_2\kappa_4 + \kappa_1\kappa_3\kappa_4 + \kappa_2\kappa_3\kappa_4$$

holds. The leading coefficients for the current densities are

$$J_{n0} = \kappa_1\kappa_3(\kappa_2 + \kappa_4)/\Sigma, \qquad J_{p0} = \kappa_2\kappa_4(\kappa_1 + \kappa_3)/\Sigma. \tag{4.5.11}$$

With the assumption that A is positive (which has been found to be necessary for a thyristor) $w_0(x_2)$ is negative which obviously is unacceptable for an approximation of the product of the carrier densities. For w taking negative values the problem (4.5.9) for the potential does not have a solution.

Let us reconsider the result we are aiming at. The problem is expected not to have a solution for values of the current between the saturation currents of the middle and the blocking branch. Since the saturation current of the blocking branch is $O(\delta^4)$ and the approximation $\delta^4 = 0$ was used in this Paragraph, the blocking branch reduces to $I = 0$. Thus, it is to be expected that there is no solution for currents below a certain threshold. This is in agreement with our results so far.

A further analysis requires the computation of higher order terms in the Taylor expansion of w. The first order term w_1 solves a linear problem just as w_0, but with different inhomogeneities. In the following only its value $w_1(x_2)$ at the middle junction will be used. Our second assumption on the device (besides $A > 0$) is that this value is positive. The approximation

$$w(x_2) \sim w_0(x_2) + I w_1(x_2) \tag{4.5.12}$$

implies that the problem has a solution if

$$I > I_m = -w_0(x_2)/w_1(x_2)$$

holds. This argument is justified as long as I_m is small enough for the Taylor polynomial (4.5.12) to be a good approximation at $I = I_m$. This condition is satisfied if the parameter A is sufficiently small, i.e. if the device is close to the critical case $A = 0$ where the shape of the characteristic changes qualitatively.

With the boundary condition

$$V(1) = V_{bi}(1) - U$$

for the potential at the right contact the voltage current characteristic is obtained by integrating (4.5.9):

$$U = U_{bi} + I \int_0^1 \frac{J_n/\mu_n + J_p/\mu_p}{\sqrt{C^2 + 4Iw}}\,dx - \sum_{i=1}^{3} [\ln(C + \sqrt{C^2 + 4Iw})]_{x_i}$$

where $U_{bi} = V_{bi}(1) - V_{bi}(0)$ denotes the built-in voltage. Using the Taylor expansions computed above this equation can be simplified considerably. The integral will be replaced by

$$\alpha = \int_0^1 (J_{n0}/\mu_n + J_{p0}/\mu_p)|C|\,dx$$

and the arguments of the logarithms by

$$\begin{array}{lll} 2C & \text{for} & C > 0, \\ 2Iw_0/|C| & \text{for} & C < 0 \quad \text{and} \quad i = 1, 3, \\ 2I(w_0 + Iw_1)/|C| & \text{for} & C < 0 \quad \text{and} \quad i = 2. \end{array}$$

The resulting approximation for the voltage-current characteristic reads

$$U = \alpha I + \ln \frac{I}{w_0(x_2) + Iw_1(x_2)} + U_{bi} + \ln \frac{w_0(x_1)w_0(x_3)}{C_1 C_4}.$$

This formula shows that I_m can be interpreted as the saturation current on the middle branch. In the limit $I \to I_m$ the voltage tends to infinity. The expression

$$I_h = (-\alpha w_0(x_2) + \sqrt{\alpha^2 w_0(x_2)^2 - 4\alpha w_0(x_2) w_1(x_2)})/2\alpha w_1(x_2)$$

for the holding current can be found by solving the equation

$$\frac{\partial U}{\partial I}(I_h) = 0.$$

The above procedure can be justified a posteriori if the computed approximation for the holding current is small. A mathematical justification by means of bifurcation theory can be found in [4.20]. For larger values of the current the full problem (4.5.8), (4.5.9) has to be solved. In [4.14] it was shown that the differential equations in (4.5.8), (4.5.9) can be integrated explicitely reducing the problem to a set of algebraic equations. For a more general model including recombination-generation effects the problem was solved numerically in [4.20] resulting in very satisfactory approximations of the characteristics.

Break Over Voltage

The aim of this Paragraph is to show the existence and to compute an approximation of a break over voltage. A branch of the characteristic which connects the two branches discussed in the preceding Paragraphs will be constructed. As mentioned above the break over voltage is large compared to the thermal voltage (the reference voltage used until now). Thus, the potential has to be rescaled. Since the main part of the potential drop occurs at the middle junction, a depletion region of significant width around that junction is to be expected. As in the case of the reverse biased diode the edges of this depletion region can be obtained by solving a free boundary problem, arising as a singular limit of the rescaled drift diffusion equations.
We consider a one-dimensional device for which the drift-diffusion model with a rescaled potential $\varphi = \lambda^2(V - V_{bi}(0))$ reads

$$\begin{aligned} \varphi'' &= n - p - C, \\ \lambda^2 J_n &= \mu_n(\lambda^2 n' - n\varphi'), \\ \lambda^2 J_p &= -\mu_p(\lambda^2 p' + p\varphi') \end{aligned}$$

where φ satisfies the boundary conditions

$$\varphi(0) = 0, \qquad \varphi(1) = -\varphi_1 := \lambda^2(U_{bi} - U).$$

For the discussion of the limit $\lambda \to 0$ we refer to the Paragraph on reverse biased diodes. In this limit the device splits into depletion and zero space charge regions. The analysis above suggests that a depletion region occurs around x_2. A more rigorous justification of this proposition by the method of matched asymptotic expansions can be found in [4.21]. A straightforward computation leads to a limiting potential which varies quadratically within the depletion region (x_l, x_r) and is piecewise constant outside of this interval. The edges of the depletion region are given by

$$x_l = x_2 - \sqrt{\frac{2C_3\varphi_1}{C_2(C_2 + C_3)}}, \qquad x_r = x_2 + \sqrt{\frac{2C_2\varphi_1}{C_3(C_2 + C_3)}}.$$

This result can be justified by the method of matched asymptotic expansions as long as $x_l > x_1$ and $x_r < x_3$ holds, i.e. for voltages small enough such that punch through does not occur.
For the computation of the carrier densities we introduce the variable w as above. Since w vanishes within the depletion region it solves the problem

$$\begin{aligned} w' &= \frac{J_n}{2\mu_n}(-C + \sqrt{C^2 + 4Iw}) - \frac{J_p}{2\mu_p}(C + \sqrt{C^2 + 4Iw}) \\ &\quad \text{in} \quad (0, 1)\backslash(x_l, x_r), \qquad (4.5.13) \\ w(0) &= w(1) = 0, \qquad w(x_l) = w(x_r), \qquad J_n + J_p = 1 \end{aligned}$$

subject to

$$w(x_l) = 0. \tag{4.5.14}$$

The reason for separating the condition (4.5.14) from the rest of the problem is that (4.5.13) is a modified version of (4.5.8). It reduces to (4.5.8) for $\varphi_1 = 0$. For given current I and voltage φ_1 the problem (4.5.13) can be solved. Then the voltage-current characteristic is obtained by substituting the result in (4.5.14). As above, the solution of (4.5.13) can be approximated by Taylor expansion in powers of I. The leading term evaluated at x_l has a representation similar to $w_0(x_2)$ above:

$$w_0(x_l) = -\tilde{A}/\tilde{\Sigma},$$

where $\tilde{A}$ and in $\tilde{\Sigma}$ are defined like A and Σ with the parameters κ_2 and κ_3 replaced by

$$\tilde{\kappa}_2 = \frac{\mu_p}{C_2(x_l - x_1)}, \qquad \tilde{\kappa}_3 = \frac{\mu_n}{C_3(x_3 - x_r)}.$$

Since in the preceding Paragraph the first order term evaluated at x_2 was assumed to be positive, $w_1(x_l)$ is positive if φ_1 is not too large. As an approximation of the characteristic, (4.5.14) yields

$$I = -w_0(x_l)/w_1(x_l). \tag{4.5.15}$$

By setting $\delta^4 = 0$ the blocking branch of the characteristic was reduced to $I = 0$. This solution was eliminated from (4.5.13) by the specific scaling used. Thus, (4.5.15) can only be an approximation of the middle branch. For $\varphi_1 = 0$ the current takes the value I_m of the saturation current of the middle branch as expected. The break-over voltage is determined by the intersection point of the middle branch (4.5.15) and the lower branch $I = 0$ of the characteristic. Consequently, the break over voltage φ_{1b} satisfies the equation

$$\tilde{A}(\varphi_{1b}) = 0,$$

which can be rewritten as

$$z^2 - \left(\frac{\mu_p}{\kappa_2} + \frac{\mu_n}{\kappa_3}\right) z + \frac{\mu_n \mu_p}{\kappa_2 \kappa_3} - \frac{\mu_n \mu_p}{\kappa_1 \kappa_4} = 0, \tag{4.5.16}$$

where $z = C_2(x_2 - x_l) = C_3(x_r - x_2)$ holds. An expression for the break over voltage is given by

$$\varphi_{1b} = z^2(C_2 + C_3)/2C_2C_3.$$

The variable z is proportional to the width of the depletion region which implies that punch through does not occur as long as

$$z < \min\left\{\frac{\mu_p}{\kappa_2}, \frac{\mu_n}{\kappa_3}\right\}$$

holds. The smaller solution of (4.5.16) is given by

$$z = \frac{1}{2}\left(\frac{\mu_p}{\kappa_2} + \frac{\mu_n}{\kappa_3} - \sqrt{\left(\frac{\mu_p}{\kappa_2} - \frac{\mu_n}{\kappa_3}\right)^2 + 4\frac{\mu_n \mu_p}{\kappa_1 \kappa_4}}\right)$$

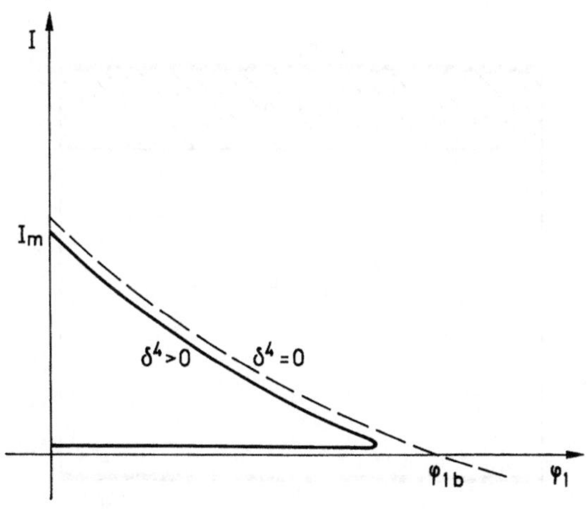

Fig. 4.5.3 Approximate voltage-current characteristic

which can be estimated from above:

$$z < \frac{1}{2}\left(\frac{\mu_p}{\kappa_2} + \frac{\mu_n}{\kappa_3} - \left|\frac{\mu_p}{\kappa_2} - \frac{\mu_n}{\kappa_3}\right|\right) = \min\left\{\frac{\mu_p}{\kappa_2}, \frac{\mu_n}{\kappa_3}\right\}.$$

By combining our results an approximation of the characteristic is given by the segments of (4.5.15) and $I = 0$ which lie between $\varphi_1 = 0$ and $\varphi_1 = \varphi_{1b}$. Again the results can only be expected to be quantitatively correct if the involved currents are not too large. Otherwise the problem (4.5.13), (4.5.14) has to be solved exactly.

4.6 MIS Diode

The MIS (*M*etal *I*nsulator *S*emiconductor) diode has applications as a capacitor with voltage dependent capacitance. In the context of this book an understanding of the performance of this device is an important preparatory step for the analysis of the MOSFET (see the following Section).
The MIS diode consists of a uniformly doped piece of semiconductor coated with a thin layer of insulating material which carries a metal contact called the *gate* (see Fig. 4.6.1). The semiconductor which has an Ohmic contact, is assumed to be of *p*-type (The analysis of this Section also applies to MIS diodes with *n*-type semiconductor after the obvious changes).
In order to simplify the presentation we assume that the insulator is free of charges and that no trapped charges at the interface between the semiconductor and the insulator occur (See [4.22] for a justification of these assumptions). Because of the simple device geometry a one-dimensional model is certainly sufficient for the analysis of the relevant effects. Since no current flow through the insulator is possible the MIS diode is always in thermal equilibrium.

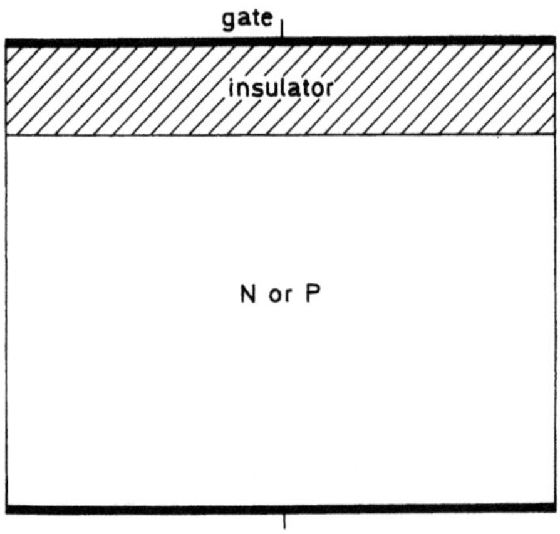

Fig. 4.6.1 Cross section of a MIS diode

The uniform concentration of acceptors in the semiconductor is introduced as a reference value for the scaling of the doping profile and the thickness of the semiconductor as the reference length. Thus, the Poisson equation for the scaled potential V

$$\lambda^2 V'' = \delta^2 e^V - \delta^2 e^{-V} + 1$$

holds in the interval (0, 1) representing the semiconductor part of the device. At the bottom of the device we impose the Ohmic contact boundary condition

$$V(1) = V_{bi} = -\ln\left(\frac{1 + \sqrt{1 + 4\delta^4}}{2\delta^2}\right).$$

The insulator is located in the interval $(-d, 0)$ where d denotes the scaled thickness of the insulating layer. Since no charges are present in the insulator the Laplace equation

$$V'' = 0$$

holds in $(-d, 0)$. At the interface between semiconductor and insulator continuity of the potential and the electric displacement is required:

$$V(0+) = V(0-), \qquad \varepsilon_s V'(0+) = \varepsilon_{\text{ins}} V'(0-),$$

where ε_s and ε_{ins} denote the permittivities of the semiconductor and the insulator, respectively. At the gate contact the potential is prescribed by

$$V(-d) = V_{bi} + \tilde{U}.$$

In the case of an ideal MIS diode the contact voltage is given by $\tilde{U}$. An ideal MIS diode is characterized by a vanishing metal-semiconductor work-function difference φ_{ms} (see [4.22] for details). In realistic cases the contact voltage is given by $\tilde{U} + \varphi_{ms}$.
The first step in the analysis is the computation of the potential and the electric field in the insulator:

$$V(x) = V(0) + (V(0) - V_{bi} - \tilde{U})x/d,$$
$$V'(x) = (V(0) - V_{bi} - \tilde{U})/d \qquad \text{in} \quad (-d, 0).$$

This result is substituted in the interface conditions at $x = 0$:

$$\frac{\varepsilon_s d}{\varepsilon_{\text{ins}}} V'(0) = V(0) - V_{bi} - \tilde{U},$$

which reduces the problem to a boundary value problem on the interval (0, 1) with a mixed boundary condition at $x = 0$.
By introducing the change of variables

$$W = \gamma V, \qquad \xi = x\sqrt{\gamma}/\lambda$$

—where $\gamma = (\ln \delta^{-2})^{-1}$ is a small parameter—the problem is transformed to

$$\partial_\xi^2 W = \exp\left(\frac{W-1}{\gamma}\right) - \exp\left(\frac{-W-1)}{\gamma}\right) + 1 \quad \text{in} \quad (0, \sqrt{\gamma}/\lambda),$$

$$\alpha\partial_\xi W(0) = W(0) - \gamma V_{bi} - U, \qquad W(\sqrt{\gamma}/\lambda) = \gamma V_{bi}. \tag{4.6.1}$$

The new parameters in (4.6.1) are defined by

$$\alpha = \frac{\varepsilon_s d\sqrt{\gamma}}{\varepsilon_{\text{ins}}\lambda}, \qquad U = \gamma\tilde{U}.$$

In Section 4.2 it was shown that the scaled width of the depletion region of a P-N-junction in thermal equilibrium is $O(\lambda/\sqrt{\gamma})$. If the thickness of the insulator is of that order of magnitude and the semiconductor is much thicker, $\lambda/\sqrt{\gamma}$ is a small parameter but α is $O(1)$. For significant doping levels, γ is also a small parameter. The scaling of the potential has been chosen such that the scaled built-in potential is $O(1)$:

$$\gamma V_{bi} = -1 + \text{TST}.$$

It will be assumed that U is also $O(1)$ which means that the applied voltage is of the order of magnitude of the built-in potential.

The solution of (4.6.1) will be approximated by letting $\lambda/\sqrt{\gamma}$ tend to zero. The approximating problem is posed on the infinite interval $(0, \infty)$. After multiplication by $\partial_\xi W$ the differential equation can be integrated, which gives

$$(\partial_\xi W)^2/2 = \gamma \exp\left(\frac{W-1}{\gamma}\right) + \gamma \exp\left(\frac{-W-1}{\gamma}\right) + W + k, \tag{4.6.2}$$

where the constant of integration k is computed by evaluating (4.6.2) at $\xi = \infty$ (where, obviously, $\partial_\xi W = 0$ holds):

$$k = 1 - \gamma + \text{TST}.$$

Since for $U = 0$ the solution is constant and equal to the value γV_{bi} prescribed at infinity, we expect the solution to be decreasing (increasing) for $U > 0$ $(U < 0)$. This observation determines the sign when taking the square root of (4.6.2):

$$\partial_\xi W/\sqrt{2} = -\text{sign}(U)\sqrt{\gamma \exp\left(\frac{W-1}{\gamma}\right) + \gamma \exp\left(\frac{-W-1}{\gamma}\right) + W + k}\,. \tag{4.6.3}$$

Evaluation of (4.6.3) at $\xi = 0$ and substitution for $\partial_\xi W(0)$ (by using the initial condition) leads to the equation

$$(W(0) - \gamma V_{bi} - U)/\alpha\sqrt{2}$$
$$= -\text{sign}(U)\sqrt{\gamma \exp\left(\frac{W(0)-1}{\gamma}\right) + \gamma \exp\left(\frac{-W(0)-1}{\gamma}\right) + W(0) + k} \tag{4.6.4}$$

for $W(0)$. It is easy to show that this equation is uniquely solvable (see [4.9]).

This reduces the problem to an initial value problem for the first order equation (4.6.3). Because of the similarity of the differential equations (4.6.3) and (4.2.14) parts of the analysis below are very similar to that of the Paragraph "Strongly asymmetric junctions" of Section 4.2. In particular, we refer the reader to the detailed explanation of the methods of singular perturbation theory there.
The aim of our analysis is to obtain an approximation for the charge

$$Q = \int_0^1 (n - p - C)\, dx$$

in the semiconductor, where the integrand (the space charge) is given by the right hand side of (4.6.1). The *capacitance* of the MIS diode is defined by $C = \partial Q/\partial U$. The case $U = 0$, where the potential W is constant and equal to the built-in potential (~ -1), is commonly referred to as *flat band condition.* The charge Q vanishes in this situation. The density of the majority carriers, i.e. the holes, is approximately equal to the doping concentration and the electron density is much smaller.
For negative U the potential is smaller than in the flat band case in the vicinity of the interface and, accordingly, an *accumulation* of holes occurs. The situation where U is positive but small enough for $W(0)$ to be between -1 and 0, is called *depletion* because both carrier densities are so small that the space charge is essentially equal to the density of the fixed charges. This is also true for $W(0)$ between 0 and 1 but in this case the additional phenomenon occurs that the minority carrier density at the interface is larger than the majority carrier density. Therefore this situation is called *weak inversion.* Finally, the case $W(0) > 1$ is referred to as *strong inversion.*

Accumulation

The equation (4.6.4) cannot be solved explicitely, but it is amenable to an asymptotic analysis as $\gamma \to 0$. For the case of accumulation we make the ansatz

$$W = -1 - \gamma \ln \gamma^{-1} + \gamma y.$$

Substitution in (4.6.4) and letting $\gamma \to 0$ gives an equation for $y(0)$ with the solution:

$$y(0) = \ln(2\alpha^2/U^2).$$

Similarly an initial layer equation for y in terms of the fast variable $\tau = \xi/\gamma$ is determined from (4.6.3):

$$\partial_\tau y = \sqrt{2} e^{-y/2}$$

The solution $y = 2 \ln(\tau/\sqrt{2} - \alpha\sqrt{2}/U)$ of the initial value problem can be used to compute an approximation for the charge by noting that in the initial layer the space charge is dominated by the hole density

$$p = \exp\left(\frac{-W-1}{\gamma}\right) = \frac{1}{\gamma}e^{-y} = \frac{1}{\gamma}(\tau/\sqrt{2} - \alpha\sqrt{2}/U)^{-2}.$$

A simple integration yields

$$Q = \frac{\lambda}{\sqrt{\gamma}} U/\alpha.$$

In terms of unscaled variables this can be written as

$$Q = \varepsilon_{\text{ins}} U/d$$

which implies that in the case of accumulation the (unscaled) capacitance can be approximated by the insulator capacitance

$$C_{\text{ins}} = \varepsilon_{\text{ins}}/d.$$

The approximation of the potential computed above does not converge to the equilibrium value as the fast variable τ tends to infinity. This disturbing situation can be eliminated by introducing a transition layer for values of W close to -1 (see Problem 4.11). The transition layer solution tends to -1 as the corresponding layer variable tends to infinity and it can be matched to the inner layer solution. The contribution of the transition layer to the charge is small compared to the computed approximation, and therfore neglected.

Depletion—Weak Inversion

When $W(0) \in (-1, 1)$ holds, both exponential terms (4.6.4) can be neglected. The resulting equation has the solution

$$W(0) = -1 + (\sqrt{\alpha^2 + 2U} - \alpha)^2/2$$

which is in $(-1, 1)$ if

$$U \in (0, 2 + 2\alpha).$$

The upper bound marks the limit between weak and strong inversion. In terms of unscaled variables it is given by

$$U = 2V_{bi} + 2\sqrt{U_{\text{ref}} V_{bi}},$$

where $U_{\text{ref}} = \varepsilon_s q\tilde{C}/C_{\text{ins}}^2$ is a reference voltage and $\tilde{C}$ denotes the concentration of acceptors used for scaling the doping profile. The built-in potential is approximated by $V_{bi} = U_T/\gamma = U_T \ln(\tilde{C}/n_i)$ in the above equation.
Returning to scaled variables, the onset of weak inversion is determined by requiring $W(0) = 0$, which leads to the voltage

$$U = 1 + \alpha\sqrt{2}.$$

Since W can be expected to be between -1 and 1, the problem (4.6.1) reduces to a double obstacle problem in the limit $\gamma \to 0$ (compare to the thermal equilibrium problem (4.2.7), (4.2.8), (4.2.9) for the P-N diode). The limiting solution is given by

$$W = \begin{cases} -1 + (\xi - \xi_d)^2/2 & \text{for} \quad \xi \leqslant \xi_d\,, \\ -1 & \text{for} \quad \xi > \xi_d\,, \end{cases}$$

where the width of the depletion region is

$$\xi_d = \sqrt{\alpha^2 + 2U} - \alpha\,.$$

For later reference we note that the depletion width at the onset of strong inversion (for $U = 2 + 2\alpha$) is 2.
The computation of the charge reduces to the multiplication of the acceptor concentration by the depletion width. In terms of unscaled variables it is given by

$$Q = C_{\mathrm{ins}} U_{\mathrm{ref}} (\sqrt{1 + 2U/U_{\mathrm{ref}}} - 1)\,.$$

The capacitance depends on the voltage in this regime and is given by

$$C = C_{\mathrm{ins}} / \sqrt{1 + 2U/U_{\mathrm{ref}}}\,,$$

which reduces to the insulator capacitance for $U = 0$.

Strong Inversion

For analyzing the case of strong inversion we make the ansatz

$$W_{\mathrm{inv}} = 1 + \gamma \ln \gamma^{-1} + \gamma z$$

for the potential in an inversion layer. An equation for $z(0)$ is obtained by going to the limit $\gamma \to 0$ in (4.6.4). It has the solution

$$z(0) = \ln \frac{(U - 2 - 2\alpha)(U - 2 + 2\alpha)}{2\alpha^2}\,.$$

Obviously, we only consider voltages larger than the above obtained threshold $2 + 2\alpha$ for strong inversion. With the fast variable $\tau = \xi/\gamma$, the layer equation

$$\partial_\tau z = -\sqrt{2(e^z + 2)}$$

is obtained from (4.6.3). The solution of the initial value problem for z is given by

$$z = \ln(2 \sinh^{-2}(\tau + c)) \qquad \text{with} \qquad c = \frac{1}{2} \ln \frac{U - 2 + 2\alpha}{U - 2 - 2\alpha}\,.$$

Since the solution decreases rapidly (as a function of the fast variable τ), strong inversion only takes place within an *inversion layer* of thickness $O(\gamma)$. Outside this layer, there is a depletion region where the potential satisfies the reduced equation

$$\partial_\xi^2 W_{\text{depl}} = 1\,.$$

The general solution

$$W_{\text{depl}} = a + b\xi + \xi^2/2$$

can be matched to the inversion layer solution W_{inv} by expressing W_{inv} in terms of the slow variable ξ:

$$W_{\text{inv}} = 1 - 2\xi + o(1)\,.$$

Matching by equating coefficients implies $a = 1$, $b = -2$ and, thus,

$$W_{\text{depl}} = -1 + (\xi - 2)^2/2\,.$$

Similarly to above this solution is valid until W_{depl} takes the value -1 at $\xi = 2$ which is the depletion width. For $\xi > 2$, W is approximately equal to -1. Note that the depletion width is equal to its maximal value obtained in the case of weak inversion and does not depend on the applied voltage any more.

For the computation of the charge, the contributions from the depletion and inversion layers have to be added. Within the inversion layer the electron concentration

$$n = \exp\left(\frac{W_{\text{inv}} - 1}{\gamma}\right) = \frac{1}{\gamma} e^{z} = \frac{2}{\gamma} \sinh^{-2}(\tau + c)$$

dominates and integration gives the inversion layer contribution Q_{inv} of the total charge:

$$Q_{\text{inv}} = 4\lambda/\sqrt{\gamma}(e^{2c} - 1)^{-1} = \frac{\lambda}{\alpha\sqrt{\gamma}}(U - 2 - 2\alpha)\,.$$

Adding the depletion layer charge we end up with the unscaled total charge

$$Q = C_{\text{ins}}(U - 2V_{bi})\,.$$

As in the case of accumulation, this is a linear relation between the voltage and the charge which means that the capacitance is independent of the voltage and equal to the insulator capacitance C_{ins}.

Our results are summarized in Fig. 4.6.2. Depending on the value of U different cases occur. We have

accumulation	for	$U < 0$,
depletion	for	$0 < U < 1 + \alpha\sqrt{2}$,
weak inversion	for	$1 + \alpha\sqrt{2} < U < 2 + 2\alpha$,
strong inversion	for	$2 + 2\alpha < U$.

We have seen that the capacitance of the MIS-diode depends on the voltage. In the accumulation and strong inversion regimes it is approximately constant and equal to the insulator capacitance. In depletion and weak inver-

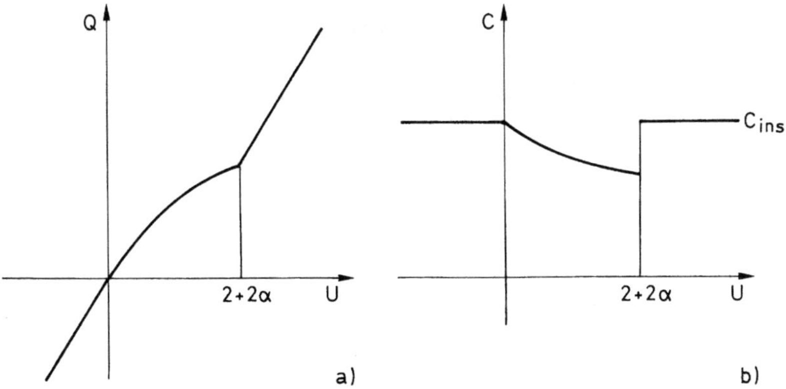

Fig. 4.6.2 (a) Charge and (b) capacitance vs. voltage

sion it is a decreasing function of U and takes its minimum value close to the threshold voltage separating weak and strong inversion. The jump discontinuity in the computed approximated for the capacitance indicates fast variation close to $U = 2 + 2\alpha$, which we have not analyzed in detail.

4.7 MOSFET

The MOSFET (*M*etal *O*xide *S*emiconductor *F*ield *E*ffect *T*ransistor) is one of the most important semiconductor devices. In general, it is used as a switch. Its importance is due to the fact that no power is consumed by the switching (as opposed to the bipolar transistor). The MOSFET is a *unipolar* device, i.e. charge transport is due to only one type of charge carrier. Since the mobility of the electrons is usually higher than that of the holes, most MOSFETs are so called *n-channel* devices (this term will be explained below). Although the following discussion will be restricted to this group of devices the results carry over to *p*-channel devices after obvious changes.
Like the bipolar transistor, the MOSFET is a three layer device with two highly doped *n*-regions called *source* and *drain* and a *p*-region called *bulk* with lower doping. Source and drain always have Ohmic contacts whereas the bulk might be a floating region (no contact), e.g. in the case of SOI (*S*ilicon *O*n *I*nsulator) technology. Between the source and drain contacts the semiconductor material is coated with a thin layer of oxide with a metal contact, the so called *gate* (see Fig. 4.7.1).
The analysis of the preceding Section applies to the situation close to the semiconductor oxide interface BC away from the endpoints B and C. Application of a sufficiently large voltage at the gate generates an inversion layer in the so-called *channel*, i.e. the part of the bulk region close to the interface. The fact that the electron density dominates the hole density in this region has led to the term *n*-channel. This *n*-channel is able to carry a significant

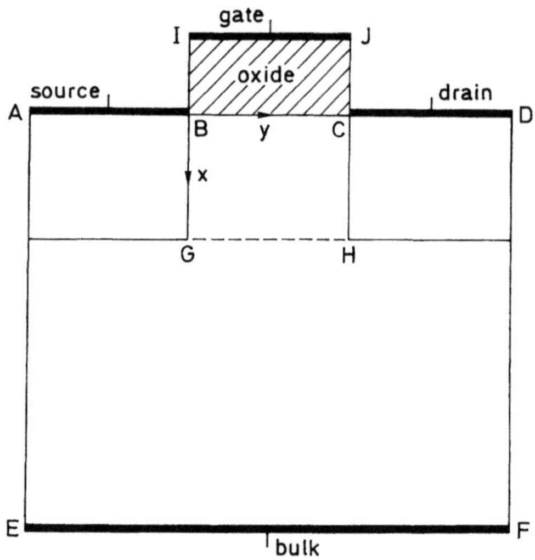

Fig. 4.7.1 Cross section of a simplified MOSFET geometry

current from source to drain as soon as there is a potential difference between these two contacts. Thus, the source-drain current can be switched on and off by applying different voltages to the gate.

An important parameter for the performance of the MOSFET is the channel length L (the length of the segment BC). *Long channel* devices are characterized by the requirement that the Debye length in the channel is significantly smaller than the channel length. This implies that the depletion layers corresponding to the source and drain junctions are well separated, and perturbation arguments lead to locally one-dimensional problems. Analytic formulas for the static characteristics of long channel MOSFETs have been obtained by models of different degrees of sophistication (see [4.22] for an overview). Recently, Ward [4.25] analyzed long channel MOSFETs by using methods of asymptotic analysis. The following presentation is based on his work ([4.25], [4.26]). We point out that no comparable results for short channel devices are available because they are intrinsically dominated by two-dimensional behaviour.

Our mathematical model is a two-dimensional version of the scaled stationary drift-diffusion equations written in terms of the quasi Fermi potentials:

$$\begin{aligned} &\lambda^2 \Delta V = \delta^2 e^{V-\varphi_n} - \delta^2 e^{\varphi_p - V} - C, \\ &J_n = -\mu_n \delta^2 e^{V-\varphi_n} \operatorname{grad} \varphi_n, \qquad \operatorname{div} J_n = 0, \\ &J_p = -\mu_p \delta^2 e^{\varphi_p - V} \operatorname{grad} \varphi_p, \qquad \operatorname{div} J_p = 0. \end{aligned} \tag{4.7.1}$$

These equations are posed on the rectangle $AEFD$ (see Fig. 4.7.1) and a coordinate system with the origin in the point B has been introduced. The validity of the two-dimensional model is restricted to devices with sufficient *channel width*, i.e. the width of the active region of the device in the direction perpendicular to the x-y-plane.

In (4.7.1) the effects of recombination and generation are neglected and the mobilities are assumed to be constant. Although these assumptions cannot be justified in general the simplified model produces qualitatively correct voltage-current characteristics. In the scaling, the *channel length* L (the length of the segment BC in Fig. 4.7.1) has been chosen as the characteristic length. The doping concentration has been scaled by the doping in the bulk region which we assume to be constant. This is a rather stringent assumption which does not hold for many practical situations. However, as demonstrated in [4.25] the principal ideas of the analysis of this Section can be carried over to the case of a doping profile which varies in the x-direction (see Fig. 4.7.1). The analysis below will show that the performance of the MOSFET can be explained by concentrating on the p-region. This has motivated our choice of the scaling where the maximal doping concentration is not scaled to 1 as usual. The parameter λ denotes the scaled Debye length. In the following, smallness of λ will be assumed which means, that we are dealing with long channel devices.

The rectangle $BCJI$ represents the oxide which we asume to be free of charges such that the Laplace equation

$$\Delta V = 0$$

holds there.

The quasi Fermi potentials satisfy Dirichlet boundary conditions along the contact segments AB, CD, and EF. We have

$$\varphi_n = \varphi_p = 0 \quad \text{on} \quad AB,$$
$$\varphi_n = \varphi_p = \tilde{U}_D \quad \text{on} \quad CD,$$
$$\varphi_n = \varphi_p = \tilde{U}_B \quad \text{on} \quad EF.$$

The voltages are referenced with respect to the source, i.e. $\tilde{U}_D$ is the drain-source voltage and $\tilde{U}_B$ the bulk-source voltage. Along the artificial boundaries AE and DF and along the interface BC homogeneous Neumann conditions for the quasi Fermi potentials hold. So currents can leave or enter the rectangle under consideration $AEFD$ only through the source, drain, and bulk contacts.

The potential satisfies the usual Ohmic contact boundary conditions at source, bulk, and drain as well as homogeneous Neumann conditions along the artificial boundary segments AE, DF, BI, and CJ. Along the interface, continuity of the potential and the vertical component of the electric displacement are required:

$$V(0-, y) = V(0+, y), \qquad \varepsilon_{ox}\partial_x V(0-, y) = \varepsilon_s \partial_x V(0+, y),$$

where ε_{ox} and ε_s are the permittivities of the oxide and the semiconductor, respectively. At the gate contact IJ the boundary condition

$$V(-d, y) = V_{bi}(0, y) + \tilde{U}_G$$

holds, where d denotes the oxide thickness and $\tilde{U}_G$ is related to the gate-source voltage as explained in the preceding Section.

Derivation of a Simplified Model

The aim of this Paragraph is to reduce the model to a boundary value problem posed on the smaller rectangle $BGHC$ by introducing several simplifying assumptions.
A singular perturbation analysis with $\lambda \to 0$ combined with the arguments from the beginning of this Chapter leads to approximations of the quasi Fermi level corresponding to the majority carriers by a constant, in each p- and n-region. Thus, as a first simplification we set

$$\begin{aligned} \varphi_n &= 0 && \text{in the source region},\\ \varphi_n &= \tilde{U}_D && \text{in the drain region},\\ \varphi_p &= \tilde{U}_B && \text{in the bulk region}. \end{aligned} \tag{4.7.2}$$

Next we consider the potential in the oxide, and assume that the oxide thickness is small compared to the channel length, i.e. $d \ll 1$. By introducing the independent variable $\xi = x/d$ the insulator region is transformed to a square and the potential satisfies

$$\partial_\xi^2 V + d^2 \partial_y^2 V = 0 .$$

In the limit $d \to 0$ the potential in the insulator is the solution of a one-dimensional problem. The resulting approximation for the potential violates the Neumann conditions at the artificial boundary segments. However, it is easy to see that $O(d)$-boundary layer correctors are sufficient as a remedy. Thus, the approximation obtained by solving the one-dimensional equation $\partial_\xi^2 V = 0$ is uniformly valid in the oxide region. As in the case of the MIS diode it leads to a mixed boundary condition at the interface:

$$\frac{\varepsilon_s d}{\varepsilon_{ox}} \partial_x V = V - V_{bi} - \tilde{U}_G \qquad \text{at} \quad BC$$

Finally, the problem is reduced to the rectangle $BGHC$. Since the hole quasi Fermi potential is assumed to be constant in this region we only consider the Poisson equation and the electron continuity equation. With the rescaled variables

$$W = \gamma V, \qquad \Phi_n = \gamma \varphi_n, \qquad \xi = x/\tilde{\lambda} \qquad (\tilde{\lambda} = \lambda/\sqrt{\gamma})$$

we have

$$\begin{aligned} &\partial_\xi^2 W + \tilde{\lambda}^2 \partial_y^2 W = \exp\left(\frac{W - \Phi_n - 1}{\gamma}\right) - \exp\left(\frac{U_B - W - 1}{\gamma}\right) + 1,\\ &\partial_\xi\left(\mu_n \exp\left(\frac{W - \Phi_n - 1}{\gamma}\right)\partial_\xi \Phi_n\right) + \tilde{\lambda}^2 \partial_y\left(\mu_n \exp\left(\frac{W - \Phi_n - 1}{\gamma}\right)\partial_y \Phi_n\right) = 0, \end{aligned} \tag{4.7.3}$$

where U_B and, in the sequel, U_D and U_G denote rescaled versions of the applied voltages. Note that the approximate built-in potential $U_T/\gamma = U_T \ln(\tilde{C}/n_i)$ is the scaling factor for voltages in (4.7.3).
By restricting our attention to the smaller rectangle $BGHC$ we ignore the electron current in the rest of the device and totally ignore the hole current. The critique of the simplified model given below also applies to these simplifications.
The problem is completed by prescribing boundary conditions. In terms of the rescaled quantities the mixed boundary condition at the interface BC reads

$$\alpha \partial_\xi W = W - \gamma V_{bi} - U_G \qquad \text{at} \quad BC \tag{4.7.4}$$

where α is defined as in the preceding Section by

$$\alpha = \frac{\varepsilon_s d}{\varepsilon_{ox} \tilde{\lambda}} .$$

The analysis below will show that the values of the potential along the *pn*-junctions BG and CH do not affect the results, and so we leave them unspecified. In order to avoid nonuniformities along the artificial boundary GH, zero space charge is required there:

$$W = -1 + U_B - \gamma \ln \frac{1}{2}\left(1 + \sqrt{1 + 4\exp\left(\frac{U_B - \Phi_n - 2}{\gamma}\right)}\right) \quad \text{at } GH \tag{4.7.5}$$

Recalling the results for the MIS diode, this condition is satisfactory as long as the edge of the depletion region stays away from GH. A quantitative formulation of this requirement will be given below.
Prescribing boundary conditions for the electron quasi Fermi level is a more subtle problem. Starting with the easy parts, we pose a homogeneous Neumann condition along the interface and the following Dirichlet conditions at the *pn*-junctions:

$$\Phi_n = 0 \quad \text{at} \quad BG, \qquad \Phi_n = U_D \quad \text{at} \quad CH, \tag{4.7.6}$$

which are motivated by (4.7.2).
The previous discussion fails to give a handle on the behavior of Φ_n at the artificial boundary GH. Therefore a boundary condition is used which reflects favourable operating conditions. The above description of the device behaviour raises the expectation that current flow only takes place in the direction tangential to the interface. This motivates the assumption that no current flow occurs across GH, resulting in the boundary condition

$$\partial_\xi \Phi_n = 0 \qquad \text{at} \quad BC \text{ and } GH. \tag{4.7.7}$$

The results of numerical simulations [4.17, Chap. 9]. Show that these assumptions are not necessarily justified. In [4.17] a parasitic effect is analyzed which can be explained by interpreting the MOSFET as a bipolar transistor where the source, bulk, and drain are identified with emitter, base, and collector, respectively. Depending on the details of the geometry and the doping profile a small bulk current can be the reason for a significant current from source to drain without an inversion layer being present. As the numerical results show, this current might very well have a significant component in the direction perpendicular to GH. According to these observations, certain parasitic effects are a priorily precluded by the conditions (4.7.7). Nonetheless, the model (4.7.3)–(4.7.7) will be sufficient for the computation of qualitatively correct device characteristics. The total current from source to drain will be approximated by the electron current across the P-N junction BG (or CH).

A Quasi One-Dimensional Model

In this Paragraph the smallness of the parameter $\tilde{\lambda}$ will be exploited for a further simplification of the model (4.7.3)–(4.7.7). As $\tilde{\lambda} \to 0$ the Poisson equation reduces to the ordinary differential equation

$$\partial_\xi^2 W = \exp\left(\frac{W - \Phi_n - 1}{\gamma}\right) - \exp\left(\frac{U_B - W - 1}{\gamma}\right) + 1. \qquad (4.7.8)$$

Assuming Φ_n to be given, this equation subject to the boundary conditions (4.7.4), (4.7.5) constitutes a one-dimensional boundary value problem for each value of y, which has a unique solution. As mentioned above the approximation of W computed in this way is independent of the boundary values at the P-N junctions. Since layer corrections along BG and CH do not significantly affect the final results for the current, they will not be considered here.

The asymptotic analysis of the continuity equation is less straightforward. The reduced problem consisting of the differential equation

$$\partial_\xi\left(\mu_n \exp\left(\frac{W - \Phi_n - 1}{\gamma}\right)\partial_\xi \Phi_n\right) = 0$$

and homogeneous Neumann conditions does not have a unique solution. It only allows the conclusion that the quasi Fermi level is independent of ξ, i.e. $\Phi_n = \Phi_n(y)$. A singular perturbation problem of this kind has been dealt with in [4.5, Section 4.3] where a formal approximation of the solution is derived. A justification for the formal approach can be found in [4.3].

Additional information on the limiting solution can be obtained by integrating the original differential equation in (4.7.3) in the ξ-direction. Using the boundary conditions, the result is

$$\int_0^{\xi^*} \partial_y\left(\mu_n \exp\left(\frac{W - \Phi_n - 1}{\gamma}\right)\partial_y \Phi_n\right) d\xi = 0$$

where ξ^* denotes the ξ-value corresponding to the boundary GH. Denoting the solution of (4.7.8) for given Φ_n by $W(\Phi_n)$, we introduce the one-dimensional electron density

$$N(\Phi_n) = \int_0^{\xi^*} \exp\left(\frac{W(\Phi_n) - \Phi_n - 1}{\gamma}\right) d\xi .$$

Since, in the limit $\tilde{\lambda} \to 0$, Φ_n only depends on y, the limit of the above equation can be written as

$$\partial_y(\mu_n N(\Phi_n)\partial_y \Phi_n) = 0 .$$

Subject to (4.7.6), this is a one-dimensional boundary value problem in the y-direction. Since its formulation involves the solution of a problem in the ξ-direction it might be called quasi one-dimensional.
The current from source to drain is equal to the current across an arbitrary vertical cross section of the rectangle $BGHC$ and is given by

$$I = \mu_n N(\Phi_n)\partial_y \Phi_n . \tag{4.7.9}$$

Since the only y-dependence in the problem for the potential originates from Φ_n, the one-dimensional electron density does not depend on y explicitely. Therefore the above equation can be integrated from $y = 0$ to 1:

$$I = \mu_n \int_0^{U_D} N(\Phi_n)\, d\Phi_n . \tag{4.7.10}$$

Computation of the One-Dimensional Electron Density

Noting the similarity of the one-dimensional problem for the potential to the MIS diode problem, we expect that the results for the MIS diode essentially carry over to the present situation. The only difference between the two problems is the occurence of the parameters Φ_n and U_B in (4.7.8). Leaving the computational details to the reader, we only summarize the results.
In general, MOSFETs are not operated in the accumulation regime. Thus, we restrict ourselves to the case $U_G > U_B$. Depletion or weak inversion occurs for values of U_G in the interval

$$[U_B, 2 + \Phi_n + \alpha\sqrt{2(2 + \Phi_n - U_B)}] . \tag{4.7.11}$$

The condition $U_B < 2$ which is necessary for the validity of this analysis corresponds to requiring that a not too large forward bias is applied to the source-bulk junction. The onset of strong inversion corresponds to the right end of the above interval or to the condition $W(0) = 1 + \Phi_n$.
The approximate solution in the case of depletion or weak inversion is determined by

$$W(0) = U_G - 1 + \alpha^2 - \alpha\sqrt{\alpha^2 + 2(U_G - U_B)},$$

$$W(\xi) = \begin{cases} -1 + U_B + (\xi - \xi_d)^2/2 & \text{for} \quad \xi \leqslant \xi_d, \\ -1 + U_B & \text{for} \quad \xi > \xi_d, \end{cases} \tag{4.7.12}$$

where the depletion width is given by

$$\xi_d = \sqrt{\alpha^2 + 2(U_G - U_B)} - \alpha .$$

Note that the potential is independent of the electron quasi Fermi level in this case. The maximal depletion width, occurring at the onset of strong inversion, is $\sqrt{2(2 + \Phi_n - U_B)}$.

In strong inversion the potential in the inversion layer is approximated by

$$W_{\text{inv}} = 1 + \Phi_n + \gamma \ln \gamma^{-1} + \gamma z$$

with

$$z = \ln\left((2 + \Phi_n - U_B) \sinh^{-2}\left(\sqrt{1 + (\Phi_n - U_B)/2}\,\frac{\xi}{\gamma} + c\right)\right) \tag{4.7.13}$$

and the constant of integration c being determined from the initial condition

$$z(0) = \ln((2 + \Phi_n - U_G)^2/2\alpha^2 - 2 - \Phi_n + U_B).$$

In the depletion layer the potential is given by (4.7.12) with the depletion width ξ_d replaced by the maximal depletion width given above. Assuming the channel close to the drain to be in the strong inversion regime, we arrive at the condition

$$\sqrt{2(2 + U_D - U_B)} < \xi^*$$

for the validity of our analysis. This inequality means that the maximal depletion width along the channel is smaller than the depth of the source and drain regions.

In the depletion/weak inversion regime the electron density is given by

$$n = \exp\left(\frac{W(0) - \Phi_n - 1 - \xi\xi_d + \xi^2/2}{\gamma}\right).$$

An approximation of $N(\Phi_n)$ can be computed by dropping the quadratic term and replacing ξ^* by ∞ in the integration:

$$N(\Phi_n) = \frac{\gamma}{\xi_d} \exp\left(\frac{W(0) - \Phi_n - 1}{\gamma}\right). \tag{4.7.14}$$

In strong inversion $N(\Phi_n)$ is the sum of contributions from the inversion layer and the depletion region. The electron density in the inversion layer is

$$n_{\text{inv}} = \gamma^{-1} e^z$$

with z given in (4.7.13). Integrating this and adding the depletion layer contribution gives

$$\begin{aligned} N(\Phi_n) = {} & (U_G - 2 - \Phi_n)/\alpha - \sqrt{2(2 + \Phi_n - U_B)} \\ & + \gamma/\sqrt{2(2 + \Phi_n - U_B)}. \end{aligned} \tag{4.7.15}$$

Computation of the Current

Depending on the biasing situation, different cases occur. If the gate voltage U_G is below the *threshold voltage*

$$U_t = 2 + \alpha\sqrt{2(2 - U_B)}$$

depletion/weak inversion prevails throughout the channel. For the computation of the current the one-dimensional electron density given by (4.7.14) can be used:

$$I = \mu_n \frac{\gamma^2}{\xi_d} \exp\left(\frac{W(0) - 1}{\gamma}\right)(1 - e^{-U_D/\gamma}).$$

In this so called *subthreshold region* the current saturates for large drain-source voltages at a value which is transcendentally small in terms of γ. The threshold voltage and the subthreshold characteristic in terms of unscaled quantities are given by

$$U_t = 2V_{bi} + \sqrt{2U_{\text{ref}}(2V_{bi} - U_B)},$$

$$I = \frac{\mu_n \varepsilon_s U_T^2 n_{\text{so}}}{L\tilde{C}x_d}(1 - e^{-U_D/U_T}),$$

where U_{ref} is defined as in the preceding Section,

$$n_{\text{so}} = n_i \exp\left(\frac{1}{U_T}(U_G - V_{bi} + U_{\text{ref}} - \sqrt{U_{\text{ref}}(U_{\text{ref}} + 2U_G - 2U_B)})\right)$$

denotes the surface electron concentration close to the source and

$$x_d = \sqrt{\varepsilon_s/(q\tilde{C})}(\sqrt{U_{\text{ref}} + 2U_G - 2U_B} - \sqrt{U_{\text{ref}}})$$

is the depletion width. Note that $U_G = U_t$ implies $n_{\text{so}} = \tilde{C}$, i.e. the threshold voltage marks the onset of strong inversion close to the source.

For $U_G > U_t$ two different possibilities have to be accounted for. In the case

$$U_G > 2 + U_D + \alpha\sqrt{2(2 + U_D - U_B)}, \tag{4.7.16}$$

called *non-saturation region*, the whole channel is in strong inversion. Thus, the formula (4.7.15) for $N(\Phi_n)$ applies and the current is given by

$$I = \mu_n\left(\frac{1}{\alpha}(U_G - 2 - U_D/2)U_D - \frac{2\sqrt{2}}{3}(2 + U_D - U_B)^{3/2} + \frac{2\sqrt{2}}{3}(2 - U_B)^{3/2} + \gamma\sqrt{2(2 + U_D - U_B)} - \gamma\sqrt{2(2 - U_B)}\right) \tag{4.7.17}$$

which—in terms of unscaled variables—reads

$$I = \frac{\mu_n}{L} C_{ox}(U_G - 2V_{bi} - U_D/2)U_D$$
$$- \frac{\mu_n}{L}\sqrt{\varepsilon_s q \tilde{C}}\left(\frac{2\sqrt{2}}{3}(2V_{bi} + U_D - U_B)^{3/2} - \frac{2\sqrt{2}}{3}(2V_{bi} - U_B)^{3/2}\right.$$
$$\left. - U_T\sqrt{2(2V_{bi} + U_D - U_B)} + U_T\sqrt{2(2V_{bi} - U_B)}\right).$$

Considering the dependence of the current on the drain voltage for a fixed gate voltage $U_G > U_t$ the formula (4.7.17) holds as long as

$$U_D < U_{Dsat} = U_G - 2 + \alpha^2 - \alpha\sqrt{\alpha^2 + 2U_G - 2U_B}$$

is satisfied. The saturation voltage is determined by assuming equality in (4.7.16). For larger values of U_D a phenomenon called *pinch-off* occurs. A transition from strong inversion to weak inversion takes place at the *pinch-off point* where the quasi Fermi level takes the value U_{Dsat}. In this case the one-dimensional electron density is given by (4.7.15) for $0 < \Phi_n < U_{Dsat}$ and by (4.7.14) for $U_{Dsat} < \Phi_n < U_D$. The current is given by

$$I = I_{sat} + \mu_n \frac{\gamma^2}{\xi_d} \exp\left(\frac{W(0) - 1}{\gamma}\right)(e^{-U_{Dsat}/\gamma} - e^{-U_D/\gamma})$$

which is essentially equal to the saturation current I_{sat} obtained by substituting $U_D = U_{Dsat}$ in (4.7.17). Due to this behavior of the characteristic the set of operating points defined by

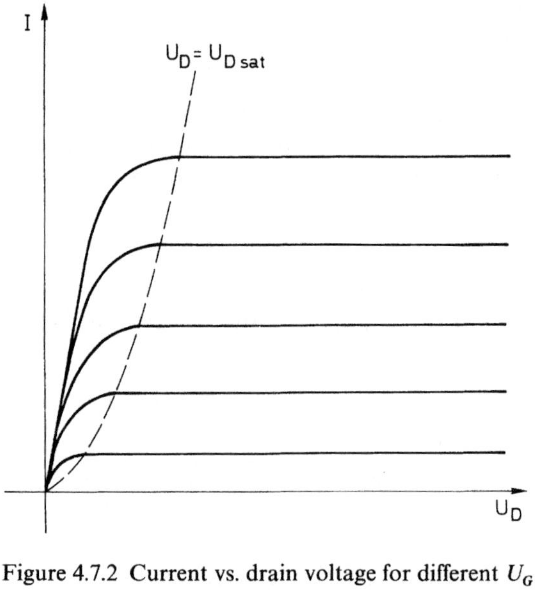

Figure 4.7.2 Current vs. drain voltage for different U_G

$$U_G < 2 + U_D + \alpha\sqrt{2(2 + U_D - U_B)}$$

is called the *saturation region.*
An approximation of the pinch-off point y^* can be computed from (4.7.9) by integration from $y = 0$ to $y = y^*$:

$$y^* = I_{sat}/I.$$

Since in the saturation region the current is only insignificantly larger than I_{sat} we obtain

$$1 - y^* \ll 1.$$

The distance of the pinch-off point to the drain is very small compared to the channel length.

4.8 Gunn Diode

The Gunn diode is an important microwave device. Its performance is based on the *transferred-electron effect* described in Chapter 2 which is responsible for a nonmonotonic velocity-field relation. A typical device consists of a homogeneously doped piece of a semiconductor whose energy-band structure supports the transferred-electron effect (e.g. gallium arsenide (GaAs) or indium phosphide (InP)). This Section is concerned with an explanation of the *Gunn effect* [4.4]: A microwave output can be generated by applying a large enough constant voltage to an n-type piece of GaAs or InP. The presentation will mostly be based on the work of Szmolyan [4.23], [4.24] who put the classical analysis (see [4.22] for references) on a mathematically sound basis. The results of the final Paragraph are new.
Consider a homogeneously doped piece of semiconductor of length L with constant donor concentration C. A one-dimensional unipolar model is given by the differential equations

$$\varepsilon_s \partial_x E = q(n - C),$$
$$\partial_t n = \partial_x (D \partial_x n - n v_{sat} v(E/E_T)),$$

where E denotes the negative electric field, D is the diffusivity and the qualitative behaviour of the velocity $v_{sat} v(E/E_T)$ is given in Fig. 4.8.1 b which also explains the meaning of the saturation velocity v_{sat} and the threshold field E_T. The graph of the scaled function v goes through the origin, has a maximum at 1 and saturates at $v = 1$ for large arguments. For negative values of the argument, v is defined by odd extension. The differential equations hold for x in the interval $(0, L)$, representing the device.
Since large fields are to be expected, the Einstein relation between the diffusivity and the mobility is dropped here (see Chapter 2). In addition the field dependence of the diffusivity will be ignored for simplicity. However, most of the arguments below go through if the field dependence is such that the diffusivity is bounded from above and away from zero (see [4.23], [4.24]), although the computations are more involved.

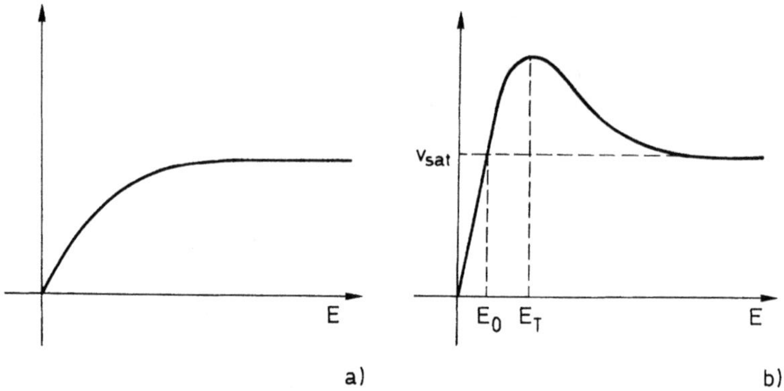

Fig. 4.8.1 Velocity vs. field for (a) Si and (b) GaAs

Ohmic contacts are modelled by the boundary conditions

$$n(0) = n(L) = C$$

and the application of a voltage U is described by an integral condition for the field:

$$\int_0^L E\,dx = U\,.$$

The problem formulation would have to be completed by imposing initial conditions for the electron density. However, our main interest will lie in the study of special solutions of the differential equations rather than in the general initial value problem.

A scaling is introduced where the device length L and the characteristic time L/v_{sat} are the reference quantities for length and time. Carrier densities, electric fields, and voltages are scaled by C, E_T, and $E_T L$, respectively. The scaled problem reads

$$\begin{aligned} &\lambda^2 \partial_x E = n - 1\,, \\ &\partial_t n = \partial_x(\gamma \partial_x n - n v(E))\,, \\ &n(0) = n(1) = 1\,, \qquad \int_0^1 E\,dx = U\,, \end{aligned} \tag{4.8.1}$$

where

$$\lambda^2 = \frac{\varepsilon_s E_T}{qCL}\,, \qquad \gamma = \frac{D}{v_{\text{sat}} L}$$

are the square of the scaled Debye length and the relative strength of diffusive and convective terms. Note, that the same symbols have been used for scaled and unscaled quantities.

Considering typical values for the material dependent parameters (see [4.22]), a device with a length of 10 μm or more, and a doping concentration of about 10^{15} cm^{-3}, both λ^2 and γ are small parameters of the same order of magnitude. Assuming the ratio $\alpha = \gamma/\lambda^2$ to take moderate values, only one small parameter appears in the differential equations:

$$\begin{aligned} \lambda^2 \partial_x E &= n - 1, \\ \partial_t n &= \partial_x(\lambda^2 \alpha \partial_x n - nv(E)). \end{aligned} \tag{4.8.2}$$

In the following two important properties of the system (4.8.1), (4.8.2) are discussed: The loss of stability of homogeneous steady states due to *bulk negative differential conductivity* (NDC) and the existence of *traveling wave* solutions. A combination of these properties will be used for the asymptotic analysis of the Gunn effect.

Bulk Negative Differential Conductivity

Consider a piece of semiconductor with homogeneous carrier density. In this case the current density is given by

$$J = nv(E)$$

and the *bulk differential conductivity* by

$$\partial J/\partial E = nv'(E).$$

Whereas in Si $\partial J/\partial E$ is always positive, bulk NDC occurs in GaAs and InP. Apart from the transferred-electron effect, other physical mechanisms can be responsible for bulk NDC. We only mention impact ionization induced bulk NDC which is used in another microwave device, the IMPATT (*imp*act ionization *a*valanche *t*ransit *t*ime) diode (see [4.22]). For a detailed discussion of bulk NDC-effects caused by recombination-generation we refer to [4.16]. Note that a global form of NDC has been observed in Section 4.5 in connection with the middle branch of the voltage-current characteristic of a thyristor. As opposed to bulk NDC, which is due to microscopic material properties, this effect is caused by the interaction of P-N junctions.
Carrying out the differentiation in the right hand side of the continuity equation leads to

$$\partial_t n = \lambda^2 \alpha \partial_x^2 n - \partial_x n v(E) - nv'(E)\frac{n-1}{\lambda^2},$$

which shows (by the smallness of λ^2) that for values of n away from the equilibrium value 1, the dynamics of the system are dominated by the ordinary differential equation

$$\partial_t n = -nv'(E)\frac{n-1}{\lambda^2}.$$

Obviously the stability of the equilibrium solution $n = 1$ is determined by the sign of $v'(E)$ with stability for $v'(E) > 0$. This heuristic argument has been made rigorous in [4.24]. A stationary solution of (4.8.1), (4.8.2) is given by

$$n = 1, \qquad E = U$$

which is called the trivial solution from now on. The stability of this solution was examined in [4.24] by linearization. It can be shown that for $U > 1$ (which implies $v'(U) < 0$) and λ^2 small enough the trivial solution is unstable whereas it is stable for $U \leqslant 1$. Furthermore a stable nontrivial solution bifurcates from the trivial solution at the critical voltage where the trivial solution looses its stability.

Traveling Waves

An important property of the equations (4.8.2) is the existence of traveling wave solutions, i.e. solutions which only depend on $x - v_0 t$ where v_0 denotes the velocity of the wave. These solutions are strictly valid only for an idealized device of infinite length. However, since the active region of the solutions will be shown to be very small with fast decay at $\pm\infty$, they can be used in the singular perturbation analysis of the following Paragraph as layer terms in a moving internal layer.
The smallness of the active region is reflected in the choice of the variable

$$s = \frac{x - v_0 t}{\lambda^2}.$$

Assuming n and E to be functions of s only, the differential equations can be written as

$$\begin{aligned} \partial_s E &= n - 1, \\ -v_0 \partial_s n &= \partial_s(\alpha \partial_s n - n v(E)). \end{aligned} \tag{4.8.3}$$

Only solutions which converge as $s \to \pm\infty$ can be used as layer terms. Besides, it will be shown in the following Paragraph that the limits of E as $s \to \pm\infty$ have to be equal. Thus, integration of the second equation gives

$$\alpha \partial_s n = n(v(E) - v_0) + v_0 - v(E_\infty) \tag{4.8.4}$$

where E_∞ is the common limit of E as s tends to $\pm\infty$. The further analysis proceeds by studying the phase portraits of (4.8.3), (4.8.4) for various choices of v_0 and E_∞. The above requirements imply that we are looking for a homoclinic orbit of the system with respect to a stationary point $(n, E) = (1, E_\infty)$. In [4.23] it is shown that such a solution can only exist if $v_0 = v(E_\infty)$ holds, which means that the velocity of the wave is equal to the drift velocity of the electrons at infinity. With this assumption the equations read

$$\begin{aligned} \partial_s E &= n - 1, \\ \alpha \partial_s n &= n(v(E) - v_0). \end{aligned} \tag{4.8.5}$$

The number of stationary points of (4.8.5) is equal to the number of solutions

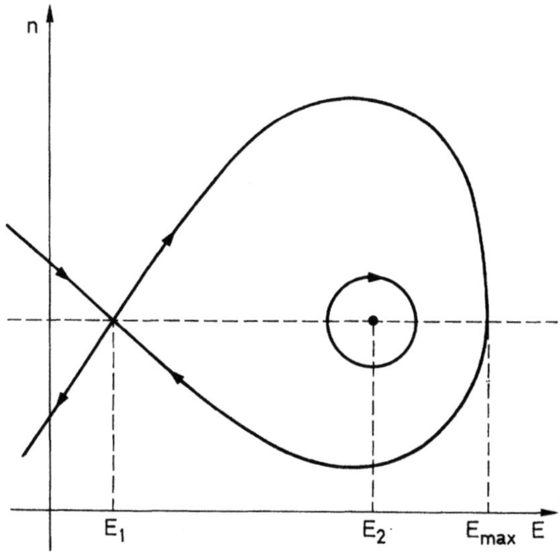

Fig. 4.8.2 Phase portrait

of the equation $v(E) = v_0$. For $v_0 \leqslant 1$ there is only one stationary point which implies that a homoclinic orbit cannot exist. Therefore we assume that v_0 lies between 1 and $v(1)$ from now on. In this case there are two stationary solutions $E_1 < 1 < E_2$. A stability analysis shows that the point $(1, E_1)$ is a saddle and that the eigenvalues of the Jacobian of the right hand side of (4.8.5) evaluated at $(1, E_2)$ are imaginary.
Separation of variables and integration gives

$$\alpha(n - \ln n - 1) = \int_{E_{ref}}^{E} (v(y) - v_0)\, dy \tag{4.8.6}$$

which can be used for drawing a picture of the phase portrait of (4.8.5), Fig. 4.8.2.
For $E_{ref} = E_1$, (4.8.6) describes the stable and unstable manifolds of the stationary point $(1, E_1)$. It is easily seen that the part of the curve with $E \geqslant E_1$ is closed which means that the stable and unstable manifolds meet and a homoclinic orbit exists. On this orbit the maximal value E_{max} of the field satisfies the equation

$$\int_{E_1}^{E_{max}} (v(y) - v_0)\, dy = 0$$

which is known under the name *equal area rule* (see Fig. 4.8.3).
For E_{ref} between E_1 and E_{max}, (4.8.6) is the equation of a closed curve around $(1, E_2)$ corresponding to a periodic solution. Thus, the stationary point $(1, E_2)$ is a center.
The homoclinic orbit is the traveling wave solution we have been looking

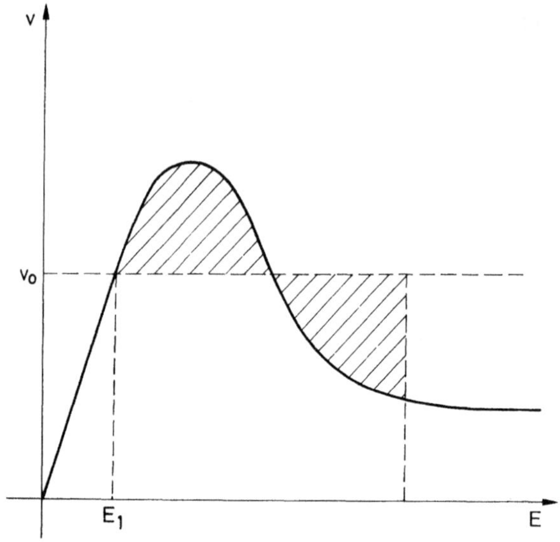

Fig. 4.8.3 Equal area rule

for. The qualitive shape of the wave can be determined from Fig. 4.8.2. The field has the form of a single pulse whereas the electron density forms a *dipole* with a depleted region ($n < 1$) followed by a region of accumulation ($n > 1$).

The Gunn Effect

It was observed in [4.4] that for sufficiently large applied voltages small perturbations of the homogeneous steady state grow, until a stable configuration (called *domain*) is reached which then travels through the semiconductor without changing its form. The electric field outside the domain is lower than the threshold field E_T but it takes values in the region of bulk NDC inside the domain. As soon as the domain leaves the device the electric field grows to a value above E_T throughout the device and a new domain is built. The aim of this Paragraph is to relate the shape and velocity of the domain to the applied voltage.
Applying the methods of singular perturbation theory to (4.8.2), we try to obtain a solution which can be approximated by a solution of the reduced equations

$$\bar{n} - 1 = 0, \qquad \partial_x v(\bar{E}) = 0$$

away from layers. We are interested in the case that the only layer is given by a traveling wave. The above equations would allow for a jump of $\bar{E}$ across the wave if the values at the left and at the right give the same velocity. This would imply that the integral of E changes with time because the

integral of $\bar{E}$ changes as the wave travels through the device and the integral of the contribution from the wave is constant. This contradicts the integral condition in (4.8.1). Thus, $\bar{E}$ has to be constant and the travelling wave is given by the homoclinic orbit constructed above. $\bar{E}$ lies between E_0 and 1 where $v(E_0) = 1$ holds (see Fig. 4.8.1 b).
The integral condition on the field implies $\bar{E} = U$ because the width of the domain is $O(\lambda^2)$. Since we are interested in applied voltages larger than 1 this would make a domain solution impossible. The reason for this problem is that the contribution of the domain to the integral of E is too small. Therefore we shall try to construct a wider domain with larger values of the electric field. By the equal area rule large values of the maximal field imply that the velocity of the wave is close to the saturation velocity (see Fig. 4.8.3). This in turn implies that $\bar{E}$ is close to E_0. These observations motivate the following transformations in the traveling wave problem:

$$E = E_\infty + e/\lambda, \qquad E_\infty = E_0 + \lambda\bar{e}, \qquad \sigma = \lambda s,$$

where e and $\bar{e}$ remain to be determined. Substitution in (4.8.6) gives

$$n - \ln n - 1 \sim \frac{1}{\alpha}\int_{E_\infty}^{E_\infty + e/\lambda} (v(y) - 1 - \lambda v'(E_0)\bar{e})\, dy$$

$$\sim \frac{1}{\alpha}\int_{E_0}^{\infty} (v(y) - 1)\, dy - \frac{1}{\alpha} v'(E_0)\bar{e}e = A - B\bar{e}e,$$

where we assumed that the improper integral on the right-hand side converges. This is an assumption on the speed of convergence of the velocity to its saturation value as the field tends to infinity. Assuming knowledge of $\bar{e}$ the rescaled field e in the domain can be computed in terms of n:

$$e = -(B\bar{e})^{-1}(n - \ln n - 1 - A)$$

and takes the maximal value $A(B\bar{e})^{-1}$. Obviously this equation is valid as long as e remains positive which holds for n between the zeros $0 < n_2 < 1 < n_1$ of the right-hand side.
Introducing the transformation in the differential equation for n implies

$$\lambda\alpha\partial_\sigma n \sim n(v(E_\infty + e/\lambda) - 1 - \lambda v'(E_0)\bar{e}) \sim n(o(\lambda) - \lambda v'(E_0)\bar{e}).$$

In the limit $\lambda \to 0$ we obtain

$$\partial_\sigma n = -B\bar{e}n$$

with the solution

$$n = \exp(-B\bar{e}\sigma),$$

where a different choice for the constant of integration corresponds to a shift in the σ-direction which obviously does not change the results. This solution is valid for σ between the values σ_1 and σ_2 where n takes the values n_1 and n_2, respectively. The construction of the asymptotic form of the traveling wave solution would be completed by considering layers in neighbourhoods

of σ_1 and σ_2 which smooth the jumps in the electron density from 1 to n_1 and from n_2 to 1, respectively.
It remains to determine the value of $\bar{e}$ from the integral condition on the electric field. Asymptotically the integral is given by

$$E_0 + \int_{\sigma_1}^{\sigma_2} e\, d\sigma = E_0 + \kappa/\bar{e}^2$$

with

$$\kappa = B^{-2}\left(n_2 - n_1 - \ln\frac{n_2}{n_1}\left(\frac{1}{2}\ln(n_2 n_1) + A + 1\right)\right).$$

Since $\bar{e}$ has to be positive, we obtain

$$\bar{e} = \sqrt{\kappa/(U - E_0)}.$$

Summarizing the results of this Paragraph we note that the dipole formed by the electrons has sharp boundaries represented by σ_1 and σ_2. The velocity of the wave is close to the saturation velocity and, thus, together with the microwave frequency essentially independent from the applied voltage.

Problems

4.1 Instead of the SRH-term consider a more general recombination-generation model of the form

$$R = Q(n, p, x)(np - \delta^4) \qquad \text{with} \qquad Q > 0$$

in (4.1.4). This includes the band-band recombination term (2.2.13) and the Auger term (2.6.1). Carry over the discussion of the close-to-equilibrium case to this model.

4.2 Solve the double obstacle problem (4.2.10) by patching together solution pieces with $W = -1$, $\partial_\eta^2 W + \hat{C} = 0$ and $W = 1$, respectively, such that W is continuously differentiable. Convince yourself that the solution is unique. Verify (4.2.11).

4.3 a) Verify the results in the Paragraph "Strongly asymmetric junctions" of Section 4.2.
b) Singular perturbation theory leads to different approximations of a function in different regions. Here these approximations are $W_{in}(\sigma)$, $W_{tr}(\tau)$ and $W_{depl}(\eta)$. The question arises if an approximation can be found which is uniformly valid in the full region of interest. In general the answer is positive. Consider the example

$$f(x, \varepsilon) = \cos(e^{-x/\varepsilon} + x) \qquad \text{with} \qquad \varepsilon \ll 1.$$

Away from the boundary layer at $x = 0$ the approximation

$$\bar{f}(x) = \cos x$$

is valid, whereas

$$\hat{f}(\xi) = \cos(e^{-\xi}), \qquad \xi = \frac{x}{\varepsilon}$$

approximates f within the layer. The matching condition

$$\bar{f}(0) = \lim_{\xi\to\infty} \hat{f}(\xi) = 1$$

holds. A uniformly valid approximation for f can be obtained by adding the individual approximations $\bar{f}$ and $\hat{f}$ and subtracting their "common part" which is 1 in our example. Thus, we have

$$f(x, \varepsilon) \sim \cos(e^{-x/\varepsilon}) + \cos x - 1 .$$

Use these ideas and the common parts (4.2.19), (4.2.21) for obtaining an approximation of W which is uniformly valid in the η-interval $[-2, 0]$.

4.4 a) Verify the formula (4.2.26) for the saturation current by explicitely solving the problems for φ_1 and φ_2 in the one-dimensional case (4.2.1), (4.2.2).
b) Extend this result to the recombination-generation model of Problem 4.1.

4.5 a) Verify (4.2.29).
b) Rescale the current in (4.2.30) by $I = \delta^4 \tilde{I}$. Obtain the Shockley equation by letting $\delta^4 \to 0$ in the equation for $\tilde{I}$.

4.6 a) Solve the double obstacle problem (4.2.33) and obtain the noncoincidence set (4.2.34).
b) Compute w, I_n and I_p and verify the formula (4.2.38) for the characteristic.

4.7 Compute the common-emitter current gain (4.3.3) of a bipolar transistor under the following, simplifying assumptions:
a) According to (4.3.4), the second term in the formula for a_3 can be neglected.
b) In the computation of φ_1 and φ_2 the emitter and base regions are represented by intervals whose lengths are the distance between the emitter contact and the emitter junction and between the emitter and the collector junctions, respectively.
c) In the computation of φ_2 the base contact can be ignored.
d) The mobilities are constant and the doping profile is piecewise constant.

4.8 Solve the double obstacle problem (4.4.2) for a short PIN-diode in thermal equilibrium in the one-dimensional case (4.4.7) and with the assumption of a piecewise constant doping profile. Verify that (4.4.6) holds in the limiting case $\tilde{\lambda} \to 0$.

4.9 a) Analyze the thyristor close to thermal equilibrium considering the Shockley-Read-Hall term for recombination-generation.
b) In the case of the one-dimensional model (4.5.5), (4.5.6) compute the constant A in (4.5.4).

4.10 Verify (4.5.10), (4.5.11) by solving the problem for w_0, J_{n0}, J_{p0}.

4.11 For the MIS diode problem (4.6.3), (4.6.4) in the accumulation regime an approximation of the potential has been obtained in Section 4.6. However, this approximation cannot be uniformly valid because it does not converge as $\xi \to \infty$.
a) Introduce a transition layer solution of the form

$$W_{\text{trans}}(\sigma) = -1 + \gamma z(\sigma), \qquad \sigma = \frac{\xi}{\sqrt{\gamma}}$$

which connects the initial approximation to the prescribed value for W at $\xi = \infty$.
b) Construct a uniformly valid approximation for the potential by the method introduced in Problem 4.3.
c) Show that the contribution of W_{trans} to the total charge is small compared to the approximation (4.6.5).

References

[4.1] W. Eckhaus: Asymptotic Analysis of Singular Perturbations. North-Holland, Amsterdam (1979).

[4.2] A. Friedman: Variational Principles and Free-Boundary Problems. John Wiley & Sons, New York (1982).

[4.3] P. Grandits, C. Schmeiser: A Mixed BVP for Flow in Strongly Anisotropic Media. Applicable Analysis (to appear).
[4.4] J. B. Gunn: Microwave Oscillations of Current in III-V Semiconductors. Solid State Comm. *1*, 88 (1963).
[4.5] J. Kevorkian, J. D. Cole: Perturbation Methods in Applied Mathematics. Springer, New York (1981).
[4.6] D. Kinderlehrer, G. Stampacchia: An Introduction to Variational Inequalities and Their Applications. Academic Press, New York (1980).
[4.7] C. C. Lin, L. A. Segel, Mathematics Applied to Deterministic Problems in the Natural Sciences. Macmillan, New York (1974).
[4.8] P. A. Markowich: A Nonlinear Eigenvalue Problem Modelling the Avalanche Effect in Semiconductor Diodes. SIAM J. Math. Anal. *16*, 1268–1283 (1985).
[4.9] P. A. Markowich, C. A. Ringhofer, C. Schmeiser: An Asymptotic Analysis of One-Dimensional Semiconductor Device Models. IMA J. Appl. Math. *37*, 1–24 (1986).
[4.10] P. A. Markowich, C. Schmeiser: Uniform Asymptotic Representation of Solutions of the Basic Semiconductor Device Equations. IMA J. Appl. Math. *36*, 43–57 (1986).
[4.11] C. P. Please: An Analysis of Semiconductor *P-N* Junctions. IMA J. Appl. Math. *28*, 301–318 (1982).
[4.12] A. Porst: Halbleiter. Siemens AG, Berlin-München (1973).
[4.13] M. H. Protter, H. F. Weinberger: Maximum Principles in Differential Equations. Prentice Hall, Englewood Cliffs, NJ (1967).
[4.14] I. Rubinstein: Multiple Steady States in One-Dimensional Electrodiffusion with Local Electroneutrality. SIAM J. Appl. Math. *47*, 1076–1093 (1987).
[4.15] C. Schmeiser: A Singular Perturbation Analysis of Reverse Biased *PN* Junctions. SIAM J. Math. Anal. *21* (1990).
[4.16] E. Schöll: Nonequilibrium Phase Transitions in Semiconductors. Springer, Berlin (1987).
[4.17] S. Selberherr: Analysis and Simulation of Semiconductor Devices. Springer, Wien-New York (1984).
[4.18] W. Shockley: The Theory of *p-n* Junctions in Semiconductors and *p-n* Junction Transistors. Bell Syst. Tech. J. *28*, 435 (1949).
[4.19] J. W. Slotboom: Iterative Scheme for 1- and 2-Dimensional D.C.-Transistor Simulation. Electron. Lett. *5*, 677–678 (1969).
[4.20] H. Steinrück: A Bifurcation Analysis of the Steady State Semiconductor Device Equations. SIAM J. Appl. Math. *49*, 1102–1121 (1989).
[4.21] H. Steinrück: Asymptotic Analysis of the Current-Voltage Curve of a PNPN Semiconductor Device. IMA J. Appl. Math. (1989) (to appear).
[4.22] S. M. Sze: Physics of Semiconductor Devices. John Wiley & Sons, New York (1969).
[4.23] P. Szmolyan: Traveling Waves in GaAs-Semiconductors. Physica D (1989) (to appear).
[4.24] P. Szmolyan: An Asymptotic Analysis of the Gunn Effect. Preprint, IMA, Univ. of Minnesota (1989).
[4.25] M. J. Ward: Asymptotic Methods in Semiconductor Device Modeling. Thesis, California Inst. of Techn. (1988).
[4.26] M. J. Ward, D. S. Cohen, F. M. Odeh: Asymptotic Methods for MOSFET Modeling. Preprint, California Inst. of Techn. (1988).

Appendix

Physical Constants

Quantity	Symbol	Value
Boltzmann constant	k_B	1.38×10^{-23} VAs/K
Electron rest mass	m_0	0.91×10^{-30} kg
Electron volt	eV	1.6×10^{-19} VAs
Elementary charge	q	1.6×10^{-19} As
Permittivity in vacuum	ε_0	8.85×10^{-14} AsV^{-1} cm^{-1}
Reduced Planck constant	$\hbar$	1.05×10^{-34} VAs^2

Properties of Si at Room Temperature

Permittivity: $\varepsilon_s = 11.9\varepsilon_0$
Bandgap: $E_g = 1.12$ eV
Low field mobilities: $\mu_n = 1500$ cm^2 V^{-1} s^{-1}, $\mu_p = 450$ cm^2 V^{-1} s^{-1}
Typical values for recombination-generation parameters:

$$C_n = 2.8 \times 10^{-31} \text{ cm}^6/\text{s}, \quad C_p = 9.9 \times 10^{-32} \text{ cm}^6/\text{s}$$
$$\tau_n = 10^{-6} \text{ s}, \quad \tau_p = 10^{-5} \text{ s}$$
$$\alpha_n^\infty = 10^6 \text{ cm}^{-1}, \quad \alpha_p^\infty = 2 \times 10^6 \text{ cm}^{-1}$$
$$E_n^{\text{crit}} = 1.66 \times 10^6 \text{ V/cm}, \quad E_p^{\text{crit}} = 2 \times 10^6 \text{ V/cm}$$

Subject Index

Springer-Verlag Wien New York

Computational Microelectronics
Editor:
S. Selberherr

P. A. Markowich

The Stationary Semiconductor Device Equations

1986. 40 figs. IX, 193 pages.
Cloth DM 98,—, öS 686,—
ISBN 3-211-81892-8

Prices are subject to change without notice.

The static semiconductor device problem is treated in an "applied mathematics" way. The device equations are derived from physical principles; qualitative properties like existence, uniqueness, continuous dependence-on-data, and regularity of solutions are analysed by means of the modern theory of elliptic boundary value problems, and singular perturbation methods are employed to investigate the structure of solutions. This analysis describes depletion layer phenomena qualitatively and quantitatively. Physical interpretations of the mathematical results are given and pitfalls of the physical model are discussed. The structural results obtained by the singular perturbation analysis serve as a basis for the derivation and convergence analysis of numerical discretisation techniques, i.e. of finite element and finite difference methods.
The monograph provides device modelers with the basic mathematical results and techniques necessary for designing efficient simulation programs and for a profound interpretation of simulation results. Also, it serves as a source of mathematically challenging problems for researchers in numerical analysis, singular perturbation theory, and p.d.e. analysis.

Springer-Verlag Wien New York

Moelkerbastei 5, P.O. Box 367, A-1011 Wien
Heidelberger Platz 3, D-1000 Berlin 33
175 Fifth Avenue, New York, NY 10010, USA
37-3, Hongo 3-chome, Bunkyo-ku, Tokyo 113, Japan

9 783211 821572

GPSR Compliance

The European Union's (EU) General Product Safety Regulation (GPSR) is a set of rules that requires consumer products to be safe and our obligations to ensure this.

If you have any concerns about our products, you can contact us on ProductSafety@springernature.com

In case Publisher is established outside the EU, the EU authorized representative is:

Springer Nature Customer Service Center GmbH
Europaplatz 3
69115 Heidelberg, Germany

Batch number: 10577884

Printed by Printforce, the Netherlands